Italy and the Suez Canal, from the Mid-nineteenth Century to the Cold War

Barbara Curli
Editor

Italy and the Suez Canal, from the Mid-nineteenth Century to the Cold War

A Mediterranean History

palgrave
macmillan

Editor
Barbara Curli
Dipartimento Culture, Politica e Società
Università di Torino
Turin, Italy

ISBN 978-3-030-88254-9 ISBN 978-3-030-88255-6 (eBook)
https://doi.org/10.1007/978-3-030-88255-6

This Palgrave Macmillan imprint is published by the registered company Springer Nature Switzerland AG.
The registered company address is: Gewerbestrasse 11, 6330 Cham, Switzerland

Passage through the Suez Canal of the AGIP (ENI Group) oil tanker, recognisable from the typical six-legged dog, the symbol of AGIP. Frame from the documentary film *Il Canale* by Bernardo Bertolucci (1966, commissioned by ENI, Ente nazionale idrocarburi)

ACKNOWLEDGEMENTS

I wish to thank Gianmaria Gros-Pietro, chairman of Intesa Sanpaolo, for the generous sponsorship of the international conference from which this book originates, and Giovanna Paladini for her liaison work; Romano Prodi and Alessandro Ovi for friendly support of the project; the Dipartimento Culture Politica e Società of the Università di Torino; Massimo Mori, president of the Accademia delle Scienze in Turin, for having hosted the conference in the Accademia's beautiful rooms and Elena Borgi for collaboration on the documentary material of the Accademia's Library. I also thank for having variously contributed to the conference Francesca Biancani, Lucia Carminati, Giovanni Costenaro, Paolo Gallarati, Bahar Gursel, Nicola Labanca, Gaetano La Nave, Peppino Ortoleva, Lucy Riall, Marcello Tadini, Alessandra Venturini. Attempts to involve Egyptian scholars in this project met with difficulties of various kind. I hope there will be other opportunities for collaboration.

A special thank to Lucia Nardi, head of ENI's cultural activities, Elena Testa of Archivio nazionale del cinema di impresa, and Sergio Toffetti, for generously making available fotograms from *Il Canale* by Bernardo Bertolucci; and to the following institutions for permission on the use of the illustrations: Archivio Corriere della Sera, Milan; Fondazione Giorgio La Pira, Florence; Archivio Negrelli, Biblioteca Intercomunale Primiero, Trento; Museo Collezione Nando Salce, Treviso; Archivio Museo Revoltella, Trieste.

Contents

Notes on Contributors

Carlo Bovolo is a postdoc fellow at the Università del Piemonte Orientale (Vercelli, Italy) and collaborator with the Museum of Criminal Anthropology 'Cesare Lombroso' of the Università di Torino. In 2017 he obtained a PhD in history at the Università del Piemonte Orientale with a thesis dealing with the apologetical uses of science in Italian Catholic press. Then, he was Fellow of the Fondazione Filippo Burzio and Visiting Fellow at the KU Leuven, Belgium. The volume *I cattolici italiani e la scienza: il discorso apologetico sulla stampa clericale nell'età del Positivismo (1848–1814)*, was published in 2017.

Luca Castiglioni was awarded his PhD in History by the Università di Pavia in 2016, defending his dissertation 'Assab. Alle origini del colonialismo italiano'. He is currently a lecturer (docente a contratto) at the Department of Scienze della Mediazione Linguistica e Culturale, Università degli Studi di Milano. He has been employed as a post-doc researcher at the Italo-German Historical Institute (FBK-ISIG) in 2020/2021 and is currently working on the history of Italian colonial rule in the Dodecanese and its impact on traditional polities. He is also a subject expert in Contemporary History at the Università degli Studi di Milano.

Giampaolo Conte is Research and Teaching Fellow at Università Roma Tre where he teaches Economic History and History of Capitalism. He has been Visiting Fellow at the LSE and Visiting Scholar at Cambridge University. His research articles appeared in scholarly journals such as *Oriente Moderno, The Journal of European Economic History, Eurasian Studies, The International History Review, and Capital & Class.* He is co-author and author of three monographs: *L'Odissea del Debito. Le Crisi finanziarie in Grecia dal 1821 ad oggi (2015); Il Tesoro del Sultano. L'Italia, le Grandi potenze e le finanze ottomane 1881–1914 (2018); Il Credito di una Nazione. Politica, diplomazia e società di fronte al problema del debito pubblico italiano 1861-1876 (2021).*

Frederick Cooper is Professor of History at New York University. Among his books are *Colonialism in Question: Theory, Knowledge, History* (2005), *Empires in World History: Power and the Politics of Difference* (with Jane Burbank, 2010), *Citizenship between Empire and Nation: Remaking France and French Africa, 1945–1960* (2014), *Africa in the World: Capitalism, Empire, Nation-State* (2014), and *Citizenship, Inequality, and Difference: Historical Perspectives* (2018).

Giovanni Cristina is a researcher in Contemporary History at the Università di Genova, and was post-doc researcher at the Università di Catania. Previously a Marie Curie Fellow at EHESS (Paris), his research deals with public housing districts in postwar Italy and the processes of modernisation in the port cities of Mediterranean Europe between the nineteenth and twentieth centuries. His recently published work include: *Il porto di Catania nel lungo Ottocento: Infrastrutture, traffici, territorio (1770–1920)*, 2019; *Persistence and Evolution in the Eastern Sicilian Coastal Corridor: The Mobility of Goods and People at the Port of Catania (1817–1860)*, *The Urban Logistic Network: Cities, Transport and Distribution in Europe from the Middle Ages to Modern Times*, Palgrave Macmillan, Basingstoke, 2019.

Barbara Curli is Professor of Contemporary History and Global History at the Università di Torino where she is the coordinator of the Ph.D. programme in Global history. She has extensively researched on the history of European integration, nuclear energy history, and history of women's work. She is member of the editorial board of *Italia contemporanea* and of the Scientific Board of Fondazione Leonardo-Civiltà delle Macchine. Her recently published work includes *Discourses and Counter-Discourses on Europe. From the Enlightenment to the EU* (edited with M. Ceretta), 2017, and on the history of the Suez Canal, Il Piemonte e il Canale di Suez, 1855–1856. *Studi Piemontesi*, XLIV, 2015, 2; Dames employées at the Suez Canal Company: The "Egyptianization" of Female Office Workers, 1941–1956. *International Journal of Middle East Studies*, 2014, 46, 3.

Angelos Dalachanis is a researcher at the French National Centre for Scientific Research (CNRS). He is based at the Institute of Early Modern and Modern History (IHMC—UMR 8066) in Paris. He holds a PhD in history from the European University Institute, Florence, and was a postdoctoral research fellow at the Seeger Center for Hellenic Studies at Princeton University. His research interests include the history of migration, business, and labor in the Eastern Mediterranean in the modern period. He is the author of *The Greek Exodus from Egypt: Diaspora Politics and Emigration, 1937–1962* (2017) and co-editor of *Ordinary Jerusalem, 1840–1940: Opening New Archives, Revisiting a Global City* (2018).

Massimo De Giuseppe is Associated Professor at Università IULM Milano, where he teaches Contemporary History. His main research interests include the study of social, religious, and political movements, transnational networks, pacifism, and the impact of conflicts in Italy and Latin America during the Twentieth Century. He is member of the editorial staff of "Contemporanea", "Modernism", "Glocalism" and of the scientific committee of "Historia de las mentalidades" and "Annali di Storia dell'educazione". Since 2019 he is international member of Academia mexicana de la historia, corripondent of Academia Real de España. Among his most recently published work are, *Italian Catholic Intellectuals and Indigenous Latin Americans: Transnational Networks and Violence at the End of the Cold War*,, *Italian Intellectuals and International Politics, 1945–1992*, Palgrave Macmillan, London 2019, *Italian Catholics and Latin America During the Long 1968*, *The Revolutionary Sixties: "1968" in Europe and Latin America*, 2020, and the monograph (with G. La Bella), *Storia dell'America latina contemporanea*, 2019.

Fabio De Ninno is Assistant Professor at the Università di Siena. His research deals with the military and naval history of contemporary Italy. He is author of the books *Civili nella guerra totale. 1940–1945. Una storia complessa* (2020); *Piero Pieri. Il pensiero e lo storico militare* (2019) and *Fascisti sul Mare. La Marina e gli ammiragli di Mussolini* (2017), in addition to various chapters and articles in international journals like *War in History* and *The Journal of Military History*. He is Editorial Secretary and Member of *Italia contemporanea* board of editors.

Giorgio Ennas is a Postdoctoral Fellow at Franklin University Switzerland, Lugano, Switzerland. He graduated with a Master of Arts in Languages and Civilisation of Asia and Mediterranean Africa from

Università Ca' Foscari in Venice with a thesis on 'Ottoman *élites* and modernization actors. Between integration and exclusion: the controversial path of Ottoman Empire to Modern times'. Between 2017 and 2021, he was a researcher at the European University Institute in Fiesole, where under the supervision of Lucy Riall and Pieter Judson his work focused on the cultural aspects of diplomatic history of the Italian and Ottoman relations during the long nineteenth century. In June 2021, he successfully defended his dissertation entitled '*The Mediterranean Mirror. Italo-Ottoman Relations in an Age of Transition, 1856–1871*'. Among his published work are, 'The Birth of the Ottoman Colonial Space. The Libyan Case (1835–1918)', *Proposte e ricerche*, 77, 2016; and the book *Reports of Cesare Durando Italian Vice-Consul in Sarajevo (1863–1867)*, 2020.

Andrea Giuntini is Professor of Economic History, Labour History and History of Economic Thought at Università di Modena e Reggio Emilia. He has extensively researched on the history of railways, transport, and infrastructure and has been president of the Istituto di Studi Storici Postali in Prato. He is member of the editorial boards of *Memoria e Ricerca*, *Storia economica*, *Ricerche Storiche*, *Revue d'histoire des chemins de fer*, *Transportes, servicios y telecomunicaciones.* He devoted a chapter on the Suez Canal in his book *Le meraviglie del mondo. Il sistema internazionale delle comunicazioni nell'Ottocento*, Prato, Istituto di Studi Storici Postali, 2011.

Andreas Guidi is a postdoctoral research and teaching assistant (wissenschaftlicher Mitarbeiter) at the University of Konstanz and currently a visiting fellow at the German Historical Institute in Washington, DC. His forthcoming monograph entitled "*Generations of Empire: Youth between Ottoman and Italian Rule in the Mediterranean*" (Toronto University Press) discusses changing interactions between the state, religious communities, and families in twentieth- century Rhodes with a focus on generational dynamics. His current project explores the notion of illegality through the history of smuggling in the modern Mediterranean. Andreas is also the creator and editor of 'The Southeast Passage', a podcast about history in the making.

Andrea Leonardi is Professor of Economic History at the Università di Trento and was a Visiting Professor at the University of Innsbruck and the Università Cattolica in Milan. His research interests include the trajectory of economic development in the Alps and the diversification of economic growth in the Habsburg Empire. His most recent studies are devoted to the financial history of Italy in the twentieth century, as well as to business history and the economic repercussions of tourism in the Alpine macro-region. He has coordinated the recovery of the Negrelli Nachlaß in Vienna, which consists of 2345 documents that are now being systematically investigated. He has organized several international conferences and authored over 220 books and articles, many of which are monographs.

Giovanni Modaffari geographer and research fellow at the Università di Milano-Bicocca, holds a PhD in Cultural Heritage and Geography from the Università di Roma Tor Vergata. His research focuses on the geography of explorations in the modern age and cultural exchanges between Europe and the Ottoman Empire in the fifteenth–nineteenth centuries. He has co-edited *Sir Richard F. Burton, Trieste e l'esplorazione*, 2019.

Steven Morewood is Associate Professor in International history at the University of Birmingham. He is the author of *The British Defence of Egypt 1935–1940* (2014), 'Prelude to Suez: The Rise and Fall of British Dominance over the Suez Canal, 1869–1956', in *Reassessing Suez 1956* (ed. Simon C. Smith), and numerous chapters and articles on the British role in the Eastern Mediterranean from 1935–1941. He is working on a new book about the first British desert offensive which destroyed the Italian 10th Army.

Giuseppe Natale is a professor at the University of Nevada, Las Vegas, where he teaches Italian Studies, Classical Studies, and Translation Studies. He has published on Italian literature (Foscolo, Leopardi, Pavese), cinema (Fellini and the Felliniesque), and translation theory and pedagogy. He is the founding editor of *TransScribe: The Teaching Translation Journal.* Natale has also translated several major American novels such as Herman Melville's *Moby Dick*, Toni Morrison's *Beloved*, and Thomas Pynchon's *Gravity's Rainbow* to Italian.

Alice Riegler is a research fellow at the Department of Economics and Management at the Università di Trento, where she researches technological transformations in the Alpine area during the nineteenth and twentieth centuries as part of a wider project funded by the Euroregion Tyrol- South Tyrol- Trentino. She completed her undergraduate studies at the University of Florence and her MA and PhD at University College London's History Department. She has been working on the Nachlaß of Luigi Negrelli since 2018.

Leonardo Scavino is a post-doctoral Fellow at the Università di Genova, and a member of NavLab. In 2020, he was awarded the European PhD Label, after having discussed his dissertation conducted in the context of the ERC SG project SeaLiT (Seafaring Lives in Transition. Mediterranean Maritime Labour and Shipping during Globalization, 1850s–1920s), coordinator Apostolos Delis (IMS-FORTH, Crete). His main research interests deal with Mediterranean maritime history and the economic and social relations between Western and Eastern Mediterranean (including the Black Sea). He is focusing on a specific seafaring Ligurian community, Camogli, in the age of transition from sail to steam shipping.

Paola Schellenbaum is an independent scholar. She is a trained psychologist and received her PhD in cultural anthropology and ethnology from the Università di Torino with a thesis on a multi-sited ethnography between Cairo and Milan on transnational migrant families in the 1990s. Before this, she was at UCLA and at Stanford University in California for a research on Italian American families, which was published as 'Stereotypes as Cultural Constructs' in *Studies in Italian American Folklore*, Luisa Del Giudice (ed), 1993. She worked as research assistant of Prof. Sylvia Yanagisako in a research on gender roles in family firms in Italy, which was published as *Producing Culture and Capital. Family Firms in Italy*, 2002. She was a consultant for ISMU Foundation (Milan) on intercultural education and international migrations and for IOM (Rome) on refugees and trauma studies. In 2017, she was in the Peacemaking Program of the Presbyterian Church USA for an itinerant seminar with conferences on humanitarian corridors. She is currently a member of the Waldensian Studies Society (Società di Studi Valdesi).

Filippo Gattai Tacchi is completing his PhD at the Scuola Normale Superiore, Pisa. In 2016 he obtained a Master's degree in History and Civilization from the Università di Pisa with a thesis on 'Britons rule the waves? La Gran Bretagna, la Guerra Anglo-Boera e la prospettiva italiana (1899–1902)', with Alberto Mario Banti as his supervisor. His research interests are the cultural, political, diplomatic, and social history of Great Britain and its Empire, in particular during the late Victorian age and the brief Edwardian period, and its political and cultural relations with Liberal Italy. In 2018, his book *L'impero britannico in 100 date* was published.

Kevin A Tang is a research associate at the Smith School of Enterprise and the Environment, School of Geography, University of Oxford, and a research fellow at the Skoll Centre for Social Entrepreneurship, Said Business School, University of Oxford. His research is focused on spatial and geographic analysis of the effects of infrastructure investments on economic development, the economic history of infrastructure developments and its economic and social impacts, economic policy for climate change, and finance and economics for sustainability.

Olga Verlato is a Doctoral candidate in History and Middle Eastern and Islamic Studies at New York University. Her research explores the social and cultural history of language and multilingualism in nineteenth and early-twentieth century Egypt, focusing on education and journalism. Towards her dissertation, she is conducting archival research in Egypt, Italy, France, and the United Kingdom.

Joseph John Viscomi is a lecturer in European History at Birkbeck, University of London. He held a faculty fellowship at NYU''s Center for European and Mediterranean Studies (2017–2019) after completing his doctoral studies in Anthropology and History at the University of Michigan, Ann Arbor, in 2016. He was a researcher on the ERC project, 'Crosslocations,' at the University of Helsinki from 2017–2021. Viscomi serves on the executive committee for the Association for the Study of Modern Italy and on the convening committee for the IHR's Modern Italian History seminar. He has published in *The Journal of Modern History, History and Anthropology*, and *Modern Italy*. He is currently completing a book on the departure of Italians from Egypt and has begun new research on depopulation in Southern Italy after 1783.

List of Figures

List of Tables

Statue commemorating Pietro Paleocapa as architect of the Suez Canal and the Fréjus tunnel, piazza Paleocapa, Torino
The statue, by Odoardo Tabacchi, was erected in 1871. The inscription reads: "To Pietro Paleocapa, illustrious engineer, eminent statesman, the Italians from all provinces"
(Pictures by Barbara Curli)

Luigi (Alois) Negrelli
@ Courtesy Archivio Negrelli, Biblioteca Intercomunale di Primiero, Trento (Italy)

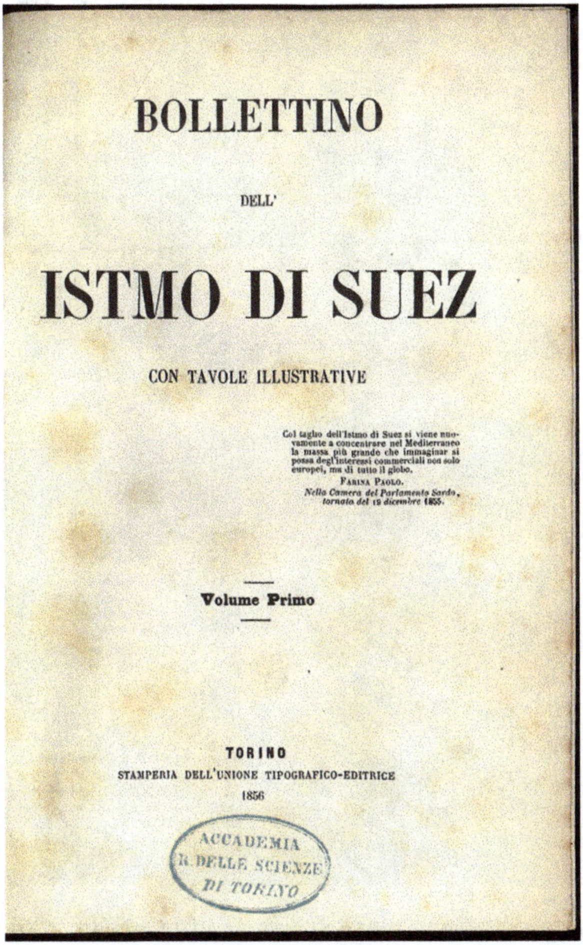

BOLLETTINO

DELL'

ISTMO DI SUEZ

CON TAVOLE ILLUSTRATIVE

Col taglio dell'Istmo di Suez si viene nuovamente a concentrare nel Mediterraneo la massa più grande che immaginar si possa degl'interessi commerciali non solo europei, ma di tutto il globo.

FARINA PAOLO.

Nella Camera del Parlamento Sardo, tornata del 12 dicembre 1855.

Volume Primo

TORINO

STAMPERIA DELL'UNIONE TIPOGRAFICO-EDITRICE

1856

The *Bollettino dell'Istmo di Suez* started publication in Torino in 1856
Courtesy: Library of *Accademia delle Scienze*, Torino

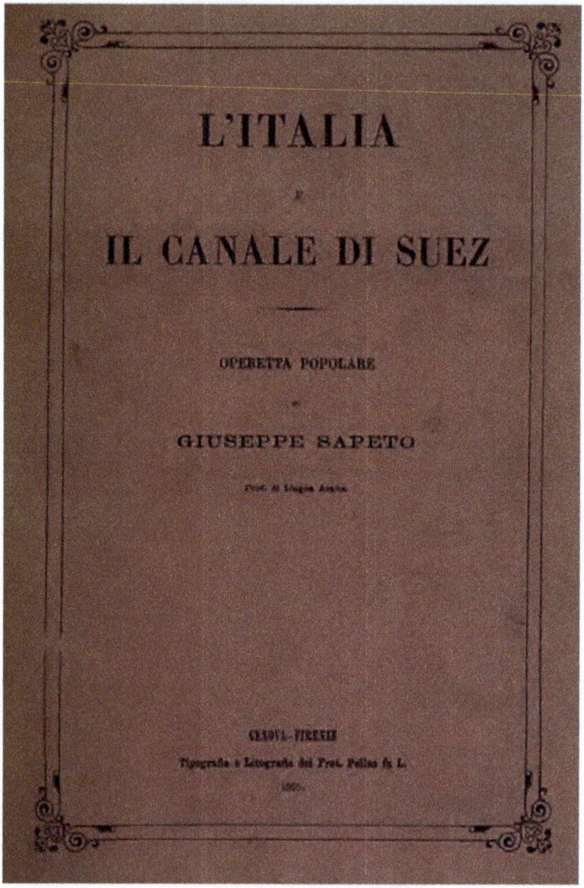

Giuseppe Sapeto, *L'Italia e il canale di Suez. Operetta popolare*, 1865

Il taglio dell'istmo di Suez (The cutting of the isthmus of Suez)
Allegorical marble group by Pietro Magni (1863), depicting Europe uniting the Mediterranean and the Red Sea, while the geniuses of Commerce and Shipping behind them give their blessing

Poster of the allegorical *Ballo Excelsior*, La Scala Milano (originally featured 1881), the Suez Canal scene on top right. In 1913 the Ballo was transformed in a movie by Luca Comerio, to which this poster is related.

LA DOMENICA DEL CORRIERE

Si pubblica a Milano ogni settimana

Supplemento illustrato del "Corriere della Sera"

Uffici del giornale: Via Solferino, 28 - Milano

Anno XXXVII — N. 39

29 Settembre 1935 - Anno XIII

Centesimi 30 la copia

Le entusiastiche dimostrazioni degli Italiani residenti in Egitto al passaggio nel canale di Suez delle navi che trasportano le nostre truppe nell'Africa Orientale.

(Disegno di A. Beltrame)

Front cover of *La Domenica del Corriere*, 29 September 1935: Passage through the Suez Canal of Italian troops heading to Ethiopia and welcomed by the Italian residents in Egypt

Cairo, 1968. Gamal Abdel Nasser and Giorgio La Pira

Frames from the documentary film *Il Canale* by Bernardo Bertolucci (1966, commissioned by ENI, Ente nazionale idrocarburi).

CHAPTER 1

Italy and the Suez Canal: Historical and Historiographical Passages from a Euro-Mediterranean Perspective

Barbara Curli

'The most absolute, the most pure, the most river of all is not a river: it is an artificial channel dreamed up by a visionary Italian and realized by a determined French, Mr. de Lesseps.' Such was the voiceover introducing *Il canale* (*The Canal*), Bernardo Bertolucci's 12-minute companion piece to his better-known *La via del petrolio* (*The Oil Route*), a documentary commissioned in 1966 by Italy's state energy company ENI and discussed by Giuseppe Natale in Chap. 22. While shooting *The Oil Route*, the young Bertolucci was so fascinated by the Suez Canal that he celebrated it in a short film of its own. Taking as leitmotif the canal's historic ties with Italy (ties that began with the Italian-speaking Austrian engineer Luigi Negrelli, who had the idea how to cut the isthmus in the mid-1800s), he put to the fore a sort of rediscovered Euro-Mediterranean unity in the years of the economic boom and of decolonisation. While the camera dwelt on the

B. Curli (✉)
Dipartimento Culture, Politica e Società, Università di Torino, Turin, Italy
e-mail: barbara.curli@unito.it

B. Curli (ed.), *Italy and the Suez Canal, from the Mid-nineteenth Century to the Cold War*,
https://doi.org/10.1007/978-3-030-88255-6_1

venerable colonial houses, the text—by Bertolucci himself—voiced the rhetorical optimism of the sixties' economic miracle and its belief in a modernisation that would bring the Mediterranean's two shores closer, levelling the differences between late-comers (at bottom, the Suez shoe-shine boy was just 'the younger brother of the Neapolitan sciuscià') in a return to a Mediterranean togetherness rooted in a common culture: 'Port Said: the smells, the colours, the sounds here are the same as in Crete, Marseilles, Barcelona; they are the smells, the colours, the sounds of our civilization.'

That the traditional and the new can be both seamless and separate is a recurrent theme of Mediterranean histories and rhetorics. Ancient and modern, or rather, age-old and ultra-modern crop up in the findings of the Italian National Research Council's recent Egyptological Mission to Tell el-Maskhuta, a few kilometres west of Ismailia, in search of the legendary 'canal of the Pharaohs'. Mentioned by Herodotus, Strabo, and Pliny the Elder, the canal was said to have connected the Nile Delta to the Red Sea by way of the Wadi Tumilat, once a major artery between the Mediterranean and the East. Traces of the ancient waterway were found during the mid-nineteenth-century excavations for Suez, and were later sought by numerous international archaeological expeditions. In 2016, aided by radar satellite remote sensing data from ASI, the Italian Space Agency, and ESA, the European Space Agency, the Italian archaeologists found the ruins of a fortress and a wall guarding the ancient canal. The most futuristic techniques were used to locate the oldest-known canal.[1]

Bertolucci's fascination and the rediscovery of the fortress of Tell el-Maskhuta are only two of the many episodes in the long and multifaceted history of the relationship between Italy and the Suez Canal, the artificial waterway that crosses some 190 kilometres of Egypt to join the Mediterranean to the Red Sea. The canal was conceived in the 1840s and 1850s, built between 1859 and 1869, and officially opened to shipping on 17 November 1869. The years of the canal's political and engineering genesis were also the years of Italy's Risorgimento and unification, and of the country's first foray into nineteenth-century globalisation. While Italian unification was the Euro-Mediterranean area's greatest geopolitical transformation in centuries, the opening of the canal—which had long figured in the European imagination—was not only the start of a new seaway, but a passage that contributed to reshaping the geopolitics of the

[1] Capriotti et al. (2019).

Mediterranean, colonial and imperial spaces and international networks of transportation, communication and trade, as well as cultures of mobility and forms of international governance.[2] Liberal Italy's state-building, budding foreign and colonial policy, and industrial take-off were intimately connected with this new global modernity.[3] United Italy was thus built in the interaction between two territorialities, that of the national space and that of the new Euro-Mediterranean spatial order, that was to be the main—though not the only—frame of Italy's turn-of-the-century modernisation, and then in the Fascist and Republican periods.

The canal's recent expansion, the so-called new Suez Canal inaugurated in 2015 with double the traffic capacity, has raised a host of issues—the Mediterranean's position in world trade routes and the new silk roads; the emergence of new actors (such as China) and the apparent 'decline' of others (Europe?); the role played by Egypt—that are surprisingly similar to those current at the time of the canal's construction in the mid-1800s, though the stage and the actors on it have changed. Once again, debate has centred on Italy's economic prospects and its ability to exert political, economic and cultural influence in the post-Cold War world of globalisation driven by the third/fourth industrial revolution. The canal's doubling, for example, has again sparked discussion on the role of the Italian ports—Genoa, Trieste, Naples—and of the Italian economy's place in the Mediterranean, in an animated debate that is in some ways reminiscent of that of the 1800s as reviewed in several chapters of this book.[4]

The opening of the canal in 1869—like its doubling today—was the outcome and not the cause, as is often maintained, of sweeping changes in the economy, technology and culture, and in war and the tools of war, while the resulting shifts in global power relationships made themselves felt in the conduct of diplomacy, as did the pursuit of new forms of international governance.[5] Little wonder, then, that the Suez Canal, though covered by an extensive albeit somewhat repetitive bibliography,[6] has again attracted scholars' interest in recent years, benefitting from innovative perspectives from global, imperial, and maritime history and from

[2] Osterhammel (2014), Ballantyne Burton (2013), Headrick (2009), Giuntini (2011), Wenzlhuemer (2012), Carreras et al. (1995).

[3] Middel Naumann (2010).

[4] For a recent assessment of this debate see SRM (2017, 2019).

[5] Buzan Lawson (2015).

[6] Among traditional and still fundamental accounts, see Charles-Roux (1901), Siegfried (1945), Farnie (1969).

wide-ranging studies on the Mediterranean. International business, labour, and social history has been shedding new light on the experience of the *Compagnie universelle du canal maritime de Suez* (the Suez Canal Company, henceforth the Company) and its relationship with modern Egypt, from its creation by Ferdinand de Lesseps in 1858, to the entrance of the British government as the single main shareholder in 1875, to nationalisation in 1956.[7] Even the had-the-canal-not been-opened counterfactual exercise has contributed to a new historiography of international trade.[8] These approaches have introduced new actors, topics and periodisation, thus helping supplement and revise traditional—if not stereotyped or even mythical—narratives of the canal.

Historiographical Passages

The Suez Canal has thus never stopped raising questions about the past and the present, at the intersection between the history of the Mediterranean, of Europe, and of Italy on the global scene. Nevertheless, the links between Italy and the canal have long been unfairly neglected by both international and Italian historiography, despite Italy's fundamental contribution to the canal's construction and later development.[9] There was, however, a flurry of publications at the time of the canal's opening—the 'isthmian literature', as it was called with a tinge of irony in 1869 by Cesare Correnti, then vice-president of the *Società geografica italiana*, founded in 1867.[10] Apart from its rhetoric and certain interpretative shortcomings, the isthmian literature offered important reflections on Italy's role in the Mediterranean and the infrastructures for its entry in the new geography of growth. The theme of Italy's relationship with the canal will return in the 1930s with an effort to reassess the country's contribution to constructing the canal, both as part of a more general refocusing on the 'Mediterranean ideals' of the Risorgimento, and to assemble a historical rationale for the so-called Italian claims on the canal in the years of the Ethiopian war, when Italy became (for two years, between 1936 and 1937) the Suez Company's second-largest customer after Great Britain.[11] The works of Monti and

[7] Huber (2013), Bonin (2010), Piquet (2008, 2009).
[8] Findlay and O'Rourke (2007).
[9] Bono (2006).
[10] Correnti (1869).
[11] Monti (1937a, 1937b), Sammarco (1940), Curcio (1941).

Sammarco in particular are generally solid, though they carry a clear ideological bias and are sprinkled with inaccuracies. Above all, much of the original documentation and correspondence on which they are based can no longer be found, making verification impossible. The literature of this period, moreover, is marked by more of the endless skirmishes about just how 'Italian' Luigi Negrelli—to whom we owe the project of the lock-free canal—can be considered to be. After a few investigations in the 1980s,[12] this question was finally resolved by recent work by Andrea Leonardi based on a thorough survey of the original sources in the Negrelli archives at the Technisches Museum in Vienna.[13] As Andrea Leonardi and Alice Riegler discuss in Chap. 7, this work has shed light on Negrelli's role as a cosmopolitan engineer and leading expert in nineteenth-century transport systems and communication networks.

Garnering little attention from economic historiography,[14] the relationship between the canal and Italy has been addressed in connection with Italy's post-war foreign policy and the 1956 crisis,[15] and has recently been the subject of a long-term review of the secondary literature.[16] The substantial historiography on the relations between Italy and Egypt and on the Italian communities there has rarely focused specifically on the canal.[17] By contrast, as this volume demonstrates, Italian communities on the isthmus were affected in many ways by the role and use of the canal in peace and war, while Italian investment, trade, banking, and maritime interests have long revolved around the canal. Even the recent historiography on Italy's ties to the sea mentions the Suez Canal only in passing,[18] and the 2018 *Storia mondiale dell'Italia*, though purporting to emphasise mobility, the Mediterranean, migrations, voyages, spaces, and empires, neglected it entirely.[19]

To fill at least some of these gaps, an international conference was held in Turin in 2019, the 150th anniversary of the canal's opening. The conference was attended by scholars engaged in innovative research based on original archival sources that offers new perspectives. Some of the papers

[12] Algardi (1989), Leonardi (1990).
[13] Leonardi (2020).
[14] One of the few exceptions is Caizzi (1985).
[15] Calchi Novati (1998).
[16] Valle (2018).
[17] Petricioli (2007), Cresti (2008).
[18] Frascani (2001), Fanfani (1993).
[19] Giardina (2018).

presented in Turin have been reworked as chapters in this book, which covers a story that unspools from the 1850s to the 1960s. The book does not seek simply to foreground a presence—of Italian work, finance, diplomacy, cultures, politics, migration, and so on—in the history of the canal. While it is hoped that this will be one of the book's outcomes, it is not the objective that inspired it. Nor is it our intention to restore a national narrative to primacy, or to perpetuate an Italy-seen-from-abroad approach. Rather, this book is an attempt to go beyond traditional narratives of the canal, the 'epic' of Suez (as fascinating as it was, as it still is). The book's intent is more limited but also more ambitious, namely, to reconstruct a number of passages—historical and historiographical—of a history of Italy that runs through Suez. The canal, literally, is a passage but it is also a mirror that reflects several major issues of the history of Italy in a Euro-Mediterranean frame, which a series of historiographic developments have brought to the fore in recent years.

The first of these developments is related to the end of the Cold War and the resulting need to redefine European regionalism and its flexible physical and thematic borders from a historical perspective. Unsurprisingly, the historiography of major infrastructures and Europe's 'material integration' was among the first to look at the construction of a European 'identity' (and the discursive rhetorics supporting it) under a new lens, in support of, rather than in opposition to, the national states' economic and political interests. In the second half of the 1800s—the era of railroads, bridges, and roads, of canals, tunnels, telegraphs, and undersea cables—creating communication networks and material interconnections called for new forms of technical and legal governance and was thus fertile ground for the rise of transnational technocracies that at the same time embodied national modernisation projects. The dense dialectic between nation, Europe, and modernity has thus problematised the relationship between technological change and European integration in a long-term perspective. In particular, scholars have noted the contribution made by 'small states' and their expert élites in driving the push towards new forms of international regulation.[20] As this book shows, the case of Italy and the Suez Canal can be seen in this light: Suez has been both an infrastructure for projecting the Italian economy onto the international stage—and thus a means of state-building—and a 'European' infrastructure, one of those

[20] Van der Vleuten and Kaijser (2006, 10–11); Badenoch and Fickers (2010), Kohlrausch and Trischler (2014).

portals of globalisation, those critical junctures that reframe hierarchies of spaces and forms of regulating modernity, fuelling a continual dialectic between the national, the regional and the global.[21]

The second historiographic development is connected with the recent resurgence of Mediterranean studies, likewise inspired by the end of the Cold War and the redefinition of spaces and mobility resulting from globalisation, new wars and migrations, and likewise the outcome of a re-regionalisation of historical research. This has spurred original work on the contemporary Mediterranean between the 1800s and the 1900s, long overshadowed by the traditional (largely Braudelian) historiography on earlier centuries.[22] In its newly recognised capacity for developing *national* narratives in terms of interconnections and entangled modernisations, the Mediterranean offers valuable insights for Italian history.[23] Such approaches have also drawn on the new historiography of empires, as the history of global dynamics, interdependence, cultural interactions and hybridisation, and frames of international mobility.[24] Suez has traditionally been seen as central to the history of imperialism and its periodisation. The opening of the canal is usually considered the beginning of a new 'multipolar', 'progressive' imperialism associated with the scramble for Africa, the emergence of aspiring colonial powers (Italy and Germany), Britain's occupation of Egypt and nascent hegemony in the Eastern Mediterranean, and the opening of new colonial spaces (the Horn of Africa) as Ottoman rule eroded. Rooted in powerful cultural and racial assumptions, this phase is said to end with the 1956 Suez crisis, the symbolic sunset for that type of British and European colonialism.[25]

However, as recent studies of the intersecting histories of the Mediterranean and imperialism have shown, Suez is more than a strategic passageway to the empires. The canal's opening also triggered a 'colonial reappropriation' of the Mediterranean as a 'European space' freighted with a new ideological burthen, and as a sort of 'rhetorical device serving to reinforce political hegemonies' and to redefine racial hierarchies as well as national projects of modernisation and expansion, particularly revealing

[21] Middel and Naumann (2010).

[22] Harris (2004), Horden and Purcell (2000, 2006, 2019), Horden 2005 Sant Cassia and Schäfer (2005), Abulafia (2011), Burke (2012).

[23] Benigno (2009), Verga (2017), Labanca (2019).

[24] Burbank and Cooper (2010), Fischer-Tiné and Gehrmann (2008).

[25] Darwin (2009, 66–67), Holland (2012).

in the process of Italian nation-building.[26] Somewhat paradoxically a common imperial European identity was constructed in the Mediterranean just when colonial rivalries and fractures and modern notions of the nation—on both shores—brought fractures and divisions among European states.[27] Uncoincidentally, this identity was reframed in the 1930s, with the resurrection of a 'Latin' Mediterranean and a new stage of Italian colonial expansion that also ushered in a new season for Italy's Mediterranean imaginings.[28]

In these passages, the canal zone took on an identity of its own. Suez is not just a passageway, but a place where community is constructed: from the 1860s onwards, the economic activities centring on it (primarily work for the Company) attracted streams of migrants from virtually all sides of the Mediterranean. Gradually, the canal became a place of 'international colonisation',[29] whose transformation over the decades coincides with Egypt's formation as a modern state as well as sweeping changes in the migrants' home countries. Suez is thus also an unconventional vantage point for observing Italian emigration in the Mediterranean, 'from North to South'.[30]

The third set of developments concerns Italian history. Italy was the European country most affected by the Cold War framework, and the one that had to adapt most to the post-Cold War global reconfiguration. This momentous passage was also a historiographic passage, which provided an opportunity for rethinking modern Italian history and intersected the debate on Italy's so-called economic 'decline' over the last 30 years.[31] Our contemporary period thus approaches Italian history from new angles, and calls for interpretative categories that are broader than consolidated national narratives with their stereotypical dichotomies of modernity and backwardness, and that put greater emphasis on Italy's place in the world, its growth dynamics, and the periodisations of its entry into the streams of global modernity.[32] This book hopes to contribute to this historiographic renewal. Italy's relationship with the Suez Canal provides a view of Italy's modern experience in the frame of shifting definitions of Europe and the

[26] Isabella and Zanou (2016), McGuire (2020), Fogu (2020).
[27] Burton (2003), Borutta and Gekas (2012), Gekas (2012).
[28] Consolo Cassano (2000); Chambers (2007), Bouchard and Ferme (2013), Fogu (2020).
[29] Huber (2012).
[30] Clancy-Smith (2010), Choate (2008), Gabaccia (2003).
[31] On the question of Italy's 'decline', Ciocca (2020), Amatori (2016), Toniolo (2014).
[32] For a recent discussion of these issues, see Benigno Mineo (2020).

Euro-Mediterranean area, while crossing the chronological boundaries between pre-unification Italy and the united nation.

Italian State-Building and the Canal Moment

In November 1846, when the Saint-Simonian *polytechnicien* Barthélemy Prosper Enfantin and the Italian-speaking Austrian engineer Luigi Negrelli founded the *Société d'Études du Canal de Suez* in Paris, Europe was on the brink of the revolutionary explosion that also marked the beginnings of Italy's Risorgimento. As several chapters in Parts I and II of this book note, 1848 was a turning-point in the history of Europe, Italy and the canal. Though short-lived, the *Société* was an important step in forming a European technical-scientific identity. As Negrelli wrote, 'we will sacrifice our services on the altar of Europe and Europe will be grateful to us only if the canal is built' (quoted in Leonardi and Riegler, Chap. 7). The events of 1848 helped frame the whole Suez affair in a new economic-diplomatic context, which included reflections on the nature of change in European societies and on the role of economic growth in shaping both the social question and international relations. The early activity of the *Société* intersected with the initiatives of French diplomat and entrepreneur Ferdinand de Lesseps, who in 1854—while the Crimean War raged—procured a firman of concession from the Khedive Muhammad Said. The firman called for the formation of a *Compagnie universelle du canal maritime de Suez* tasked with 'cutting through the Isthmus of Suez, the construction of a passage suitable for extensive navigation, the foundation or appropriation of two sufficient entrances, one from the Mediterranean and the other from the Red Sea, and the establishment of one or two ports'.[33]

Consequently, an *International Commission for the piercing of the isthmus of Suez* consisting of major experts in naval and civil engineering and large-scale public works from France, the United Kingdom, Prussia, the Habsburg Empire, the Netherlands, Spain, and the kingdom of Piedmont-Sardinia was set up in October 1855. With the creation of the Commission—called on to express 'la décision scientifique de l'Europe'—the Suez affair logged a political breakthrough towards high-level diplomacy and establishing a European identity based on technical capacity, while the Khedive

[33] The English text of early original documents (the firman and subsequent conventions) can be found in Harlow and Carter (2003).

took advantage of Europe's interest in the canal to advance his goals of independence from the Sublime Porte.

The Commission's work coincided with the opening in February 1856 of the Congress of Paris, held to re-establish order in Europe and settle the 'Eastern question' in the aftermath of the Crimean War, while also addressing the issue of major growth infrastructures in an industrialising Europe where the revolutions of 1848 re-echoed still.

The construction of a European technical-scientific identity—at Suez, Cesare Correnti wrote, 'the old world is re-made and re-concentrated'[34]—and the canal's early diplomacy paralleled the new diplomatic strategy forged by Cavour, present in Paris as the prime minister of a small but important European state that had taken the lead in the gigantic enterprise that was Italian unification. The Kingdom of Sardinia's representative in the *International Commission* was Pietro Paleocapa, Minister of Public Works, who in 1857 also began work on the Fréjus tunnel, which was to connect Piedmont to France. Barbara Curli (Chap. 2) reconstructs this passage, drawing on the Piedmontese archives and the correspondence between Cavour and Paleocapa, and showing how Piedmont's—and Italy's—inclusion in the concert of great European powers and in the infrastructures of global modernity laid the groundwork for Cavourian policy in the 'decade of preparation' for unification.

The Suez debate that began at that time in all Italian States involved many different actors on the political, diplomatic, economic, and cultural scene. As Andrea Giuntini notes (Chap. 3), it was primarily a discourse about the infrastructures needed for Italy's modern growth, which helped define the meaning of 'Italy' and its position in the Mediterranean well before unification. Indeed, as Giovanni Cristina (Chap. 4) writes in relation to Italian port cities: 'The time of geo-economic strategies and technological progress anticipated that of political change.' The Suez issue calls traditional views of the backwardness of pre-unification states versus the modernity of united Italy into question, forcing us, as recently suggested, to rethink them in the light of that 'set of powers, institutions and cultures that transcend the bounds of official policies'.[35] Faced with the prospect of the canal's construction, many Italian ports, far from being pre-modern, parochially self-interested monads, as often portrayed, were—as shown in Part I of this book, whose contributors draw exten-

[34] Correnti (1869).
[35] Benigno Mineo (2020, 29–30).

sively from mostly unpublished local archival sources—places where leading merchants and shipowners hammered out strategies for joining the new networks of global trade and navigation. Giovanni Cristina emphasises (Chap. 4) how different strategies were implemented in southern ports such as Messina, Catania, or Siracusa, in mid-sized ports (Livorno, Ancona) and in Genoa. Here, for example, precisely because Genoa had sought to extend its reach across the Atlantic and Pacific since the early 1800s, there was considerable scepticism about Suez's real potential for Italian trade. In Trieste, by contrast, as Giovanni Modaffari shows (Chap. 9), the canal was optimistically expected—especially by Metternich—to relaunch Austria's position in the Mediterranean, thanks to a dynamic local business and maritime community. This included Baron Pasquale Revoltella, entrepreneur and merchant, who was appointed vice-president of the Suez Canal Company and in 1862 led a mission to Egypt on behalf of Trieste's City Council and Austrian Lloyd, in order to see how works on the canal were progressing, leaving a journal of his mission. Revoltella's former home in Trieste, now a museum, displays the allegorical marble group by Pietro Magni, *Il taglio dell'istmo di Suez* (1863), depicting Europe uniting the Mediterranean and the Red Sea, while the geniuses of Commerce and Shipping behind them give their blessing. 'When the canal was at last opened—writes Pedrag Matvejević—the internal sea became a strait, a passageway between two oceans. The human hand had altered the map of the Mediterranean and our conception of it.'[36]

Clearly, the opening of the Suez Canal sped up processes that were already under way. Contemporary sources emphasise that it was not the opening of the canal that revolutionised transportation and spurred the shift from sail to steam, but the opposite: booming world trade made the canal necessary. In particular, increased trade from the Indian Ocean was clogging the port of Suez, where goods for the Mediterranean had to be offloaded onto caravans as far as the first rail lines laid by the British.[37] Had the isthmus not been cut, Cesare Correnti observed, the 'irresistible infiltrations of trade would have opened other routes'. Now, however, Italy needed 'merchants, adventurers, scholars' and daring leaders to not risk being a 'stumbling block tossed in mid-Mediterranean'.[38] Indeed, as Andrea Giuntini shows (Chap. 3), the debate over Suez was a catalyst for

[36] Matvejević (1999, 32).
[37] Garelli (1864).
[38] Correnti (1869).

a more general political and economic discourse about the infrastructures of the newly unified state (ports, railways, tunnels and telegraph networks) and their connections to global markets.

The 'isthmian literature' that sprang up in those years and is discussed in Parts I and II of this book is thus—despite its rhetoric and even because of its rhetoric—an invaluable source for historians delving into the cultures that contributed to redefining Italy's place in the new Euro-Mediterranean modernity. Filippo Gattai Tacchi (Chap. 8) looks at a classic of this literature, the 'popular operetta' *L'Italia e il canale di Suez* by Giuseppe Sapeto, where Catholic missionary spirit, colonialist drive and nationalistic discourse were combined. Sapeto provides an intriguing mixture of orientalist imagery, narratives of the revival of the Roman Empire and a new age for the Mare Nostrum and the Maritime Republics in Italy's hoped-for political and commercial role in the Mediterranean, and the country's potential as a great power on the international and colonial stage. The Suez issue raised political questions about the prospects of the new Kingdom, and brought to the fore areas of civil society and the business world that had long remained in the background, including new players such as the Italian Geographic Society.[39] According to Carlo Bovolo (Chap. 10), it gave Catholic circles the opportunity to take part in a broader public debate on the nature of modern growth and the hopes that the Ottoman Empire's decline held out for gaining converts in majority-Muslim areas. Similarly, Paola Schellenbaum (Chap. 17) discusses the interest the canal's opening sparked in Protestant circles, whose missionary work in the Mediterranean had begun in Malta in the early nineteenth century and was closely tied to the slavery abolition movement, especially by Presbyterians and the Bible societies. Operations intensified in Egypt after the canal opened, and Waldensians from Piedmont were present in Egypt both as missionaries and as merchants and entrepreneurs.

As Fabio De Ninno notes (Chap. 14), Italy had planned to be represented at the opening ceremony on a level reflecting the new Kingdom's interest in the canal. Prince Amedeo of Savoia, rear admiral of the Italian Navy, was scheduled to attend with six battleships, but had to hasten back to Italy because of the king's sudden illness. Nevertheless, many prominent figures from Italy's business circles were present, along with Chamber of Commerce representatives and a number of merchant ships. An opera was commissioned from Giuseppe Verdi to commemorate the event:

[39] Natili (2009, 9).

Aida, to a libretto by Antonio Ghislanzoni following a scenario by the French Egyptologist Auguste Mariette, premiered under the baton of Giovanni Bottesini in December 1871 at the new Cairo Opera House designed by the Italian architect Pietro Avoscani.[40]

Italy's hopes as a great maritime nation soon clashed with a more complicated and at times frustrating reality, given the country's technological and organisational lag, slowness in moving to steam, and difficulties in making itself felt in ocean-going trade.[41] But rather than clinging to the well-worn paradigm of Italian backwardness and the late-comer's expectations and disappointments, the book frames these issues in the broader dynamics of nineteenth-century globalisation and the resulting marginalisation of the Eastern Mediterranean. Although the canal contributed to reducing travel times and transportation costs, it did not revitalise the economies of Mediterranean ports and its overall effects on the Middle East were negative.[42] As Kevin Tang's findings on the Indian trade demonstrate (Chap. 6), while the Mediterranean was previously central to trade between Europe and Asia, the Suez Canal's opening made the region more isolated and less integrated, and although it rerouted international trade and traffic through the Mediterranean, ports across the region were not the main beneficiaries of this dynamic. Trade to Mediterranean ports did not grow by as much as trade to northern European ports, which witnessed the greatest increase in trading opportunities with India. The relative marginalisation of Italian ports should be assessed in this more global framework. However, as Leonardo Scavino shows (Chap. 5), different actors reacted to this marginalisation in different ways. The alternatives to Suez continued to be used by actors on the Italian trading scene, many of whom still preferred the Cape route: this was the case of sailing vessels carrying heavy low-value cargo such as coal, demand for which rose precisely because of the increase in steamship navigation worldwide.[43] Indeed, the Cape's importance even after the canal opened enables us to take a broader global view of these issues.

The Suez passage thus continued to be marginal to Italian foreign trade for many years, not so much because of Italy's backwardness, but because

[40] Correspondence between Verdi, Mariette, and others on Aida is published in Harlow and Carter (2003, 557 *ff*).

[41] Monina (2008), Natili (2009), De Courten (1989), Spadoni (1970, 1973).

[42] Pamuk and Williamson (2000), Issawi (1982).

[43] Spadoni (1973), Doria (2001).

global growth sidelined the Mediterranean. As Fabio De Ninno (Chap. 14) points out, from the turn of the century to the First World War the bulk of Italian maritime trade was with Europe and North America via Gibraltar, while trade with India and the Far East, via Suez, accounted only for 9.1 percent of imports and 4.2 percent of exports. A global view thus provides a different problematisation of the relationship between centres and peripheries, a better grasp of where local realities fit into national growth, and in general enables us to place the Italian experience in the Mediterranean and world context.

New Colonial Spaces

Suez was one of the main physical and symbolic passages of the systemic shift which in the second half of the nineteenth century was taking place in the structure of world power, in the logic of empire, and in modern state-building. In the popular imagination of the time, Suez ends up representing a kind of boundary between Europe and non-Europe, between obscurantism and modernity. And it was thus, as the triumph of light over darkness, that the canal's opening was portrayed in the *Ballo Excelsior*, the allegorical 'gran ballo' choreographed by Luigi Manzotti to music by Romualdo Marenco, performed first at Milan's La Scala in January 1881, and then at the 1889 Paris World's Fair. The ballet celebrated the successes of modern science in a series of scenes—from Alessandro Volta to the Fréjus tunnel, and thence to the opening of the Suez Canal. Here, Italy was one of the modern nations that surrounded 'exultant Civilisation' at centre stage and extolled the 'triumph of civilized mankind', as the *Corriere della sera* commented. This picture of Suez was part of a more general ideological-discursive construction embedded, in Italy as elsewhere in Europe, in the rhetoric of nineteenth-century scientific internationalism, liberal free-trade mysticism and racial Darwinism, which provided the ideology for that systemic shift and for a new phase of colonial expansion.

The canal opened—and indeed invented—new colonial spaces in the Red Sea and the Horn of Africa for a newcomer to the imperial race like Italy, that were at the same time national spaces in the making (Egypt), spaces of global imperial strategies (Britain, the Ottoman Empire), and spaces where pre-existing local actors redefined their identity. The Genoese shipping company Rubattino was the first to pursue the geopolitical openings offered by the canal, to which Giuseppe Sapeto had drawn the Italian government's attention as early as the 1860s. In 1882 Rubattino bought the bay of Assab, which was then acquired by the Italian government and became the country's first colonial possession.[44] The chapters in Parts III and IV, mainly based on Italian diplomatic archives, revisit these spaces and the dynamics engendered by the increased traffic through the canal and Italy's new economic, political and military presence. The first impact was on pre-colonial local power structures. As Luca Castiglioni shows (Chap. 11), different actors—for example, local sultans—used the new Italian presence to settle their own scores, while also playing off the Anglo-Italian rivalries in the area. Ottoman misgivings about the canal, as a recent study describes,[45] found a new channel when Italy established its Red Sea outpost, sparking a long period of inter-imperial rivalries between Italy and the Sublime Porte which gradually developed into a new modus vivendi. These topics, long ignored by historiography, are analysed here on the basis of Ottoman diplomatic sources by Giorgio Ennas (Chap. 15), who also plumbs the Ottoman position on the controversial question of the canal's neutralisation and includes other hitherto neglected actors such as the Yemenite authorities in the picture.

Spaces thus opened were also spaces of economic and financial expansion. The 1866 financial crisis and the credit-crunch of the 1870s led to sovereign defaults in Tunisia, the Ottoman Empire and Egypt, where the financial crisis of 1875 stemmed specifically from the construction of the canal. Local financial troubles were leveraged by the creditor countries, above all Great Britain and France, to impose indirect control in an area vital to international trade, especially after the canal opened. As Giampaolo Conte argues (Chap. 12), this situation paved the way for Italy to gain seats on the boards of ad hoc financial commissions dealing with sovereign debts, which was also a way to increase Italy's political influence and promote its economic interests in the Mediterranean. This strategy helped strengthen ties with Italian-Levantine communities in the main Mediterranean port cities. For example, much Egyptian foreign and domestic debt was held by small and mid-sized bondholders from the Italian community in Egypt.

As these new spaces opened, different actors came to the fore: both Luca Castiglioni (Chap. 11) and Andreas Guidi (Chap. 13) emphasise Egypt's persistent role as a geopolitical presence. The construction of the

[44] Labanca (2002).
[45] Bilici (2019).

canal was a fundamental passage in the process of Egypt's state- and national identity building. It is in these new spaces—physical and political—that the quintessentially modern question of national sovereignty arises in connection with smuggling during the Italo-Ottoman war of 1911–1912, which, as Guidi notes (Chap. 13) relying on both Italian and Ottoman sources, shifts the rivalry between Italy and the Ottoman Empire to the Red Sea, historically rife with contraband, and reveals the 'densification' of Mediterranean connections gravitating around Egypt and its fragile and ambivalent territorial status. According to Giorgio Ennas (Chap. 15), with the opening of Suez, the Eastern Mediterranean as a geopolitical area was gradually enlarged to include the Red Sea, and was peopled by a host of new actors moving in scenes of shifting sovereignties and extraterritorialities. Moreover, the Italian settlements and the increased European presence in general was a threat to the area's slave trade, long the most profitable activity in the region and the basis for local leaders' wealth.

Against this backdrop, which raises the 'humanitarian' dimension of this new stage of European imperialism,[46] the United Kingdom as the dominant power in the Mediterranean continued to be ambiguous towards Italy's presence in the area, seeing it at times as an opportunity to fill the power vacuum left by waning Ottoman rule, at others as a military threat on the route to India. As Fabio De Ninno shows (Chap. 14) in a study based on Italian military and naval archives, unified Italy arrived on the scene as a new major sea power that affected the military balance of power in the Mediterranean. While Italian trade through the canal, especially in coal and foodstuffs imported from India and Australia, rose temporarily during the First World War when submarine warfare curtailed Atlantic shipping, it dropped again in the interwar period. By contrast, Suez maintained and indeed increased its importance to Italy's naval strategies in the 1920s and 1930s. During the Fascist period, the canal became part of the regime's foreign policy objectives, largely thanks to the geopolitical thinking and war planning of the Italian Navy, which coveted control over Suez as part of Fascist Italy's ambition to escape its 'Mediterranean prison'. The canal then took on an essential role during the Ethiopian war. The question—or dilemma—of whether to blockade the canal to prevent Italian aggression against Ethiopia is discussed by Steven Morewood (Chap. 16), who introduces us to a delicate moment in European history. We cannot

[46] Darwin (2009).

know whether a British blockade (with or without the League of Nations' blessing) would have changed this history's course. Perhaps it would have been enough if the Royal Navy had sent a cruiser to guard Port Said: would Italy have risked war with Britain? Would a blockade have delivered Mussolini into Hitler's hands and sent Europe skidding towards war? Could Italy itself have blockaded the canal? All questions that entered very much into Britain's calculations, but, as Morewood concludes, the Italian troops' entry into Addis Ababa two months after the German occupation of the Rhineland put an end to the League of Nations' sanctions and to any question of blockading the canal, which from February 1935 through July 1936 carried almost 600,000 men, 634,900 tons in supplies, 10,084 vehicles, and 40,859 animals to the Italian war in Ethiopia. Nevertheless, not even appeasing Mussolini about the canal proved useful in preventing his alliance with Hitler and warding off world war.

A Transnational Social History of Suez

The Suez Canal was not only a passageway to the empires—a gateway to new trading spaces, the physical and symbolic locus of military strategies and influence between powers great and small. It has also been a place of settlement, of community-building, the outcome of flows of North-South migration, characterised by continuous negotiations over identities and by fluid notions of citizenship and of 'Europeanness' as politics and power structures changed over time (e.g. during the long-drawn-out demise of Ottoman rule and Egypt's development as a modern State after 1922). The economic and social development of the canal zone was thus part of that 'intense circulation of money, goods, steamships and human beings' that made the Mediterranean into a 'colonial' sea in the late nineteenth and early twentieth centuries, and provided for practices of racial discrimination and segregation, as well as of cooperation, coexistence and hybridisation that merged 'European' and 'non-European' cultures.[47]

As Part V shows, however, hierarchies in the canal zone were also formed within each European community. Suez had been one of the world's largest construction sites, and the largest in the Mediterranean.[48] For construction work, the Company had relied extensively on a local labour force, mostly *en corvée*. Subsequently, though most of the unskilled

[47] Borutta and Gekas (2012), Gekas (2012).
[48] Montel (1998), Karabell (2003).

labour force (largely daily or seasonal) were still Egyptians, the development of the canal and the Company's hiring opportunities attracted a more diversified personnel (sailors, navvies, skilled workmen, engineers and pilots) from Southern and Mediterranean Europe. As a result, a new multinational, multilingual, and multireligious community progressively settled along the isthmus. At the Company, which became the area's largest employer, labour was organised in a rigid distinction between *européens* and *indigènes*—to use the Company's terminology—in a racial and segregated model that facilitated social control and contributed to the construction of a 'Europeanness' that provided a community of non-Egyptian workers with a common identity. It was only in the 1930s, when the Egyptian government set quotas for Egyptian personnel and pursued other measures of 'economic nationalism', that this model of work organisation gradually changed.[49]

The Italians were the second largest group in the Company's workforce (around 15 percent in the 1920s), after the Greeks (around 30 percent in the 1920s). Though there are as yet no studies of the history of Italian manpower at the Company, Angelos Dalachanis (Chap. 19) gives us an overview in a comparative study of the Italian and Greek workers based on the Company's archives that sheds light on the divergences and rivalries among the '*européens*', especially when the Company's Italian employees—caught up in the politics and human tragedies of the Second World War—were fired *en masse* in the aftermath of 10 June 1940. Italian workers did not regain their jobs after the war, notwithstanding pressure from the diplomatic authorities as well as individual claimants. This was not so much because of the Company's policy of hiring more Egyptians, but because of the competition from other national groups, the Greeks in particular. In general, competition between the foreign communities in the labour market was particularly harsh in the immediate post-war years, a period of increasing unemployment when all communities sought to reinforce their position in Egypt.

In addition, a series of activities arose around the canal for provisioning and coaling the ships in transit, and small businesses, restaurants and bank branches were opened as the area urbanised, bringing social and economic changes. Likewise, roads, railways, schools, and hospitals were constructed, consulates were established to handle the influx of immigrants, an entire 'canal service industry' came into being, and whole new cities such as

[49] On work organization at the Suez Canal Company, see Curli (2014).

Ismailia were built. Drawing from both Italian and Egyptian archival sources, Olga Verlato (Chap. 18) enters into the social dimension of this Suez community with an analysis of elementary education. Though there were more educational opportunities in the canal region than in other parts of the country, factors such as social extraction, legal standing, religion, gender, and geographic location made students' access to elementary education highly uneven. Focussing on the schools administered and financed by the Egyptian, Italian, and French states, and private schools such as al-Tawil's, which enrolled students from different communal, linguistic, religious, and socioeconomic backgrounds, Verlato argues that their educational provision was largely shaped by the local environment, and that coexistence and competition between different local and foreign schools (and between foreign schools themselves, in particular between French and Italian schools) and the resulting uneven educational environment oblige us to reconsider the alleged insularity of different communities and to reformulate the notion of cosmopolitanism. Paola Schellenbaum (Chap. 17) also looks at education, though from a different perspective, namely, in relation to how Waldensian migrants and Protestant missionaries to the Southern Mediterranean designed a 'new political and religious geography'. After the canal opened, Egypt became strategic for the international missionary movement, which in turn influenced migratory flows and education patterns, in particular through the advocacy of anti-slavery positions, women's rights and education for women. The 'civilising' efforts of Protestant and Catholic missionary activities and the development of transnational faith-based solidarity cast doubt on the notion that national or religious affiliations—that is, Europeans versus Muslims—are homogeneous categories.

Post-war Suez

The story of the Italian community that settled along the isthmus after the canal opened is just one facet of the history of the larger Italian community in Egypt, which numbered around 47,000 people, or a quarter of all foreigners, in 1937, and whose wartime and post-war destinies intersect with the end of fascism, the birth of the new republican Italy, the beginnings of the Cold War and Egypt's transformation, and, lastly, the Suez crisis in 1956. Joseph Viscomi (Chap. 20) offers a re-reading of the Italians' departure in 1956, placing it in the longer chronology and genealogy of the 'uncertainty' that was such a dramatic part of the condition of the Italians in Egypt during the war and thereafter. When Italy entered the war, diplomatic ties between Italy and Egypt were broken off, political figures were repatriated, Italian institutions closed and Italian bank accounts were frozen. Over 5000 Italian civilians were arrested and imprisoned, and most Italians were detached from social life and from their families until 1945. After the conflict, the Italian community found itself living 'on the boundaries of destitution', emarginated in Egypt and substantially ignored in Italy, a harsh letdown after the 'inflated sense of their centrality' cultivated by the Fascist regime. In the canal zone, following the closure of the internment camps, over 90 percent of Italians were unemployed, with little chance of being hired, as jobs at the Company had been filled by Egyptians and workers from other ethnic groups. For many, leaving Egypt began to suggest itself as a way out: as early as 1947, the Italian community had decreased in size more than any other.

These human dramas coincide with the first steps in building the Italian Republic and, later, the Egyptian Republic, against the backdrop of the budding Cold War and the wider transformations in Mediterranean geopolitics. Although the Italian immigrants' departure accelerated in the late autumn of 1956 and in 1957, in Viscomi's account the Suez crisis is not a primary event in the history of Italian departures from Egypt, but the outcome of extended processes, whereby over 40,000 Italians would leave Egypt by 1960. However, because Italy and Egypt were not in a state of political conflict, Italians arriving in Italy did not qualify as refugees, but as 'poverty-stricken' (*indigenti*) to be housed temporarily in displacement camps. These events are not simply the story of the Italian community in Egypt, but are part of the history of the Italian Republic and of its arduous search for an identity both as a new state facing the challenges of citizenship in a broad sense, and in terms of its place on the post-war Mediterranean scene. And yet, the 1956 watershed also shows us the intellectual ferment and strong personalities that were involved in seeking to reformulate that identity in this period. One of the prominent figures in this effort was Enrico Mattei, the dynamic chairman of ENI, whose aggressive policies helped break the Seven Sisters' monopoly in the oil-producing countries of the Middle East. A day before the Egyptian government announced the nationalisation of the Company, Mattei had inaugurated the Suez-Cairo pipeline with Nasser.[50] Mattei was involved in the development of Italy's new Mediterranean foreign policy—neo-Atlanticism—and in a far-

[50] On Mattei's policies in Egypt, see Tonini (2003).

reaching network of other actors, including the Italian Catholic circles analysed by Massimo De Giuseppe (Chap. 21), that were particularly engaged with the emerging issues of 'Third-Worldism', decolonisation and non-alignment.

De Giuseppe scrutinises the role of the charismatic Catholic mayor of Florence, Giorgio La Pira, using, among other sources, the correspondence on the Suez crisis preserved in the La Pira archive. Organiser of the Conference of the mayors of the world's capital cities held in Florence in October 1955, La Pira pursued a sort of parallel diplomacy that emphasised the role of non-state actors in promoting international dialogue, and an original pacifism, connected to ecumenism, which hoped to eliminate the world's bipolar divisions. He was convinced that the Middle East had become a seismic fault at the intersection of East-West and North-South interests, and that without peace in the region it would be increasingly difficult to promote peace on a global level. In August 1956, in the aftermath of nationalisation, he even wrote to Nasser offering Florence as a neutral venue for a conference of the Arab peoples, in order 'to rethink a new agenda and face the problems affecting peace, the dignity of the Arab world and all other parts of the world'.

While most Italian residents were leaving Egypt, in Italy the Suez crisis contributed to drawing the general public's attention to the issues of development, non-alignment, and the concept of the Third World. ENI, however, remained in Egypt, strengthening the two countries' joint energy exploration efforts, which have continued to our own day with the discovery of the Zohr gas field in 2015. Giuseppe Natale (Chap. 22) looks at ENI's corporate propaganda and communication strategy in the Mediterranean. This strategy included the production of documentaries and 'techno films', which during the 1950s increased in parallel with ENI's expanded search for oil resources. ENI's cinema production not only was part of an innovative strategy for communicating domestically and internationally, but also contributed to constructing a narrative of Italian modernity and openness towards the oil-producing countries, with ENI representing itself (and Italy) as their like-minded interlocutors on an equal footing. Bernardo Bertolucci's *La via del petrolio*, perhaps the best-known of these films, not least because of the fame its director later achieved, starts from the Iranian mountains and wends its way to the Persian Gulf, where crude oil would be loaded on a ENI tanker that would pass through the Red Sea and the Suez Canal to reach the port of Genoa, then a refinery in Germany through a pipeline from Genoa to Ingolstadt.

While in Egypt, instead of navigating the Suez Canal on the tanker, Bertolucci disembarked to film the isthmus and its surroundings. As we indicated in the opening pages of this Introduction, the sequences shot in Suez ended up forming a separate 12-minute short, *The Canal*. Perceptively, Natale notes that *The Oil Route* and *The Canal* are the least techno of all ENI's techno films. With them, Bertolucci creates a veritable poetic document that is still one of the most sensitive testimonies to the long relationship between Italy and the Suez Canal, which we hope can inspire further research on the canal, Italy and the Mediterranean.

References

Abulafia, David. 2011. *The Mediterranean in History*. Getty.

Algardi, Zara Olivia. 1989. *Luigi Negrelli, l'Europa, il canale di Suez*. Firenze: Le Monnier.

Amatori, Franco, ed. 2016. *L'approdo mancato. Economia, politica e società in Italia dopo il miracolo economico*. Annali della Fondazione Feltrinelli, Milano, 51, 2016–2017.

Badenoch, Alexander, and Andreas Fickers. 2010. *Materializing Europe: Transnational Infrastructures and the Project of Europe*. New York: Palgrave Macmillan.

Ballantyne, Tony, and Antoinette Burton. 2013. Empires and the Reach of the Global. In *A World Connecting, 1870–1945*, ed. Emily Rosenberg. Cambridge, MA: Harvard University Press.

Benigno, Francesco. 2009. *Il Mediterraneo*. XXI secolo, Enciclopedia Treccani Online. http://www.treccani.it/enciclopedia/il-mediterraneo_(XXI-Secolo).

Benigno, Francesco, and E. Igor Mineo. 2020. *Introduzione. Discutere il canone nazionale*. In *L'Italia come storia. Primato, decadenza, eccezione*, ed. Francesco Benigno and E. Igor Mineo. Roma: Viella.

Bilici, Faruk. 2019. *Le canal de Suez et l'Empire ottoman*. Paris: Editions CNRS.

Bonin, Hubert. 2010. *History of the Suez Canal Company, 1858–2008: Between Controversy and Utility*. Geneva: Droz.

Bono, Salvatore. 2006. Il canale di Suez e l'Italia. *Mediterranea* 3: 411–422.

Borutta, Manuel, and Sakis Gekas. 2012. A Colonial Sea: The Mediterranean, 1798–1956. *European Review of History-Revue européenne d'histoire* 19 (1): 1–13.

Bouchard, Norma, and Valerio Ferme. 2013. *Italy and the Mediterranean: Words, Sounds, and Images of the Post-Cold War Era*. Palgrave Macmillan.

Burbank, Jane, and Frederick Cooper. 2010. *Empires in World History, Power and the Politics of Difference*. Princeton: Princeton University Press.

Burke, Edmund, III. 2012. Toward a Comparative History of the Modern Mediterranean, 1750–1919. *Journal of World History* 23: 4.

Burton, Antoinette, ed. 2003. *After the Imperial Turn. Thinking with and Through the Nation*. Durham: Duke University Press.

Buzan, Barry, and George Lawson. 2015. *The Global Transformation: History, Modernity and the Making of International Relations*. Cambridge University Press.

Caizzi, Bruno. 1985. *Suez e San Gottardo*. Milano: Cisalpino.

Calchi Novati, Gian Paolo. 1998. *Il canale della discordia: Suez e la politica estera italiana*. Urbino: Quattro Venti.

Capriotti, Vittozzi Giuseppina, Andrea Angelini, and Annalinda Iacoviello 2019. Dall'Egitto sulla via dell'Oriente: le campagne di scavo a Tell el-Maskhuta lungo lo wadi Tumilat. In *Atti della Pontificia Accademia romana di Archeologia, Rendiconti*, Vol. XCI. Roma: Tipografia Vaticana.

Carreras, Albert, Andrea Giuntini, Michele Merger, eds. 1995. *Les réseaux européens transnationaux: 19–20. siècles: quels enjeux?* Nantes: Ouest.

Chambers, Iain. 2007. *Mediterranean Crossings: The Politics of an Interrupted Modernity*. New York: Berghan Books.

Charles-Roux, Jules. 1901. *L'isthme et le canal de Suez: Historique—Etat actuel*. Paris: Hachette.

Choate, Mark I. 2008. *Emigrant Nation: The Making of Italy Abroad*. Cambridge and London: Harvard University Press.

Clancy-Smith, Julia. 2010. *Mediterraneans. North Africa and Europe in an Age of Migration, c. 1800–1900*. Berkeley: University of California Press.

Ciocca, Pierluigi. 2020. *Ricchi per sempre? Una storia economica d'Italia (1796–2020)*. Torino: Bollati Boringhieri.

Consolo, Vincenzo, and Franco Cassano. 2000. *Rappresentare il Mediterraneo: Lo sguardo italiano*. Mesogea.

Correnti, Cesare. 1869. *L'istmo di Suez e il commercio orientale*, Bollettino della Società Geografica Italiana, 3, Re-published in: *Scritti scelti di Cesare Correnti in parte inediti o rari*, ed. Tullo Massarani, Vol. IV, Roma, Forzani (1894), 477–490.

Cresti, Federico. 2008. Comunità proletarie italiane nell'Africa mediterranea tra XIX secolo e periodo fascista. *Mediterranea* 5: 189–214.

Curcio, Carlo. 1941. *Ideali mediterranei nel Risorgimento*. Roma: A. Urbinati.

Curli, Barbara. 2014. *Dames employées* at the Suez Canal Company: The "Egyptianization" of Female Office Workers, 1941-56. *International Journal of Middle East Studies* 46 (3): 553–576.

———. 2015. Il Piemonte e il canale di Suez, 1855–1856. *Studi Piemontesi* XLIV (2): 325–338.

Darwin, John. 2009. *The Empire Project: The Rise and Fall of the British World-System, 1830–1970*. Cambridge, UK and New York: Cambridge University Press.

De Courten, Ludovica. 1989. *La marina mercantile italiana nella politica di espansione (1860–1914): industria, finanza e trasporti marittimi*. Roma: Bulzoni.

Doria, Marco. 2001. La marina mercantile a vela in Liguria dalla metà dell'Ottocento alla prima guerra mondiale. In *A vela e a vapore*, ed. Paolo Frascani, 83–107. Roma: Donzelli.

Farnie, Douglas A. 1969. *East and West of Suez. The Suez Canal in History, 1854–1956*, Oxford University Press.

Fanfani, Tommaso, ed. 1993. *La penisola italiana e il mare. Costruzioni navali, trasporti e commerci tra XV e XX secolo*. Napoli: ESI.

Findlay, Ronald, and Kevin H. O'Rourke. 2007. *Power and Plenty: Trade, War, and the World Economy in the Second Millennium*. Princeton University Press.

Fischer-Tiné, Harald, and Susanne Gehrmann, eds. 2008. *Empires and Boundaries: Rethinking Race, Class and Gender in Colonial Settings*. New York and London: Routledge.

Fogu, Claudio. 2020. *The Fishing Net and the Spider Web: Mediterranean Imaginaries and the Making of Italians*. Palgrave Macmillan.

Frascani, Paolo, ed. 2001. *A vela e a vapore. Economie, culture e istituzioni del mare nell'Italia dell'Ottocento*. Roma: Donzelli.

Gabaccia, Donna R. 2003. *Italy's Many Diasporas*. London: Routledge.

Garelli, Felice. 1864. *La sezione dell'istmo di Suez e la perforazione delle Alpi italiane*. Torino: Tipografia scolastica di Sebastiano Franco e figli.

Gekas, Sakis. 2012. Colonial Migrants and the Making of a British Mediterranean. *European Review of History-Revue européenne d'histoire* 19 (1): 75–92.

Giardina, Andrea, ed. 2018. *Storia mondiale dell'Italia*. Bari-Roma: Laterza.

Giuntini, Andrea. 2011. *Le meraviglie del mondo: Il sistema internazionale delle comunicazioni nell'Ottocento*. Prato: Istituto di studi storici postali.

Harlow, Barbara, and Mia Carter, eds. 2003. *Archives of Empire, Vol. 1: From the East India Company to the Suez Canal*. Durham and London: Duke University Press.

Harris, William V., ed. 2004. *Rethinking the Mediterranean*. Oxford: Oxford University Press.

Headrick, Daniel R. 2009. *Power over Peoples. Technology, Environments and Western Imperialism, 1400 to the Present*. Princeton University Press.

Holland, Robert. 2012. *Blue Water Empire. The British in the Mediterranean after 1800*. London: Allen Lane.

Horden, Peregrine. 2005. Mediterranean Excuses: Historical Writing on the Mediterranean since Braudel. *History and Anthropology* 16 (1): 25–30.

Horden, Peregrine, and Nicholas Purcell. 2000. *The Corrupting Sea: A Study of Mediterranean History*. Oxford and Malden, MA: Blackwell.

———. 2006. The Mediterranean and 'The New Thalassology'. *American Historical Review* 111 (3): 722–740.

———. 2019. *The Boundless Sea, Writing Mediterranean History*. Routledge.

Huber, Valeska. 2012. Connecting Colonial Seas: The 'International Colonisation' of Port Said and the Suez Canal during and after the First World War. *European Review of History-Revue européenne d'histoire* 19: 141–161.

———. 2013. *Channelling Mobilities: Migration and Globalisation in the Suez Canal Region and Beyond*. Cambridge University Press.

Karabell, Zachary. 2003. *Parting the Desert. The Creation of the Suez Canal*. New York: Vintage.

Kohlrausch, Martin, and Helmuth Trischler. 2014. *Building Europe on Expertise: Innovators, Organizers, Networkers*. London: Palgrave Macmillan.

Isabella, Maurizio, and Konstantina Zanou, eds. 2016. *Mediterranean Diasporas: Politics and Ideas in the Long 19th Century*. London: Bloomsbury.

Issawi, Charles. 1982. *An Economic History of the Middle East and North Africa*. London: Methuen.

Labanca, Nicola. 2002. *Oltremare. Storia dell'espansione coloniale italiana*. Bologna: il Mulino.

———. 2019. La storia contemporanea del Mediterraneo: Per una discussione. *Il mestiere di storico* XI (2): 5–50.

Leonardi, Andrea, ed. 1990. *Luigi Negrelli ingegnere e il canale di Suez*. Trento: Società di studi trentini di scienze storiche.

———. 2020. A Prominent Figure in the Creation of the Suez Canal: Luigi Negrelli (1799–1858). *The Journal of European Economic History* 49 (2): 11–20.

Matvejević, Predrag. 1999. *Mediterranean: A Cultural Landscape*. Berkeley: University of California Press.

McGuire, Valerie. 2020. *Italy's Sea. Empire and Nation in the Mediterranean, 1895–1945*. Liverpool University Press.

Middel, Matthias, and Katja Naumann. 2010. Global History and the Spatial Turn: From the Impact of Area Studies to the History of Critical Junctures of Globalization. *Journal of Global History* 5: 149–170.

Monina, Giancarlo. 2008. *La Grande Italia marittima. La propaganda navalista e la Lega navale italiana (1866–1918)*. Soveria Mannelli: Rubbettino.

Montel, Nathalie. 1998. *Le chantier du canal de Suez (1859–1869). Une histoire des pratiques techniques*. Paris: Presses de l'École nationale des Ponts et Chaussées.

Monti, Antonio. 1937a. *Storia del canale di Suez. Con un diario di Luigi Torelli ed altri documenti inediti*. Milano: Ispi.

———. 1937b. *Gli italiani e il canale di Suez*. Lettere inedite di P. Paleocapa, L. Torelli, E. Gioia/ a cura di. Roma: Vittoriano.

Natili, Daniele. 2009. *Un programma coloniale. La Società geografica Italiana e le origini dell'espansione in Etiopia (1867–1884)*. Roma: Gangemi.

Osterhammel, Jurgen. 2014. *The Transformation of the World: A Global History of the Nineteenth Century*. Princeton and Oxford: Princeton University Press.

Pamuk, Sevket, and Jeffrey G. Williamson, eds. 2000. *The Mediterranean Response to Globalization before 1950*. London and New York: Routledge.

Petricioli, Marta. 2007. *Oltre il mito: L'Egitto degli italiani, 1917–1947*. Milano: Bruno Mondadori.

Piquet, Caroline. 2008. *La Compagnie du canal de Suez: Une concession française en Egypte (1888–1956)*. Paris: Presses de l'Université Paris Sorbonne.

———. 2009. *Histoire du canal de Suez*. Paris: Perrin.

Sammarco, Angelo. 1940. *Suez. Storia e problemi. Secondo documenti inediti egiziani ed europei*. Milano: Garzanti.

Sant Cassia, Paul, and Isabel Schäfer. 2005. 'Mediterranean Conundrums': Pluridisciplinary Perspectives for Research in the Social Sciences. *History and Anthropology* 16 (1): 1–23.

Siegfried, André. 1945. *Suez Panama et les routes maritimes mondiales*. Paris: Armand Colin.

Spadoni, Ugo. 1970. Il canale di Suez e l'inizio della crisi della marina mercantile italiana. *Nuova Rivista Storica* 54 (V–VI): 651–702.

———. 1973. Linee di navigazione e costruzioni navali alla vigilia dell'Inchiesta parlamentare sulla Marina mercantile italiana (1881–1882). *Nuova rivista storica* 57 (III–IV): 313–372.

SRM. 2017. *Economic Relations between Italy and the Mediterranean Area*. 7th Annual Report, Europe and the Mediterranean Area between Crisis and Opportunity. The Role of Italy and Its Economy. Napoli: Giannini.

———. 2019. *6th Italian Maritime Economy Report—New Scenarios in the Mediterranean: Suez and China, the Strategies of Big Carriers, New Technologies and Energy Routes*. Napoli: Giannini.

Tonini, Alberto. 2003. *Il sogno proibito. Mattei, il petrolio arabo e le sette sorelle*. Firenze: Polistampa.

Toniolo, Gianni, ed. 2014. *L'Italia e l'economia mondiale. Dall'Unità a oggi*. Venezia: Marsilio.

Trinchese, Stefano, ed. 2005. *Mare nostrum. Percezione ottomana e mito mediterraneo in Italia all'alba del '900*. Milano: Guerini.

Valle, Marco. 2018. *Suez: Il canale, l'Egitto e l'Italia*. Roma: Historica.

Van der Vleuten, Erik, and Arne Kaijser, eds. 2006. *Networking Europe: Transnational Infrastructures and the Shaping of Europe, 1850–2000*. Sagamore Beach: Science History Publications.

Verga, Marcello. 2017. L'Italia e la 'sua' storia del Mediterraneo: cronache di storiografia italiana del secondo Novecento. *RiMe, Rivista dell'Istituto di Storia dell'Europa Mediterranea*, 18.

Wenzlhuemer, Roland. 2012. *Connecting the Nineteenth-Century World: The Telegraph and Globalization*. Cambridge University Press.

PART I

Modern Infrastructures for a Modern State: The Suez Canal and Italian Unification

CHAPTER 2

Camillo Cavour and Pietro Paleocapa: Italy's Path to Unification and the Technoscientific Diplomacy of the Suez Canal (1855–1857)

Barbara Curli

The *International Commission for the Piercing of the Isthmus of Suez*: A Technoscientific Concert of Europe[1]

With a dispatch dated 19 November 1855, Luigi Gobbi, the Piedmontese consul in Alexandria, informed the Foreign Minister of the Kingdom of Sardinia of the arrival, 'on the French steamer', of the 'European Scientific Commission for the definitive local inspection on opening the isthmus of

[1] I wish to thank Pierangelo Gentile for kindly introducing me to Turin's archives, the personnel of Turin's state archives and of Accademia delle Scienze, and Rosanna Roccia for assistance on Cavour's correspondence.

B. Curli (✉)
Dipartimento Culture, Politica e Società, Università di Torino, Turin, Italy
e-mail: barbara.curli@unito.it

B. Curli (ed.), *Italy and the Suez Canal, from the Mid-nineteenth Century to the Cold War*,
https://doi.org/10.1007/978-3-030-88255-6_2

Suez'.[2] The idea of creating an *International Commission for the piercing of the isthmus of Suez* dated to April 1855. The November 1854 firman of concession to the French diplomat and entrepreneur Ferdinand de Lesseps had called for the formation of a *Compagnie universelle du canal maritime de Suez*, which was to carry out the work.[3] The first surveys were conducted by two French engineers in charge of Egypt's public works, Adolphe Linant de Bellefonds and Eugène Mougel. On 4 February 1855, Gobbi told Cavour of the 'fledgling plan to open the maritime canal of Suez', of which he had been informed by de Lesseps himself.[4] When the French engineers had produced a voluminous technical report in March 1855, the Egyptian Khedive Muhammad Said had asked that a body of experts be created to express 'the scientific decision of Europe' on the ways to construct the canal and the ports.[5]

Consisting of the major experts in naval and civil engineering and large-scale public works of France, the United Kingdom, Prussia, the Habsburg Empire, the Netherlands, Spain, and the Kingdom of Piedmont-Sardinia,[6] the Commission met for the first time on 30 and 31 October 1855 in Paris. On that occasion, it decided that five of its members would travel to Egypt to study the question.[7] Consul Gobbi expressed his regret that the

[2] Luigi Gobbi to the Minister of Foreign Affairs Cibrario, 19 November 1855, AST, CN, Alexandria, Mazzo 7.

[3] On the history of the Suez Canal Company, see Piquet (2008), Piquet (2009), Bonin (2010).

[4] Luigi Gobbi to Cavour, 4 February 1855, AST, CN, Alexandria, Mazzo 7.

[5] For this first stage, see de Lesseps (1855), the first volume of a series of three containing documentation on the canal's 'rise and progress' between 1855 and 1856. The extensive bibliography on Lesseps varies in quality. As an example, see Association du Souvenir de Ferdinand de Lesseps et du canal de Suez (1994).

[6] The members of the Commission were: Frederik Willem Conrad, chief engineer of Holland's Water Staat; Captain Harris of the British Royal Navy; Captain Charles Jaurès, the hydrographic engineer Lieussou and Rear Admiral Charles Rigault de Genoully, all of the French Imperial Navy; Louis Auguste Renaud, inspector-general of Ponts et Chaussées; the Prussian Carl Lentze, Chief Engineer of the Works on the Vistula; the British engineers J.R. McClean, J.M. Rendel and Charles Manby; the Director-General of Public Work at Madrid Cipriano Segundo Montesino; the engineer Luigi (Alois) Negrelli, inspector-general of the railways of the Austrian Empire; and Pietro Paleocapa, Minister of Public Works of the Kingdom of Sardinia.

[7] The five were Conrad, McClean, Negrelli, Renaud, and Lieussou. The minutes of the meetings can be found in: *Extraits des Procès-Verbaux de la Commission internationale du Canal de Suez*, published in: *Percement de l'Isthme de Suez* (1856).

group of 'distinguished personages and engineers' who disembarked at Alexandria did not include Pietro Paleocapa, the Minister of Public Works of the Kingdom of Piedmont-Sardinia, whose failing eyesight had prevented him from serving as president of the Commission—an honour which thus went to Conrad of Holland—and from facing the rigours of the journey.

The role of Paleocapa and Cavour, and, more generally, the Kingdom of Piedmont-Sardinia's involvement in this stage of the Suez enterprise have been largely neglected and are absent or barely mentioned in Cavour's biographies.[8] And yet, they are a significant passage in the history of the Italian economy's internationalisation in the decade leading to unification.

'One of the most illustrious engineers of our day', as Cesare Vimercati called him,[9] Paleocapa was a leading figure in the European technocracy of major public works. Born to a family of Greek origin that had moved to Bergamo in Austrian Lombardy-Venetia, prior to 1848 he had supervised the drainage and reclamation of the Venice lagoon and the Danube delta, and the construction of the Milan-Venice railway line. After serving as Minister of Public Construction and of the Interior in Manin's Venetian Republic, he went into exile in Piedmont in 1849, continuing to be a fervent supporter of the Venetian cause. In November 1849 he was appointed Minister of Public Works in the D'Azeglio government, a position he was to hold in the successive Cavour governments until 1857. A senator of the Subalpine Parliament, he later became a senator of the Kingdom of Italy. In 1863, he became president of the Ferrovie dell'Alta Italia, and in 1865 contributed to drafting the new Kingdom of Italy's first public works legislation.[10]

Very close to Cavour, who held him in great esteem, Paleocapa was a great moderniser of Sabaudian infrastructures, which he managed with a select group of experts with extensive international connections. At the time he was appointed to the International Commission, Paleocapa was

[8] Nada (1990); Romeo (1984, III, 848) just mentions the issue of Suez and underlines its historiographical potential. For a recent approach to the topic, Curli (2015).

[9] Vimercati (1864, 130). Historian, traveller, and patriot, Vimercati had journeyed extensively in the Middle East and the Mediterranean, the 'natural seat' of civilisation (Vimercati 1849), and visited the isthmus during the canal's construction.

[10] On Paleocapa, see Gottardi (2014), Istituto Veneto di Scienze, Lettere ed Arti (1990). A number of studies were published in the 1930s and shortly thereafter, including Sammarco (1940), Di Prima (1940), Manfredi (1932).

engaged in the Moncenisio railway project, which involved tunnelling through the Alps at Fréjus to connect Piedmont and France. Work on the tunnel began in August 1857, a few months before the first excavations at Suez. Paleocapa's participation in the Commission demonstrated the importance that the Kingdom of Sardinia attached to the project and to playing a part in European infrastructure diplomacy. Trade and communication networks (railways, tunnels, canals, telegraphs) called for new forms of technical and legal governance and favoured the rise of transnational technocracies that at the same time embodied national modernisation projects. Small states and their expert élites played an important role in driving the push towards new forms of international regulation.[11] The role of Paleocapa and of Piedmont can be set against this background.

The Commission's creation was a decisive step forward in the long-discussed European effort to construct a canal linking the Mediterranean to the Red Sea, while the outbreak of the Crimean War further contributed to redefining the major powers' grand political and military strategies, in particular to Great Britain's evolving vision of the Mediterranean as a colonial sea, opening up a new phase in Anglo-French geopolitical rivalry and cooperation in the establishment of that 'legalised hegemony' of the great powers which entailed a new 'managerial responsibility' for the international order.[12]

Between 1855 and 1856, then, Suez passed from the level of theoretical discussion to that of high-level diplomacy. The International Commission—a sort of technoscientific concert of Europe—helped forge a European identity based on technical expertise and the capacity to present a vision of modernisation: 'this Europe', as Cesare Correnti, founder of the *Società geografica italiana*, commented, 'seemingly so divided and subdivided, but living essentially a single life'.[13] 'Cutting the Suez isthmus—Cesare Vimercati wrote—is one of those battles in which all of Europe must contend.'[14] Each European power came to the Commission with its own agenda, so as not to be left out and to keep a watchful eye on the others, and at the same time be part of a 'European' project contributing to varied national priorities for modernisation. In turn, the Khedive took advantage of Europe's interest in the canal to advance his goals of

[11] Van der Vleuten and Kaijser (2006, 10–11).
[12] Holland (2012), Simpson (2004, 173).
[13] Correnti (1869).
[14] Vimercati (1864, 268).

independence from the Sublime Porte; for the latter, Europe's political involvement via a technical group served chiefly as a guarantee that the canal would be neutral.[15]

When the Commission's work began, the major technical issue was whether it would be feasible—and preferable—to cut straight across the isthmus between the Mediterranean and the Red Sea (as Negrelli maintained, with Paleocapa's support), or to route the canal from Alexandria to Suez in two sections via the Nile, which would have required a system of locks. On 9 November 1855, just before the Commission left for Egypt, Paleocapa informed Cavour of the outcome of the early meetings in Paris: 'with the Suez isthmus Commission, things have gone quite expeditiously. In general, all members seem persuaded that the direct sea-to-sea cut is preferable.' However, for 'the general opinion of the sure and fruitful success of this great enterprise to be truly favourable, tests that are more certain and more concrete than those carried out so far are needed'.[16]

Paleocapa was referring to the studies by Paulin Talabot[17] and the diplomat Georges Napoléon Baude, formerly attaché at the French embassy in Rome (and who would become ambassador to the Holy See in 1876), who in 1854 had been sent on a mission to Vienna by the Foreign Minister Drouyn de Lhuys to prepare a study on the prospects for shipping and trade in Europe and the Mediterranean. Thenceforth an enthusiastic supporter of opening the Suez Canal, Baude had published lengthy considerations on the subject in March 1855, emphasising Piedmont's interest in the enterprise.[18]

The Commission was thus expected 'with great desire' at Alexandria, where it was welcomed with full honours before leaving 'with the railway for Cairo, and thence on a Government steamboat for Upper Egypt' and down to Suez, to travel back up the isthmus to the bay of Pelusium, 'in order to study matters concerning the mouth of the canal at the Mediterranean' and 'enable the canal to be navigable by large vessels'.[19]

[15] Bilici (2019).

[16] Letter from Paleocapa to Cavour, 9 November 1855, AST, AC, Mazzo 12 (Correspondence).

[17] Paulin Talabot, French engineer, politician, and banker, had been part of the 1846 *Société d'Études du Canal de Suez*. In May 1855 he had published an article defending the indirect Alexandria-to-Suez route; see Talabot (1855).

[18] Baude 1855. On Baude's mission to Vienna, Bruley (2012, 354).

[19] Luigi Gobbi to the Minister of Foreign Affairs Cibrario, 29 December 1855 and 5 January 1856, AST, CN, Alexandria, Mazzo 7.

The Commission was accompanied by Ferdinand de Lesseps and Barthélemy Saint-Hilaire, philosopher and scholar of Oriental cultures and languages, a man 'who had been through seven revolutions' in the turbulent French politics of the time. He was close to Piedmontese circles, as Cavour himself had attended some of his conferences in Paris in the early 1840s.[20] Saint-Hilaire was to be secretary of the Suez Canal Company and de Lesseps' right-hand man until 1858. The 1855 trip was the basis for his *Lettres sur l'Egypte*, which became an international sensation. Anticipating some of the themes of the cultural/racial/religious/geopolitical discourse on the meeting of East and West symbolised by the cutting of the isthmus, the book was to influence much of the later literature on the canal.[21] Saint-Hilaire also handled public relations on the Suez question during the 1856 Paris Peace Conference, keeping in constant touch with Paleocapa partly as a means of reaching Cavour.[22]

On 2 January 1856, the Commission submitted a long and detailed report to the Khedive, running through the range of geological, technical, and hydraulic issues, the ports, the rates of flow and the estimated costs. The Commission concluded that the work was not 'difficult', and was in any case urgent 'in the present state of international relations'.[23]

Cavour, the Suez Question, and the Congress of Paris

After submitting the report, the Commission left Alexandria on 5 January 1856, taking 'the Austrian steamer' for Trieste, where they were to proceed to Turin to confer with Paleocapa.[24] The Commission's return from Egypt coincided with the opening in February 1856 of the Congress of Paris, held to re-establish order in Europe and settle the 'Eastern question' in the aftermath of the Crimean War, while also addressing the question of major growth infrastructures in an industrialising Europe where the revo-

[20] See Cavour's Diaries (1991, 603).

[21] Saint-Hilaire (1856, V). For a biographical profile of Saint-Hilaire, Picot (1899).

[22] These contacts are documented in Monti (1937a, 182–183). The correspondence published by Monti is supported by other sources, but as the originals have not come to light, it is used here only as a reference.

[23] *Rapport à S.A. Mohammed Said Pacha, vice-roi d'Egypte*, Alexandrie, 2 Janvier 1856, published in *Percement de l'Isthme de Suez* (1856, 191).

[24] Luigi Gobbi to the Minister of Foreign Affairs Cibrario, 5 January 1856, AST, CN, Alexandria, Mazzo 7.

lutions of 1848 re-echoed still. The Congress marked also the entry of Piedmont and the 'Italian question' to the concert of European diplomacy. Traditionally seen as a turning point in nineteenth-century diplomacy, the Congress has recently been reinterpreted in the light of globalisation history and with new emphasis on questions such as the regulation of navigation on the Danube, the neutrality of shipping and the abolition of privateering, all bundled into an appendix to the peace treaty.[25] In particular, the European Commission of the Danube that was established in Paris was the first example (after that for the Rhine created in the 1830s) of international governance of a strategic trade route and the pattern for those that followed.[26]

Long considered as the apogee of the culture and politics of free trade—'the heyday of liberal principles'[27]—the Peace of Paris in reality ushered in a modern international institutionalism, starting from the approach to major growth infrastructures.[28] Following the legacy of the Congress of Vienna—where the question of river navigation had first been addressed—the construction and management of arteries of communication was the proving ground for a new global governance that, although based on rivalries and power politics, laid also the groundwork for a collaborative European imperial political economy in the first era of globalisation. Technological innovations (the steamship), economic growth (the shipping boom) and political changes (the rise of international public opinion) called for new forms of regulating global modernity.[29]

At the margins of the Congress of Paris and in line with its political, economic, and ideological ethos, one of the issues tackled by infrastructure diplomacy was that of cutting the Suez isthmus. The question, however, was a delicate one, given the danger of irking Great Britain, which no one—starting from Cavour—wished to risk at that time. The British government was at the time the main opponent of the canal project, as it competed with the railways in Egypt, of which the first Alexandria-Cairo section had been inaugurated in 1855, while the continuation to Suez would open three years later. In the Spring of 1855, during the French engineers' first surveys, consul Gobbi emphasised that improvements had

[25] Lemnitzer (2013, 2014).

[26] Tamborra (1956), Gatejel (2017).

[27] A thesis recently revisited by Abbenhuis (2014, 67).

[28] Ameil Nathan Soutou (2009).

[29] Buzan Lawson (2015), Jarrett (2013), Mitzen (2013), Koskenniemi (2002).

been made by the British with 'charabancs and coaches and posting stations that make passengers' travel between Cairo and Suez comfortable and easy', with the goal of continuing the railway along that route, 'if the plan to open the Suez canal does not change minds'.[30] Moreover, the British government feared that the project could weaken the Ottoman central government and frustrate the efforts to bring the Porte back into the European concert to counter Russian power, that was one of the aims of the Crimean War.

The Ottoman Empire's integrity was also among France's objectives in Paris. After the firman of November 1854, the Quai d'Orsay had instructed the Vienna and Constantinople embassies to keep the Suez question 'outside diplomacy'. Likewise, the French consul in Alexandria was instructed in January 1855 to consider de Lesseps officially as an ordinary private businessman, with no ties to the government. When the British asked about de Lesseps' activities in Egypt, the Foreign Minister Drouyn de Lhuys answered that de Lesseps had private dealings with the Viceroy, 'with no connection to our government and even without our knowledge'.[31]

On the eve of the peace congress, Gobbi met de Lesseps, who suggested that, as the entire European diplomatic corps would be present in Paris, it could be advantageous to raise the question of the canal. On 21 February 1856, Gobbi thus informed Cavour:

> In the last meeting I had today with Mr. Lesseps, he told me he had been advised by Prince Metternich and Baron de Bruck that the Austrian Plenipotentiaries in the conferences that will take place in Paris [...] have been ordered to broach the question of opening the isthmus of Suez, and he bade me inform Y.E. of these provisions on the part of Austria, and to commit the R. Government to taking the same measures, authorizing the Sardinian Plenipotentiaries to broach and support the question, given the evident and immense advantage it would have for the trade of our states.[32]

Cavour was already in Paris, where, as he informed Paleocapa, had just met Saint-Hilaire, 'who asked me to advocate his project before the Congress. I promised to support him if others moved in favour, in such a way however as not to make an enemy of England, which is said to be very

[30] Luigi Gobbi to Cavour, 23 February 1855, AST, CN, Alexandria, Mazzo 7.

[31] Bruley (2012, 355–356).

[32] Luigi Gobbi to Cavour, 21 February 1856, AST, CN, Alexandria, Mazzo 7.

much against it.'[33] For Cavour it was not just a matter of pleasing the British government. The Ottoman Empire's integrity was a priority for Piedmontese diplomacy as well. As Cavour had written a few days earlier to the plenipotentiary in London Vittorio Emanuele d'Azeglio, the Piedmontese government's objective was to pursue 'whatever it takes to preserve the Turkish empire'.[34] Paleocapa, in turn, informed Cavour of the Commission's latest decisions:

> As regards the straight cut of the Suez isthmus [...] my opinion is that not only can it be done and the artificial port in the harbour of Pelusium can be established and maintained without difficulty, and at an expense that pales before the immense utility of the enterprise, but what is more, that the indirect Canal [...] could not bear the competition and the practicability of the railway in English hands.[35]

Paleocapa drew on the experience gained in works on the Adriatic, the Danube delta and the Po, whose beds were geologically similar to Pelusium. His article comparing Pelusium to the Venetian port of Malamocco, which he had designed, caused a stir in European engineering circles.[36] This new port at the canal's Mediterranean mouth was to become Port Said. The main adversary of this plan was the English engineer and railway expert Robert Stephenson (George's son), whose public clash with Paleocapa and others during 1856—an attempt to put a technical veneer on Britain's political objections to the canal—had earned him the condemnation of much of the international community of experts. In a letter dated 2 March, Paleocapa informed Cavour that de Lesseps had written him 'asking me to urge you, as I am doing, to back the request' to broach the question of Suez in Paris. For his part, Paleocapa could only add that 'the great work could not fail, either on the technical side or the economic'.[37]

In the meantime, Cavour had directed d'Azeglio in London to sound out the British government on Suez. On 23 February, d'Azeglio reported

[33] Letter from Cavour to Paleocapa, 22 February 1856 in Cavour (1992 (1856), 105–106).

[34] Letter from Cavour to Vittorio Emanuele d'Azeglio, 6 February 1856 in Cavour (1992 (1856), 70–72).

[35] Letter from Paleocapa to Cavour, 22 February 1856, in Cavour (1992 (1856), 112–113).

[36] Paleocapa (1856).

[37] Letter from Paleocapa to Cavour, 2 March 1856, AST, AC, Mazzo 12 (Correspondence).

on his long talk with Palmerston, who was 'decidedly opposed' for technical, economic, and political reasons. Palmerston felt that the enterprise was impracticable because of the winds and the shallow sandy shores, which would have involved enormous construction and maintenance costs, leading to prohibitive tolls. These were the arguments advanced by Stephenson, which the other members of the Commission disputed. Palmerston then emphasised that the goods typically of the Indian trade, such as spices and cloth, were readily transported by rail, so in reality there was no need for the canal. Politically, Palmerston saw the canal as part of the French scheme to subtract Egypt from the Porte and make the Mediterranean 'a French lake'. Moreover, 'we cannot tell France', and certainly not at that time, that should there be a war between England and France, the latter could blockade the mouth of the canal and send a fleet to India from the Red Sea: obliged to double the Cape, the British navy would take a month or two to reach the French. 'In conclusion—d'Azeglio wrote to Cavour—England wants the railway and sees the canal as very dangerous to its most important interests.'[38] While de Lesseps, Paleocapa, and Saint-Hilaire continued to exchange correspondence about the Commission's activities in preparation for the next meeting,[39] Cavour made no further official statement about the canal while at the Congress and in subsequent months.

'Public Opinion, Queen of the World': Turin and the Suez Canal

While the dictates of high diplomacy enforced silence on Suez, Piedmont continued to adapt its infrastructures to the prospects for Mediterranean navigation. The first step was the extension of the pier at the port of Genoa, which Paleocapa proposed to Cavour in May 1856. The railway from Turin to Genoa, which had become 'one of the most important lines on the European continent', had enhanced the city's role. Now, with the new Suez route on the horizon, the line's connection to the Piedmontese rail network and its 'forthcoming junction with those of Lombardy and

[38] Letter from V.E. d'Azeglio to Cavour, 23 February 1856, in Cavour (1992 (1856), 114–118).

[39] Monti (1937a, 167*ff*). On the use of this correspondence see note 22.

France' could create the infrastructural axis of a new Mediterranean and European geopolitics.[40]

After the Congress, the Commission met again in Paris on 23 and 24 June 1856, when it addressed engineering matters, the question of fresh water supplies, the canal's depth, the new Mediterranean port, winds in the Suez harbour and where lighthouses should be placed, and the perils of navigating the Red Sea. Paleocapa reported to Cavour that 'the business of the Suez isthmus' was going ahead quickly: 'the main technical points are almost entirely decided. [...] I dined at Lesseps' yesterday with Mr. Thiers who spoke of you with great interest and bade me remember him to you.'[41]

An energetic effort was then begun to sway public opinion in favour of the canal and thus attract subscribers to the Company's shares. In Turin, a periodical, the *Bollettino dell'Istmo di Suez*, was founded in 1856 with de Lesseps' financial support. The *Bollettino* was edited by Ugo Calindri, referred to at times as an engineer, at others as a professor, but undoubtedly among the enterprise's most ardent Italian advocates. For this work, Calindri was also decorated by the Khedive.[42] The *Bollettino* was the Italian counterpart of *L'Isthme de Suez, Journal de l'Union des deux mers*, published in Paris from June 1856 onwards to cover de Lesseps' febrile public relations activity and all the initiatives fielded in Europe in favour of the canal. The *Bollettino* printed translations of much of the material appearing in the French journal, starting with Saint-Hilaire's *Lettres sur l'Egypte.* The Letters were then included in a volume for which Calindri translated many of the texts presented in de Lesseps' books.[43] The *Bollettino* reviewed Italian current affairs and journalism dealing with the canal, in practice covering everything that took place and was printed in Italy on the topic. 'Public opinion, queen of the world' was now convinced of the canal's importance, and Italian public opinion in particular, given that 'the extraordinary enterprise, in changing the direction and increasing the importance of global trade, will be supremely advantageous for Italy and the nearby islands'. Paleocapa's role had been crucial, especially in connecting the enterprise with the 'progressive, rapid and coordinated growth

[40] *Bollettino dell'Istmo di Suez* (henceforth, *Bollettino*), I, 1856, 37–39. On Cavour and the port of Genoa, Tonizzi (2011).

[41] Letter from Paleocapa to Cavour, from Paris, 25 June 1856, AST, AC, Mazzo 12 (Correspondence).

[42] Dispatch of 30 April 1857, AST, CN, Alexandria, Mazzo 7.

[43] *Apertura e canalizzazione dell'Istmo di Suez* (1856).

of the great network of domestic railways'.[44] Moreover, linking the two seas would spread 'civilisation among many rough and barbarous peoples' and promote 'the material interests of the civilised nations'.[45]

Turin was also home to Giuseppe Francesco Baruffi, another enthusiastic supporter of the canal and contributor to the *Bollettino*, as well as to the *Journal des deux mers.* Baruffi was a philosopher, author, traveller, and lecturer on scientific subjects for a popular audience. A correspondent of the French *Société de géographie*, Baruffi exchanged letters with prominent scientific and political figures of the day and had also made a trip to Egypt in 1843, as recorded in his *Viaggio da Torino alle Piramidi* published in 1848.[46] Baruffi kept up a steady correspondence with Saint-Hilaire, whereby de Lesseps kept tabs on the mood prevailing in Piedmontese and Italian circles.[47] 'Deeply enamoured of real progress' and heavily engaged in the question of the isthmus, 'fertile seed of that certain and pacific social revolution that will put the European world back on the path to becoming truly civilised', Baruffi delivered a 'popular lecture' on Suez in Turin in which he stated that 'the modern civilisation' expressed in the decisions of the Paris Congress had proclaimed that 'the Isthmuses and Straits of the globe are the inalienable property of the human race'.[48]

The *Bollettino* focused considerable attention on initiatives in other Italian States, votes by the Chambers of Commerce in various Italian cities, and predictions of the canal's effect on traffic in the main Italian ports. Through the discourse on the canal, the *Bollettino* contributed to the rhetorical recovery of themes (the maritime republics, Italy's centrality in the Mediterranean) that underpinned the Italian States' common interest in the enterprise and an Italian modernity poised between Europe and the Mediterranean. Naturally, de Lesseps' visit on 1 August 1856 to Turin—whence he went to Genoa to take ship for Marseilles—was a major event. In Turin, he 'had from many Dignitaries of the country that distinguished reception he deserved'. He was met at Turin's railway station by Paleocapa himself, who introduced him to Luigi Torelli, deputy of the Sardinian Parliament.[49]

[44] *Opinione pubblica in Italia sulla canalizzazione dell'Istmo di Suez* (unsigned, but by Calindri), *Bollettino* I, 1856, 86–87.

[45] *Bollettino* I, 1856.

[46] Baruffi (1848). On Baruffi, see Nada (1970).

[47] Letter from Saint-Hilaire to Baruffi, 21 February 1857, BSP, AB.

[48] Baruffi (1857, 19).

[49] *Bollettino* I, 1856, 145.

Torelli was already one of the keenest supporters of the canal project—'a soldier myself, and no recent recruit, in the legion of the Suez canal'.[50] Born to a noble Lombard family, Catholic patriot and author of *Pensieri sull'Italia di un anonimo lombardo* (published in Paris in 1846), Torelli had joined the Piedmontese army in the first war of independence, later going into exile in Piedmont (he called himself 'Italian, Lombard by birth, Piedmontese out of appreciation'), where he was to become Minister of Agriculture. In 1854 he began to draft a monumental work, *Dell'avvenire del commercio europeo ed in modo speciale di quello degli stati italiani*, published in 1859: by his own admission, a 'daring' theme at the time. The book drew inspiration from the prospect that both the railway, this 'great instrument of civilisation and prosperity', and the Suez Canal would connect Piedmont and 'all of Italy' with the 'great European network'.[51] The 'first historiographer of the Suez Canal', as Monti called him, Torelli championed the canal 'when Italy either did not yet exist as a nation or had existed for only a few years and was still struggling to make its way among the hostile or lukewarm Powers of Europe'.[52] Senator of the new Kingdom of Italy and then Minister of Agriculture, Industry, and Commerce, Torelli travelled to Egypt in December 1866 to view the canal's construction, keeping a diary of the trip later published by Monti.[53] In 1867 he authored *Il Canale di Suez e l'Italia*, where he proudly defined himself 'one of the strongest adherents' and 'among the promoters' of the enterprise.[54] During the last stage of the canal's construction (1867–1869), which was also the last stage of work on the Moncenisio tunnel, Torelli read a quarterly series of papers (the celebrated *Paralleli*) at the *Istituto Veneto di Scienze, Lettere ed Arti* in which he described the progress of the two projects, both of which had benefitted from Cavour's farsightedness, Paleocapa's role, and the protection of Napoleon III, though both, when first announced, 'were met with scepticism and declared impossible'.[55] The parallel between digging the Suez Canal and tunnelling through the Alps is a recurrent theme in the coeval press and in historiography.[56]

[50] Torelli (1870). On Torelli, see also Manfredi 1930.
[51] Torelli (1859, VI). Part IV of the book is devoted to the Suez Canal (quote p. 245).
[52] Monti (1937b, 173).
[53] Monti (1937b).
[54] Torelli (1867, 8).
[55] Torelli (1870).
[56] Garelli (1864), Caizzi (1985).

Conclusion

In December 1856, the Commission submitted its final report to the Khedive. The first part described the direct sea-to-sea cut, which the experts thought preferable; the second assessed the necessary resources. Following the report, the firman was modified to grant a concession for the work to the *Compagnie universelle du canal maritime de Suez*. Paleocapa forwarded the report to Cavour on 26 January 1857, underscoring that 'the great influence that this work must have on the prosperity of our trade fuels hope that H.M.'s Government [...] will view it with the favour that can do so much to overcome the difficulties still standing in its way'.[57] Gobbi's dispatches from Alexandria continued to document the local community's reactions to the British government's unrelenting hostility, but also the evolving pro-Suez attitude of British trading companies' in Alexandria, while material began to arrive for the construction of a freshwater canal to bring water from the Nile to the construction sites.[58]

Work started a year later, in 1858. The history of Italy's relationships with the canal thus entered a second phase, intersecting with new national and international dynamics—the war of 1859 and the passage to unification—while Italian workers and engineers began their migration to the isthmus. The Indian Mutiny of 1857 would also contribute to reframing the British government's approach to Suez, which will progressively acquire a central position in Victorian imperial connections and in the geopolitics of the Mediterranean.[59] During the first phase we have described here, the Suez affair contributed to forming the policies furthering the Kingdom of Piedmont-Sardinia's debut in European infrastructure diplomacy, and to constructing a new narrative on Italy's role in the Mediterranean.

The debate over Suez was a catalyst for a more general political and economic discourse about the infrastructures needed to connect the Italy of the future to global markets. The politico-rhetorical content of the discourse on the canal, which originally served to sway public opinion in its favour, would be absorbed by the 'national' theme of a maritime Greater Italy, contributing to that 'lasting rhetorical schooling' that will in turn be absorbed by the colonial discourse.[60] At the same time, through the

[57] *Percement de l'Isthme* (1856).
[58] Luigi Gobbi to Cavour, 6 February 1857, AST, CN, Alexandria, Mazzo 7.
[59] Darwin (2009), Kamel (2015).
[60] Monina (2008, 32–34); see also Natili (2009).

international Commission, the Suez project had contributed to a European technoscientific identity, as the expression of a shared modern civilisation. Italy's hopes as a great maritime nation would soon clash, however, with a more complicated and at times frustrating reality, given the country's technological and organisational lag, slowness in moving to steam, and difficulties in making itself felt in ocean-going trade. Contrary to expectations, the opening of the canal would actually lead to the Mediterranean's gradual marginalisation in global trade and, as consequence, to a more realistic assessment of Italy's position.[61]

Cavour had to some extent foreseen these developments. Writing to d'Azeglio on 27 October 1857, he summarised the question, explained his official silence on Suez, and reflected on the relationship between the technical and political spheres and between rhetoric and economic reality, which to most observers would only become clear years later. Paleocapa had handled the question 'as an engineer, not a minister', and had thus—rightly—been at loggerheads with Stephenson. By contrast,

> I abstained from ever pronouncing the word Suez during the many discussions in which the question of cutting the isthmus happened to come up. I believe the English government should bear my discretion in mind, but at bottom I cannot say that my silence stemmed only from the fear of displeasing Lord Palmerston. Though I support cutting the isthmus, I believe the consequences from a commercial standpoint are being exaggerated enormously. It is for this reason I have no difficulty in resigning myself to leaving how to solve this thorny problem to our grandchildren.[62]

Cavour's 'grandchildren', that is, successive generations of Italian policymakers, would find themselves facing the issues raised by Italy's role in the Mediterranean time and time again.

References

Archival Sources

AST Archivio di Stato di Torino.
AC, Archivio Cavour, Corrispondenza, Mazzo 12.
CN, Consolati nazionali, Alessandria d'Egitto, Mazzo 7, 1851–1859.

[61] Sevket and Williamson (2000).

[62] Letter from Cavour to V.E. d'Azeglio, 27 October 1857, in Cavour 1994 (1857), 520–521.

BSP, Biblioteca storica piemontese.
AB, Archivio Baruffi.
Bollettino dell'Istmo di Suez.

References

Abbenhuis, Maartje. 2014. *An Age of Neutrals. Great Power Politics, 1815–1914.* Cambridge: Cambridge University Press.

Ameil, Gilbert, Isabelle Nathan et George Henri Soutou (sous la direction de). 2009. *Le Congrès de Paris (1856): un événement fondateur.* Bruxelles: Peter Lang.

Apertura e canalizzazione dell'Istmo di Suez 1856*: narrazione informativa e documenti officiali del Signore Ferdinand de Lesseps Ministro Plenipotenziario; col rapporto sommario fatto al Vicere d'Egitto dalla Commissione Internazionale e lettere sull'Egitto del sig. By St. Hilaire*; traduzione del prof. Ugo Calindri, Torino, Unione Tip. Editrice.

Association du Souvenir de Ferdinand de Lesseps et du canal de Suez 1994, Centenaire de la mort de Ferdinand de Lesseps. *1894–1994*, Paris.

Baron Baude. 1855. 'De l'isthme de Suez et du canal maritime à ouvrir de la Méditerranée à la mer-Rouge', *Revue des deux mondes*, 15 mars.

Baruffi, Giuseppe Francesco. 1848. *Viaggio da Torino alle Piramidi: fatto nell'autunno del 1843.* Torino: Fontana.

———. 1856. 'Lettera del prof. Cav. G.F. Baruffi all'ing. Ugo Calindri sull'apertura dell'istmo di Suez', 6 luglio 1856, *Bollettino dell'Istmo di Suez*, I, 1856.

———. 1857. *L'Istmo di Suez: lezione popolare tenuta a Torino il 1° luglio 1857.* Torino: Stamperia Reale.

Bilici, Faruk. 2019. *Le canal de Suez et l'Empire ottoman.* Paris: Editions CNRS.

Bonin, Hubert. 2010. *History of the Suez Canal Company, 1858–2008. Between Controversy and Utility.* Droz, Geneva.

Bruley, Yves. 2012. *Le Quai d'Orsay impérial: histoire du Ministère des affaires étrangères sous Napoléon III.* Paris: Pedone.

Buzan, Barry, and George Lawson. 2015. *The Global Transformation: History, Modernity and the Making of International Relations.* Cambridge University Press.

Caizzi, Bruno. 1985. *Suez e San Gottardo.* Milano: Cisalpino.

Cavour, Camillo. 1991. *Diari, 1833–1856*, vol. II, ed. Alfonso Bogge, Roma.

———. 1992. *Epistolario*, vol 13, 1856, t.1., eds Carlo Pischedda and Maria Luigia Sarcinelli, Firenze, Leo Olschki.

———. 1994. *Epistolario,* vol. 14, 1857, eds Carlo Pischedda and Rosanna Roccia, Firenze, Leo Olschki.

Correnti, Cesare. 1869. *L'istmo di Suez e il commercio orientale*, Bollettino della Società Geografica Italiana, 3, re-published in: *Scritti scelti di Cesare Correnti in parte inediti o rari*: ed. Tullo Massarani, Vol IV, Roma, Forzani (1894), 477–490.

Curli, Barbara. 2015. Il Piemonte e il canale di Suez, 1855–1856. *Studi Piemontesi*, XLIV 2: 325–338.

Darwin, John. 2009. *The Empire Project: The Rise and Fall of the British World-system, 1830–1970*. New York: Cambridge University Press.

De Lesseps, Ferdinand. 1855. *Percement de l'Isthme de Suez: exposé et documents officiels*. Paris: Plon.

Di Prima, Giuseppe. 1940. *L'opera politica e tecnica di Pietro Paleocapa alla luce di un epistolario inedito*. Milano: Vita e pensiero.

Garelli, Felice. 1864. *La sezione dell'istmo di Suez e la perforazione delle Alpi italiane: considerazioni economico-politiche*. Torino: Tip. Scolastica di Sebastiano Franco e figli.

Gatejel, Luminita. 2017. Imperial cooperation at the margins of Europe: the European Commission of the Danube, 1856–65. *European Review of History: Revue européenne d'histoire* 24 (5): 781–800.

Gottardi, Michele. 2014. *Pietro Paleocapa*, in: *Dizionario biografico degli italiani*, (online).

Holland, Robert. 2012. *Blue-Water Empire. The British in the Mediterranean since 1800*. Allen Lane.

Istituto veneto di scienze, lettere ed arti. 1990. *Ingegneria e politica nell'Italia dell'Ottocento: Pietro Paleocapa*. Venezia.

Jarrett, Mark. 2013. *The Congress of Vienna and Its Legacy: War and Great Power Diplomacy after Napoleon*. London: I.B. Taurus.

Kamel, Lorenzo. 2015. *Dalle profezie all'impero. L'espansione britannica nel Mediterraneo orientale (1798–1878)*. Roma: Carocci.

Koskenniemi, Martti. 2002. *The Gentle Civilizer of Nations: The Rise and Fall of International Law, 1870–1960*. Cambridge: Cambridge University Press.

Lemnitzer, Jan Martin. 2013. "That Moral League of Nations Against the United States": The Origins of the 1856 Declaration of Paris. *The International History Review* 35 (5): 1066–1088.

———. 2014. *Power, Law and the End of Privateering*. London: Palgrave Macmillan.

Manfredi, Silio. 1930. *Luigi Torelli ed il Canale di Suez*. Sondrio, Tip. Arti grafiche valtellinesi.

———. 1932. *I collaboratori italiani di Ferdinando Lesseps*. Mevio Washington: Sondrio.

Mitzen, Jennifer. 2013. *Power in Concert: The Nineteenth Century Origins of Global Governance*. Chicago: University of Chicago Press.

Monina, Giancarlo. 2008. *La Grande Italia marittima. La propaganda navalista e la Lega navale italiana (1866–1918)*. Soveria Mannelli: Rubbettino.

Monti, Antonio (a cura di). 1937a. *Gli italiani e il Canale di Suez. Lettere inedite di P. Paleocapa, L. Torelli, E. Gioia.* Roma: Vittoriano.

———. 1937b. *Storia del canale di Suez: Con un diario di Luigi Torelli ed altri documenti inediti.* Milano: Ispi.

Nada, Narciso. 1970. *Giuseppe Francesco Baruffi in: Dizionario Biografico degli Italiani*, Volume 7, online.

———. 1990. *Pietro Paleocapa nel gruppo cavouriano durante il decennio di preparazione.* in: Istituto veneto di scienze, lettere ed arti 1990, 209–217.

Natili, Daniele. 2009. *Un programma coloniale. La Società geografica Italiana e le origini dell'espansione in Etiopia (1867–1884).* Roma: Gangemi.

Paleocapa, Pietro. 1856. Considerazioni sul protendimento delle spiagge e sull'insabbiamento dei porti dell'Adriatico, applicate allo stabilimento di un porto nella rada di Pelusio. *Bollettino dell'Istmo di Suez* I: 105–142.

Percement de l'Isthme de Suez 1855, Ferdinand de Lesseps, *Percement de l'Isthme de Suez: exposé et documents officiels.* Paris, H. Plon.

Percement de l'Isthme de Suez 1856, *Percement de l'Isthme de Suez. Rapport et Projet de la Commission internationale. Documents publiés par M. Ferdinand De Lesseps*, troisième série, Paris, Henri Plon.

Picot, Georges. 1899. *Barthélemy Saint-Hilaire: notice historique, lue en séance publique le 3 décembre 1898.* Paris: Hachette, Institut de France.

Piquet, Caroline. 2008. *La Compagnie du canal de Suez. Une concession française en Egypte (1888–1956)* Paris: Presses de l'Université de Paris Sorbonne.

———. 2009. *Histoire du Canal de Suez.* Paris: Perrin.

Romeo, Rosario. 1984. *Cavour e il suo tempo.* Vol. III. Laterza: Roma- Bari.

Saint-Hilaire, Barthélemy. 1856. *Lettres sur l'Egypte.* Paris: Michel Lévy Frères.

Sammarco, Angelo. 1940. *Suez. Storia e problemi. Secondo documenti inediti egiziani ed europei.* Milano, Garzanti.

Sevket, Pamuk, and Jeffrey G. Williamson, eds. 2000. *The Mediterranean Response to Globalization Before 1950.* London and New York: Routledge.

Simpson, Gerry. 2004. *Great Powers and Outlaw States.* Cambridge University Press.

Talabot, Paulin. 1855. Le canal des deux mers, d'Alexandrie à Suez. Moyens d'exécution. *Revue des deux mondes*, mai.

Tamborra, Angelo. 1956. *Un siècle de collaboration internationale sur le Danube maritime: 1856–1956.* Rome: Commission européenne du Danube.

Tonizzi, Maria Elisabetta, ed. 2011. *Cavour e Genova. Economia e Politica.* Genova University Press.

Torelli, Luigi. 1859. *Dell'avvenire del commercio europeo ed in modo speciale di quello degli stati italiani.* Firenze.

———. 1867. *Il Canale di Suez e l'Italia.* Milano: Giuseppe Civelli.

———. 1870. *Dieci paralleli fra il progresso dei lavori della Galleria del Cenisio e quelli del Canal di Suez a partire dal 31 maggio 1867 al 15 novembre 1869*, Venezia, G. Antonelli.

Van der Vleuten, Erik, and Arne Kaijser, eds. 2006. *Networking Europe. Transnational Infrastructures and the Shaping of Europe, 1850–2000*. Sagamore Beach: Science History Publications.

Vimercati, Cesare. 1849. *Constantinopoli e l'Egitto; studi statistici storici politici commerciali*. Alberghetti: Prato.

———. 1864. *Il canale dell'istmo di Suez. Sua influenza di pace per l'Europa e di rigenerazione per l'Italia*. Livorno.

CHAPTER 3

Italian Infrastructures at the Time of the Opening of the Suez Canal

Andrea Giuntini

Infrastructures in the Italian Kingdom

The opening of the Suez Canal took place at the end of the first decade of Italy's unity, when the process of defining and building the infrastructural system of the new Kingdom was underway. Between 1861 and 1869, the first liberal governments made a series of fundamental decisions regarding the various networks, which for a long time to come would represent the infrastructural backbone of the country. In 1861 modern Italy began to be built physically, not only through the establishment of a national market, but also aiming at a new spatial and organizational dimension as the necessary premise for the very idea of a nation and of a real identity. Italy was a young and ambitious new nation-state, aspiring to be a powerful country that required to be properly equipped with modern infrastructures. Therefore, a physical and territorial conception of the nation,

A. Giuntini (✉)
Università di Modena e Reggio Emilia, Modena, Italy
e-mail: andrea.giuntini@unimore.it

B. Curli (ed.), *Italy and the Suez Canal, from the Mid-nineteenth Century to the Cold War*,
https://doi.org/10.1007/978-3-030-88255-6_3

defined precisely by the space shaped by networks, would contribute to the design of a new Italy while a revolution in the global communications system was unfolding.[1]

The term infrastructure can signify different things and include multiple elements. A starting point is the definition of durable public capital present in a territory, that is, a complex of physical network, equipment, and installations indispensable for economic activity, and at same time a shared system of values and objectives, where the public goal is identified with the possible use of the infrastructures, but not necessarily with their ownership. The concept covers an area ranging from roads to railways, from ports to airports, from the postal system to communications, from energy networks to urban sanitation systems and still including bridges, tunnels, inland waterways, and oil and gas pipelines. These systems take the form of a network, extending over the territory, which constitutes the physical base, capable of allowing the flow of people, goods, capital, information, and services. More recently, infrastructure studies have adopted the new concept of mobility as a cultural turn, which allows us to think of space not as rigidly and permanently configured only in political and administrative terms but rather in a dynamic perspective of change. The network connections, or lack thereof, structure the spaces, activating and deactivating their potential uses, and ultimately determining their development trajectories. The complete picture of the infrastructure system in a given territory emerges through the analysis of the complex set of social, economic, technological, cultural, and political relationships that diachronically converge, concentrate, and interconnect, thanks to the actions of private and public, individual and collective subjects.

In post-unification Italy, the first liberal governments had to face the heritage of the previous states also in terms of infrastructure networks coming from different administrative and managerial experiences and characterized by different standards, which were quite fragmented and little functional. The road system was in a very critical condition; the rail network was poor and made up of a set of regional sub-systems disconnected from each other; ports and navigation needed a decisive relaunch; the postal and telegraphic systems still left most of the population isolated.

As regards the construction of a basic infrastructural network, the Italian case is utterly typical as compared to the experience of the new nineteenth-century Western nation-states. For this reason, the path

[1] Giuntini (1999, 2012).

followed for the construction process of the main infrastructure networks in Italy was full of successful choices and acquisitions, but also suffered from faults and missteps. This far-reaching project, which was pursued in the first decade of unity, was characterized by accumulation and overlapping rather than by a coherent vision. The reason for this urgency is explained by the need to trace new arterial roads and tracks, to modernize the numerous existing ports and the organization of the postal system, and to start a dense telegraphic network. The plan lacked however a long-term view of integration and coordination. Rush, enthusiasm, and scarce unity of intents negatively affected both the change in infrastructure systems and their management, which still lagged behind the Western European norm. In addition, politics weighed heavily on the project, distorting and affecting the path of Italian infrastructures due to the tremendous amounts of interests, to the business, political, and military lobbies, and to the institution of the concessions system for the realization of the infrastructure projects, leading to extensive speculation and profiteering.

The model followed in the early years for the creation of the infrastructure network was the French *Grands Travaux*, within which the role of technicians, mostly engineers, was particularly emphasized. The early liberal era represented a formidable call to arms for all those actors, often straddling science and technology, who had already successfully operated in the pre-unification years in the preparation of the first infrastructures. The main builders of the new spatial and organizational dimension of the nation were the most varied categories of technicians, committed to the project's achievement, but at the same time also carriers of the interests of lobbies, social, and local groups. These scientists and technicians were interpreters of the new technology but still conditioned by old ties and interests. Scientists and technicians constituted a functional link between the new Kingdom and the international centers of technology, from where they successfully transferred technology to their homeland. However, more and more members of the technical-scientific community, with which they shared tools and goals, were also proud propagators of the patriotic discourse in the name of a decisive contribution to the birth of the new state.

Despite several defects, at the time of the opening of the Suez Canal, Italy could be satisfied with what had been achieved in the sector, in particular as we consider how much behind the project was at the moment of unification. By the eve of the Great War, an intense innervation of networks in the territory and in the cities, which provided irreplaceable support for economic activity, had been built, even if the articulated commitment made toward the construction of railways, ports, and road work relating to urban, postal, and telegraphic services only partially followed a network rationality.

Railways

When in 1861 the new rulers of united Italy started the reorganization of the railway system, the shortcomings inherited from the pre-unification era were heavily felt, affecting, until today in certain cases, the physiognomy of the country's railway system. The pre-unification railway network was a set of separate sections with no elements of integration, and in general there was no real organic network on the peninsula, only a series of disconnected sub-systems that had to be linked together: reliance on the railways seemed to be the best course to promote integration between the various regional economies. Since the early years of the new Kingdom, some major routes from the north and south were designed to bring distant areas of the country closer. The inconsistent bundling of a cobweb of railroads, which changed the design of the young Kingdom's map, was almost chaotic, and accentuated the longitudinal character of the Italian railway system. The logic that was followed at the time of the creation of the national railway network was marked on the one hand by the joining of the various systems just sketched prior to the unification; and on the other hand, to the writing of guidelines in a great hurry. The network that was created in the sixties, therefore, suffered from continuous interruptions and did not have mountain cross-connections. Such a policy, justified by the belief that the railways could guarantee both the physical unification and the modernization of the state, failed however to solve the obvious problem of a net imbalance in the railway supply in the various areas of the country.

Anyway, Italy entered immediately into the restricted ranks of the great European powers: within five years, from 1861 to 1866, the network doubled, going from just over 2000 km to over 4400 km (in 1861 Great Britain had 17,000 km of railways and France 9300 km). In 1863 the tracks reached Foggia following the Adriatic coast, the following year the line from Bologna to Florence was opened, and in 1866 Rome was first connected with Ancona and then in 1867 with Pisa; in 1870 the Naples-Foggia line began operating, and in 1872 the Rome-Florence and the Indian Mail lines were completed. Some of the main tunnels were also opened or started: the Giovi line, built between 1846 and 1855; the Fréjus line, built between 1857 and 1872; the Bologna-Pistoia in 1864; and the Brenner in 1867. The opening of the crossings made it possible to connect Italy with Central Europe, opening the doors to increasingly accessible markets.

Undoubtedly, it was a huge effort also in financial terms. Although the state was the promoter of the expansion of the railway network, the management of the network formally remained in the hands of private companies until 1905, the year of the nationalisation of the railway network, and the instrument used to build and rule railways was the concessions system. This established that the lines were built and managed by private companies, to which the guarantee of a minimum income per kilometer was recognized. Through the establishment of the concession, the state contracted its own prerogatives without renouncing to supervise the construction of the various sections, which often gave rise to conflicts and favored wide speculation.

Therefore, in a very limited time and without abdicating the principles of liberalism, the Historical Right-wing political party provided the country with a basic railway system. The new bourgeoisie at the head of the country was driven by a clear political inspiration rather than by the belief that the railways could play an economic and industrial role. The railways, and infrastructures in general, were ultimately assigned a more political than economic task, thus large façade projects were prioritized over the lines that could have had a more significant economic impact. Notwithstanding these great efforts, planning railway lines—as the main historiography regarding this period pointed out[2]—was hardly capable of transforming the railways into a crucial element for the economic future of the country.

Roads

The road issue was a major public discussion topic since unification, which provoked a wave of new road constructions in the same generically innovative perspective adopted for the railways, without a coherent plan.[3] In the early 1860s, the state of the national roads was better than the provincial and municipal ones, which were frequently in bad transit conditions, and the absence of bridges was a common problem. The roads were

[2] Fenoaltea (1984); Schram (1997); Maggi (2003).

[3] Badon (2011).

constantly relegated in a perspective of clear subordination with respect to the railways; so much so that whenever the routes of the national roads coincided with those of the railways, the roads were the first to be downgraded. However, the new economic and social dynamics induced by the birth of the new Kingdom significantly increased the need for mobility and therefore a more marked use of the roads. The Italian system in 1863 consisted of 13,500 km of national roads and 8997 km of provincial roads; in 1873 the total Italian patrimony amounted to 102,000 km of roads against the 556,000 km of France and 220,000 km of Great Britain.

In an attempt to rebalance a decidedly unequal endowment, the first governments of the Kingdom followed a compensatory logic, engaging in an attempt to increase the road network also in the southern provinces. This commitment is clearly demonstrated by the expenses incurred in the sector in the first years of the Kingdom's life: between 1861 and 1871 the amount spent on roads in the south was three times greater than in the north. Construction of the new roads was quickly completed and the new road trend provides continuous maintenance instead of major periodic repairs, to which, until then, every road was normally subjected.

The secondary road system was in trouble and long neglected and therefore unable to guarantee the necessary connections. The accelerated progress of the first years then gave way to a progressive slowdown by the central state authority, which was increasingly replaced by provincial and municipal roads. Ultimately, the roads were used much more as a political-patronage tool than as the infrastructure necessary for economic development. The new network was developed in the name of a widespread need to satisfy the single requests that arose everywhere—like local communities and economic lobbies—on the peninsula and not with an organic design of the territorial development of the road network as a whole in mind.

The Postal Service

The thread that sewed the very fragmented peninsula together was the postal service.[4] Each Italian pre-unification state had its own different postal organization, which however converged on some shared basic principles. These included the state privative and the tariff linked to the distance traveled, a rule that will falter with the introduction of the stamp.

[4] Giuntini (2004).

Equally common was the difficulty in organizing dispatches to widespread areas distant from cities, some of which were often reached on foot. The habits and rules adopted in the previous states were decidedly different, and therefore the standardization process that took place in 1861 was necessary and complex. Nonetheless, the unification of Italy, thanks to a progressive improvement of the postal service, which became a significant element of the modernization of the country, gives way for the first time to a wide use of mail. Post offices and postal stations were the cornerstones of a network that in 1861 required a decisive strengthening and a definitive legislative uniformity. The new national postal law of 1862, which was a synthesis of the experience of the most advanced European postal systems, marks the start of the centralization process of the postal administration in a country that required much work to complete its communication system.

Mail was mainly carried on trains and ships. In Italy too, special railway carriages—called ambulant—were specifically equipped for the transport and sorting of the mail on board by specially appointed personnel. Sea transport was much more widespread and, ultimately, important because of the capacity of coastal shipping to reach substantially any location on the Italian coasts, from where the couriers departed on foot and horseback. The introduction of steam navigation also represented a great innovation for mail transport. It facilitated its distribution and it contributed to making the transportation of mail an economically increasingly convenient, and therefore disputed, affair. The maritime postal service in Italy was granted to private companies with a subsidy system, in line with what was happening in the major European maritime powers. Italian policy was characterized in the first decade after unification by subsidies, tax breaks and bonuses to shipping companies, so as to facilitate the circulation of mail inside and outside the country as much as possible. The system of concessions entrusted to a series of private maritime companies guaranteed the state the free transport of mail, which represented the compensation for subsidies. State intervention significantly reduced the number of small shipowners and favored the formation of an oligopoly in the sector. The first postal agreement was signed in 1862 with the *Adriatico-Orientale* shipping company, which established a line between Ancona and Egypt with a stopover in Brindisi; the following year there were already 20 functioning lines of maritime postal services.

Telegraph

In the various Italian pre-unification states, the first telegraph lines were made around the mid-nineteenth century from an experimental perspective and unrelated to any programming.[5] In Tuscany in 1847 and later in other states, governments solicited the first attempts, which, however, did not lead to an integrated system. The need to equip railway traffic with strict supervision on the one hand and military needs on the other, were the main reasons for the development of an integrated system of telegraph lines, together with the prestige that the new technology provided.

At the moment of unification in 1861, the kingdom of Piedmont-Sardinia had 1400 km of telegraph lines and 60 offices; the state with the most lines was the Bourbon kingdom with 3974 km and 114 offices. It was necessary to rationalise and standardise the fragmented organizational and regulatory framework, without departing from the principle of state monopoly, and to update the equipment. The first step in creating a unitary telegraph service was the merging of the pre-existing administrations into a single regulatory framework: in February 1861, a decree was passed for this purpose, which approved the first telegraph regulation of the Italian state. The overall configuration included two main arteries, the Tyrrhenian and the Adriatic, connected by seven transversal lines. The results were excellent: between 1860 and 1863, 4059 km of lines and 12,492 km of wire were built and 180 new offices were opened; in 1865 the network consisted of 14,792 km for a total of 37,323 km of wire. The materials and technology adopted were aligned with what was used in the most advanced European telegraph systems. The training of competent and efficient staff—including many women—by the new service managers was extremely demanding. At the end of the decade, thanks also to having joined the International Telegraph Union, Italy was integrated into the European telegraph system.

Submarine telegraphy followed a different path.[6] Although unable to compete with the great European powers, Italy played a relevant role in the field of international submarine telegraphy until 1869. This was possible thanks to its geographical location in the middle of the Mediterranean, a natural crossroad for telegraphic communications. In a few years, Italy ended up accumulating an appreciable supply of submarine cables. All this

[5] Fari (2008).
[6] Giuntini (2014).

took place also thanks to the government's forethought, which tried to exploit this opportunity by stipulating a high number of agreements with the foreign companies in charge of submarine telegraphy. The original initiative goes back to 1853—two years after the world's first submarine cable between Dover and Calais—when the French government decided to support the project for submarine communication with Algeria and Malta, and hence to the Indian Ocean through Egypt. The itinerary included crossing Corsica and Sardinia, thus developing a mixed overland and submarine route. The protagonist of this event was the *Compagnie du télégraphe électrique sous-marin de la Méditerranée pour la correspondance avec l'Algerie et les Indes*, which drew up an agreement with the Turin government to generate the plan for a cable to Corsica was laid in July of 1854, which would have been the third in the world and the first international cable in the Mediterranean. The cable—completed in 1857—that presented the greatest difficulties was the one between Sardinia and Algeria, laid in a very deep stretch of the sea plagued by currents, and probably too long for the technological equipment of the time.

The Italian peninsula, with its elongated shape and central position, was an ideal landing place for the submarine cables connecting Great Britain and other European countries with the African and Asian colonies. There were two strategic points for which Italy was at the center of a busy chain of initiatives: it was the means of access to northern Africa and the East through the upcoming canal. The Italian peninsula was a sort of bridge between Europe and Asia, and as such, was particularly useful for British strategies of communication with its Indian possessions. Before the passage to Gibraltar, the English, in order to reach the East, arranged to unite Malta with Sicily in 1859, making it more convenient to cut along the Italian peninsula, due to the length and risks of the long and demanding sea crossing. In 1867 the Italian administration decided to stipulate an agreement with Telcon; thus, for a few years, Italy found itself as the main crossing point of the flow of British correspondence to and from the Asian and African colonies. When, in 1869, the Falmouth Gibraltar and Malta Telegraph Company laid a direct cable of 2281 miles from the British Isles to Lisbon, Gibraltar, and Malta—which was the cornerstone of the English telegraph strategy—and finally to Egypt, the game for Italy was over.

Expectations and Debate on the Opening of the Suez Canal

The Suez Canal project generated great expectations in Italy, stimulating an intense debate, a series of concrete initiatives, and a kaleidoscope of different positions. The expectation of an automatic growth after the opening of the canal was rather superficial, and it turned out to be naive and wrong. The euphoria was based simplistically on the historical commercial vocation of the country and on the reduced distances of the naval routes. Suez was considered a historical opportunity for Italy, which was still tentatively facing industrial modernity, rather preferring the status of commercial power. The Italian political and economic elites—and a growing part of the public opinion—were ready to make a further step toward modernity. Protagonists of the Suez issue were trade associations—mostly Chambers of Commerce—entrepreneurs, and merchants, who predicted a bright future for the country. Techno-optimism and a positivist spirit were not enough however to foster Italian commercial export through the new sea lines, as the resettlement of the political and economic balance of power following the opening of the canal did not substantially change hierarchies in the global economy.

Since before unification, the Suez story was accompanied by a certain interest in all the pre-unification states, mostly in Piedmont,[7] and there was no shortage of commentators, although, at the time when the shares of the manufacturer of the channel were put on the market, the investors' response was all in all lukewarm. The various phases of the project attracted widespread attention and even apprehension and adverse opinions, or at least reluctant, raised to moderate the spreading optimism, were readable in the country's newspapers and listened to in the Parliament. Unquestionably, the prospect of the canal hastened the creation of the national infrastructure network and made it faster.

It was common opinion of the various observers, who took part in the debate, that the novel communication would have opened up an era of great economic growth for the new Kingdom. The possibility of reaching out into the Mediterranean in the direction of the East was an available asset, which Italian rulers intended to use with prudence, as it raised significant opportunities. Detailed studies and research were started, contributing to the progressive maturation in building Italians' infrastructure.

[7] Curli (2015).

Many went personally to Egypt to visit the worksite, as did Alessandro Cialdi, who studied the decisive topic of cover-up during the excavation of the canal,[8] Fedele Lampertico,[9] Gerolamo Boccardo,[10] who warned of miraculous expectations that without the necessary infrastructure would have not been realized, and Giuseppe Sapeto, always quite prudent.[11] Others who visited the site were Jacopo Virgilio, concerned about the insufficiently advanced state of the Italian navy,[12] Cesare Vimercati[13] and Luigi Torelli, who was the most convinced partisan of the canal and author of a series of studies of great technical precision, with which he informed the reader of the state of construction of the canal and the Moncenisio railway tunnel in parallel with the progress of the works.[14]

In 1869 Italy was far from 'ready' to enter the new global economic and commercial framework. However, the debate on the canal contributed to the growth of the Italian ruling classes in the first decade after the birth of the new kingdom. Undoubtedly, progress in the infrastructural field was evident, but shortcomings and insufficient solutions given to certainly crucial problems held back the Italian take-off. The economic and entrepreneurial Italian context was still unable to compete with the most advanced European countries and Italy was only partially prepared for the new challenge of the opening of the canal. The hastily announced economic advantages due to the opening of the canal, based on the acquired data of a modern and functioning infrastructure network, inevitably did not work. The strategic position in the center of the Mediterranean alone was not enough. It is hard to say if it was a missed opportunity or rather a surrender to a manifest inferiority. Anyway, in the following decades the lesson learned from the events tied to the canal probably contributed to making Italy a more modern country.

[8] Paleocapa (1856); *Sul Portosaido* (1867).
[9] Lampertico (1859).
[10] Boccardo (1865).
[11] Sapeto (1865).
[12] Virgilio (1869).
[13] Vimercati (1864).
[14] Torelli (1870).

References

Badon, Cristina. 2011. *Le vie dell'Italia unita. La politica stradale italiana nell'epoca delle ferrovie (1850–1900)*. Firenze: Nerbini.

Boccardo, Gerolamo. 1865. *Il canale attraverso l'istmo di Suez e gli interessi commerciali dell'Italia. Relazione di un viaggio in Egitto.* Genova: Tipografia Fratelli Pellas.

Curli, Barbara. 2015. Il Piemonte e il canale di Suez, 1855–1856. *Studi piemontesi* XLIV (2): 325–338.

Fari, Simone. 2008. *Una penisola in comunicazione. Il servizio telegrafico italiano dall'Unità alla Grande guerra*. Bari: Cacucci.

Fenoaltea, Stefano. 1984. Le costruzioni ferroviarie in Italia 1861–1913. *Rivista di Storia Economica* 1: 61–94.

Giuntini, Andrea. 1999. Nascita, sviluppo e tracollo della rete infrastrutturale. In *Storia d'Italia. Annali 15. L'industria*, ed. Franco Amatori, Duccio Bigazzi, Renato Giannetti, and Luciano Segreto, 551–616. Torino: Einaudi.

———. 2004. L'eredità preunitaria. In *Le Poste in Italia. 1 Alle origini del servizio pubblico, 1861–1889*, ed. Giovanni Paoloni, 1–41. Roma-Bari: Editori Laterza.

———. 2012. Il sistema infrastrutturale italiano fra acquisizioni e vincoli, opportunità e mancanze. *Rivista di storia economica* 1: 61–77.

———. 2014. Technology Transfer, Economic Strategies and Politics in the Building of the First Italian Submarine Telegraph. *History of Technology* 32: 277–293.

Lampertico, Fedele 1859. *Sulle conseguenze che si possono presagire pel commercio in generale, e pel commercio veneto in particolare dall'apertura di un canale marittimo attraverso l'Istmo di Suez. Memoria di Fedele Lampertico premiata dall'Istituto Veneto nell'adunanza 19 maggio 1859.* Venezia: G. Antonelli.

Maggi, Stefano. 2003. *Le ferrovie*. Bologna: il Mulino.

Paleocapa, Pietro. 1856. *Considerazioni sul protendimento delle spiagge e sull'insabbiamento dei porti dell'Adriatico applicate allo stabilimento di un porto nella rada di Pelusio.* Torino: Barera.

Sapeto, Giuseppe. 1865. *L'Italia e il canale di Suez. Operetta popolare.* Genova: Pellas.

Schram, Alberto. 1997. *Railways and the Formation of the Italian State in the Nineteenth Century.* Cambridge: Cambridge University Press.

Sul Portosaido. 1867. *Al signor Ferdinando de Lesseps presidente e direttore della Compagnia universale del canale marittimo di Suez. Lettera del comm. Alessandro Cialdi azionista nella Compagnia universale di detto canale.* Roma: Tipografia delle Belle Arti.

Torelli, Luigi. 1870. *Dieci paralleli fra il progresso dei lavori della Galleria del Cenisio e quelli del Canal di Suez a partire dal 31 maggio 1867 al 15 novembre 1869.* Venezia: G. Antonelli.

Vimercati, Cesare. 1864. *Il canale dell'istmo di Suez. Sua influenza di pace per l'Europa e di rigenerazione per l'Italia.* Livorno.

Virgilio, Jacopo. 1869. *Il commercio indo-europeo e la marina mercantile italiana a vela ed a vapore.* Genova: Tipografia sociale.

CHAPTER 4

The Suez Canal and Italian Port Cities: Local Competition and Global Ambitions Within a Nation in the Making (1850s–1870s)

Giovanni Cristina

When the Suez Canal was officially inaugurated in Port Said, on 16 November 1869, Italy was a recently formed nation-state. This meant that the strategies of the pre-unification states in trade, infrastructure, and geopolitics were still ongoing issues in the national debate. The tradition of autonomy of individual cities and territories in pursuing their own interests, which was the outcome of the political division of the Italian peninsula since at least the Middle Ages, long coexisted with a progressive centralization of economic decision-making after unification.

The prospect of a possible renewed Mediterranean centrality generated by the opening of the canal had somehow awakened a sense of 'Italian' patriotism among the various territorial contexts of the peninsula. In fact, in periodicals, publications, and pamphlets on the canal which, starting as early as the 1850s, flourished in Genoa, Turin, or Venice, the term 'Italy'

G. Cristina (✉)
Università di Genova, Genoa, Italy
e-mail: giovanni.cristina@unige.it

B. Curli (ed.), *Italy and the Suez Canal, from the Mid-nineteenth Century to the Cold War*,
https://doi.org/10.1007/978-3-030-88255-6_4

takes on different meanings and does not necessarily imply the existence of a unitary political entity. Several political, scientific, and economic figures of Italian culture—such as the banker from Trieste Pasquale Revoltella; the president of the Chamber of Commerce of Venice, Giuseppe Maria Reali; the future Italian Minister of Trade and Industry, Luigi Torelli; the Austrian engineer Luigi Negrelli; or the Piedmontese Minister of Public Works, Pietro Paleocapa—participated in different ways to the planning of the Suez Canal. The time of geo-economic strategies and technological progress anticipated that of political change.

The 'Isthmian Literature': The Suez Canal and Italian Nation-Building

The patriotic fervour, inherent in the process of nation-state building, sometimes incorporated the pre-existing geo-economic strategies of the various territories and rhetorically reinterpreted them in the light of an 'Italian' project for the global expansion of local trade. More generally, as the president of *Società geografica italiana* Cesare Correnti also argued, de Lesseps's enterprise had reawakened the idea, or the illusion, that 'the ancient Mediterranean would once again become for civil geography what it had always been for physical geography, [that is] the centre of the habitable world'.[1] Once again, according to Correnti, 'ever since the fame of Mr. Lesseps and his bold design began to rise, praise, advice, information, exhortations, prophecies abounded', giving birth to 'a kind of isthmian literature'.[2] In the numerous publications inspired by the prospect of opening the Suez Canal, at least three disciplines in support of the arguments on and analysis of the effects of a new way of communicating towards the East clearly emerged: geography, political economy and a form of narrative that could be defined as 'historicist-poetic'. Most commonly, these pamphlets were cross-disciplinary. Consequently, according to Torelli, the impact of the *grandeur* of the new public works in the collective image of the time had converted 'a host of men of lively imagination' into 'public economists', creating the figure of the 'economist-poet'.[3]

[1] Correnti (1869, 5).
[2] Correnti (1869, 11).
[3] Spadoni (1970, 658).

Alongside the dichotomy between traditional local interests and new national aspirations, the study of pamphlets, journal articles and publications in general also reveals a twofold rhetorical register. On the one hand, the potential consequences of the opening of the Suez Canal created a patriotic enthusiasm, linked to the nascent ideology of *navalism*[4] and imbued with references to history—for example, greatness of the Roman Empire, the glories of Maritime Republics, and so on—and geography (i.e. the centrality of Italy in the Mediterranean). On the other hand, more pragmatic analyses related to political economy gave a rather negative outlook of the actual state of Italian port infrastructure and merchant navy and their competitiveness in global trade. The first register includes the *toast* proposed by the representative of the Chamber of Commerce of Livorno, Giacomo Giorgio Levi, to de Lesseps, defined as the 'Dante [Alighieri] of the canal', during the Ismailia Conference of 1865:[5]

> Almost everyone believed that the grand idea of M. de Lesseps was a utopia. All great things have enemies, that's why the Suez canal had and currently has many enemies [...] I compare the canal to the Italian Unification [...] Prince Metternich, who was one of the greatest ministers and diplomats, [...] said that Italy was a geographical expression and unification a utopia. [...] Well, Italy is done, and the Suez canal is done, too.[6]

To the second register a trade treaty written by Jacopo Virgilio undoubtedly belonged. The Ligurian economist stated that 'it would be sheer folly to believe that the commercial revolution that is now taking place in the Mediterranean is an absolute consequence of the cutting of the Isthmus'.[7] In fact, the opening of the Suez Canal, and therefore the strengthening of Euro-Asian trade routes, took place when a series of breakthroughs was already underway, such as the transition from sailing to steam and the construction of the railways, both in Europe and in Egypt, which made preliminary forms of intermodality possible.

[4] Monina (2008, 32–8).

[5] The conference was an initiative planned by the Italian ministers Torelli and Jacini in agreement with de Lesseps himself, and involved representatives from several Italian Chambers of Commerce, in order to verify on the spot that the works would shortly be completed and to promote a pro-canal sentiment in national public opinion.

[6] Levi (1865, 14).

[7] Virgilio (1869, 9–10).

The contrast between euphoric enthusiasm and pessimistic realism was not only due to Levi's and Virgilio's different disciplinary backgrounds or personal viewpoints. The four-year distance between the two quotations also influenced the opposed registers towards the impact of the opening of the Suez Canal on the economic and commercial panorama of the Italian peninsula. Some passages from a lecture held at the Industrial Institute of Venice by professor Gaetano Cegani in 1869 are particularly revealing:

> Since de Lesseps disclosed his project, the thunderous applause of the deluded and the dreamers stifled the calm objections of the clairvoyant expertise and skeptically wise doctrine. The Suez canal was to open up the land of Cockaigne, India was an Eldorado, stolen by Europe from England, the East was the garden of the Hesperides, whose first fruits our Italy was supposed to pick, under the auspices of resurrected Venice.
>
> As the mirage disappeared, the naked, bleak truth emerged [...] The spell was broken, little by little every illusion vanished: and if, at first, it could be believed that all the goods sent from the Orient to northern and western Europe, had to land in the Italian ports to continue their journey on the railways, very soon it was clear that the enormous tariff of these [railways] made the sea route preferable; and that for high-volume and not so sought-after goods sail could be preferred to steam. Then finally it was discovered that, despite the railways and the canal, most of the ships of the Mediterranean nations would still have had to move to the North seas; and that the ports of England would continue to be the starting-point for Asian European trade.[8]

The Suez Canal Between Local and Global

What emerges from the Suez Canal debate is a 'national-patriotic' horizon of thought, related to Italy—as a single geographical and cultural entity—before the actual existence of a political Italy. This approach can be found, for instance, in the speech given in 1856 at the Florence-based prestigious association for the studies of agricultural issues, *Accademia dei Georgofili*, by the archivist Francesco Bonaini: the maritime trade and industry of Livorno could benefit from the opening of the Suez Canal and of a new terrestrial route with Northern Italy (e.g. the *Centrale Italiana*). With a view, however, to 'restoring to Italy the prevalence over the seas, which

[8] Cegani (1869, 35–9).

unfortunately went to foreigners'[9] since the fifteenth century. The pre-unification states—especially those more sensitive to the processes of modernization, such as the Kingdom of Sardinia—had already started to support the initiative of de Lesseps or, at least, tried to evaluate the possible consequences of a new route towards the Indies for the Mediterranean trade.[10]

It was above all Cavour who had this 'national' vision of the strategic role of rail and maritime communications since before 1848.[11] Not surprisingly, the Kingdom of Sardinia—as the state that was to lead the Italian unification process—was the main promoter of the 'pro-canal' stance throughout Italy before Italy was made. The example of the journal animated by the engineer Ugo Calindri—the *Bollettino dell'Istmo di Suez*, nsatlantica whose publications began in Turin in 1856—is particularly revealing in this regard. In addition to collecting the most relevant international writings on the Suez Canal translated into Italian, it also hosted a press review of newspapers and pamphlets from various contexts of 'geographical' Italy, including Malta, which dealt with the topic.

The response by the Italian cities to de Lesseps' endeavour was not homogeneous. Indeed, it was also affected by the European geopolitical situation of the time. The issue of the Suez Canal stood at the crossroads between international diplomacy, the aspirations of individual port cities, and the interests of private capitalism. If it is true that France, in this diplomatic war, was the main supporter of the construction of the canal, it is also true that those who promoted it was a private company, in need for subscriptions not only in France, but in the rest of Europe and in Egypt. As for the Italian context, if in the 1850s Cavour did not want to lose English support for Italian unification, Venice, instead, under Austrian domination until 1866, had enthusiastically welcomed de Lesseps' crowdfunding. On the other hand, Genoa's reaction remained rather lukewarm to the stock subscription. In general, as Spadoni maintains, 'if we consider the considerable contribution by technicians and the interest of some politicians, such as Ricasoli and Cavour himself, it is certain that the stocks subscribed in the [...] cities [of the Italian peninsula], and particularly in the maritime ones, were very scarce'.[12]

[9] Bonaini (1856, 32).
[10] Curli (2015).
[11] Torelli (1859).
[12] Spadoni (1970, 658).

In a sense, the difficulty encountered by de Lesseps in reaching out to most of the entrepreneurial and commercial class of Italian maritime cities represents yet another episode of the historically conflictual relationship between Italy and the sea, including symbolic aspects, images, and mental frameworks.[13] Moreover, the opportunity generated by the Suez Canal highlighted all the ambiguities of the relationship between the central state and local contexts in defining geo-economic strategies related to infrastructure and the establishment of new routes in international trade. Often, the modernizing zeal displayed by the Kingdom of Sardinia and its activism in international geo-economic diplomacy did not find an adequate response at the local level, mostly for lack of dynamism. Genoa and Venice were the most relevant exceptions because they had been the two most important maritime-commercial cities of the Italian peninsula since the Middle Ages, following, between highs and lows, the traditional division of Italian maritime traffic between the Tyrrhenian and Adriatic sides.[14] The case of Trieste had its own peculiarity, given that most of its economic élite, although mostly of Italian culture and language, remained closely linked to the geo-economic horizons of the Habsburg Empire and Adriatic area, rather than joining the Italian national project.[15]

Venice, Trieste, and the Adriatic Space

Together with Genoa, Venice was Italy's most important port. If Genoa's main competitor in the Mediterranean was the port of Marseille, on the Adriatic side Trieste was undoubtedly Venice's main rival. Even in these inter-port and interurban competitions, there were rhetorical justifications that referred to a 'national' mission. This is particularly evident, for example, in a report from the general assembly of 26 May 1859 of the *Société des ports de Marseille*. In this report, cited in an extract in Italian in the periodical *Bollettino dell'Istmo di Suez*, it was stated that the mission of the *Compagnie des Docks et Entrepôts de Marseille* was to 'fight for France against all foreign ports of the Mediterranean and Adriatic sea'.[16]

Although in a less nationalistic way, Italy was also animated by such patriotic purposes. The national dimension was able to temper even the

[13] Frascani (2008, 24–26).
[14] Cafarelli (2014, 14).
[15] Revoltella (1864).
[16] *Bollettino*, 28 May 1859, 306.

potential inter-port rivalries transforming them into a positive emulation effect, as the Venetian Cegani shows:

> Aware of our inadequacy, ashamed of our misery, would we [the Venetians] want to withdraw from this field [i.e. international trade], in which our ancestors were always leaders?
>
> And will Venice, the ancient queen of the Adria, give way, humiliated, to rival Trieste; will it end up under its rule, will it recognize its high dominion?
>
> Oh! Also Genoa, the superb queen of Liguria, hears at her side the industrious Marseilles in full swing; nevertheless it does not collapse at the feet of the mighty rival, but it surges forward, spurred on by emulation. Now if Genoa was doing so by private impulse and against the mighty Marseilles, why will Venice sit, inert and discouraged, before rival Trieste?[17]

The identity crisis of the port of Venice, both with respect to the competition of Trieste in trade with the Levant and the Far East and with respect to its possible role in the new Italian national state, had started to emerge as early as the 1860s. The opportunity to redeem Venice from a state of subordination with respect to Trieste in Adriatic trade in connection with Northern Europe came from the definition of shipping lines—subsidized or not—to the south-eastern Mediterranean and to the Indian Ocean. Before Venice was annexed to Italy in 1866, five years after unification, the Italian government had already decided to create a railway system that would reach the ports of Ancona and Brindisi, 'effectively transforming the shipping lines into a sort of extension of the railways'.[18] The choice to exploit the land route as much as possible had led the National government to focus on the port of Brindisi which, thanks to its geographical position, was considered the starting-point for communication by sea with Egypt and the Red Sea. In 1862 it was decided to entrust the English *Palmer* with the establishment of the shipping company *Adriatico-Orientale*, which operated a subsidized line between Ancona and Alexandria. At the same time, the Italian government invested many resources in strengthening the port of Brindisi, which was opened to bigger steamers in 1866. Both decisions raised strong doubts among many Italian economists, such as Luzzatti, Boccardo, and Virgilio.[19]

[17] Cegani (1869, 43–5).
[18] Cafarelli (2014, 12).
[19] Cafarelli (2014, 20–32).

In the 1860s, Venice found itself caught between two fires: on the one hand, the Italian government supported rail transportation and Genoa as a national port; on the other, Austria, delaying the opening of the Brenner tunnel, brought most of the Central European traffic to Trieste. Even when *Adriatico-Orientale* decided to extend the Ancona-Alexandria line to Venice in 1867, the overwhelming power of *Lloyd Adriatico* was evident and Trieste's superiority over *Serenissima* could not be questioned. A last hope for Venice could come from the possibility of establishing a line with Alexandria, via Ancona and Brindisi, by the Egyptian company *Azizieh*. However, despite the favourable vote of an *ad hoc* municipal commission, in 1867 *Azizieh* withdrew the proposal and *Adriatico-Orientale* remained the only shipping company to pass through the Venetian port.[20]

In the 1860s, the possibility of transiting the Indian Mail through the Italian peninsula, defined as the 'loading and unloading dock of world maritime trade, [which] reaches out into the Mediterranean more than any other part of Europe',[21] threw a spotlight on the ports of Ancona and Brindisi. A letter dated July 1863 from the president of the Chamber of Commerce of Ancona, Pietro Farsetti, addressed to de Lesseps himself, expressed the enthusiasm from that port-city towards the opening of the Suez Canal:

> Distinguished gentleman!
>
> If Italy as a whole wishes your business well and feeds the hope of its realization, these hopes and desires are all the more profound in the inhabitants of Ancona. In fact, the powerful current of exchange that, thanks to you, will be established between the eastern oceans and Europe—dividing the navigation of the world into four great branches, that of the Danube, that of the Adriatic, that of the Mediterranean and of Gibraltar—will push towards us a considerable part of this navigation and much of the new wealth.[22]

The case of Brindisi differed from the others. Compared to Venice, Genoa, and even Ancona, its local entrepreneurial and political fabric was weaker and did not ask central government to protect its aspirations within the growing international trade. The decision to focus on Brindisi 'as a natural seaport for mail and commerce' came exclusively from the Italian

[20] Cafarelli (2014, 24–42).
[21] Romano (1869, 4).
[22] Revoltella (1864, 18–9).

government and was based on geographical considerations that went beyond the local dimension. This choice, as has been mentioned, aroused strong opposition, both because of territorial rivalries and because of the lack of infrastructure in the Apulian port, which reinforced scepticism regarding its success *ex nihilo*.[23]

On the opposite side of the Adriatic, Trieste also tried to stimulate the Austrian imperial government in order to strengthen its maritime activities. In 1864 the banker Pasquale Revoltella addressed a pamphlet to the highest authorities of the Austro-Hungarian Kingdom 'to expand its political-commercial horizon, in order to actively associate itself with transoceanic world trade'.[24] Strategically, Revoltella made Trieste's interests coincide with those of the Empire and denounced its delay with respect to other powers in international trade with the Americas, Asia, and Oceania:

> If we really want to progress, it is necessary to look frankly and spiritually towards more distant regions [...]. If we are not able to profitably export our products to the German Custom league, to England, France, Belgium and Switzerland, this does not mean that we cannot send them directly to those foreign consumer markets, where the English, the French, the Germans of the North, the Belgians and the Swiss usually send theirs, and it is even less doubtful that we will be able to withstand the competition in those countries, where the geographical position and the cheaper transport rate is more favorable for us.[25]

The antidote to backwardness, according to Revoltella, lay in a greater commitment by the central government of Vienna in the diplomatic war against Austria's two biggest competitors in the Mediterranean: France and the new Italian national state. The city of Trieste thus tried to protect itself under the imperial aegis in order to defeat its commercial rivals. The possible opening of the Suez Canal, even in the case of Trieste, reveals that rivalries between competing port cities, to be fought effectively, must necessarily refer to a state and then to the more complex perspective of international relations. On the one hand, most of the commercial élite and representatives of the Chambers of Commerce in the second half of the nineteenth century leaned towards free-trade economic doctrine also at an

[23] Jacini (1869, 141–53 [quote 141]).
[24] Revoltella (1864, 2).
[25] Revoltella (1864, 8).

international level. On the other hand, it was clear that without the support of a state with greater powers and duties—that is, the authority to conclude commercial treaties with other states, military defence of maritime trade, geopolitical strategies before the development of new transoceanic routes, construction of railway infrastructures, establishment of consular networks—it was practically impossible to develop the commercial vocations of the single territories. While Revoltella insisted on the competition with Marseille, he saw the greatest threat to the development of the maritime trade of the port of Trieste, and therefore of Austria, coming from Italy:

> Despite a financial deficit, which is increasing year by year, the Turin government is working hard with alacrity and it must be said with admirable success in order to strengthen the Italian trade, which essentially damages Austria. Without dwelling on Genoa's efforts to replace our navigation more and more in the ports of the [Aegean] Archipelago and the Black Sea [...] we must here clearly highlight that the main tendency of the Italian Government is evidently directed to overcome Austria in the Adriatic sea, and to attract our maritime trade towards its ports.[26]

Therefore, the major concern of Revoltella was Italian competition on the Adriatic Sea, embodied by the strengthening of the railways for Ancona and Brindisi, considered as possible final entrepôts for goods and mail from Northern Europe:

> The railway already leads from the slopes of the Savoy Alps and Switzerland up to Ancona, and within a year direct railway communication will be achieved up to the shores of the Ionian Sea, that is up to Brindisi. An English company has already been set up to carry out the new project to conduct the railway up to Otranto [...]; huge sums are spent annually on improving the ports of Ancona and Brindisi; in Ancona there is a steam navigation company, which is very meaningfully named Adriatica-Orientale, which competes with Lloyd Austriaco on the line between Alexandria and Trieste [...]. The Federal Council of Switzerland is already negotiating with this company for the forwarding of correspondence to the Indies via Ancona, Colico-Chur, and according to the news reports from the newspapers, the idea of creating an Italian steam navigation line to Suez and the Indies is being cultivated in Italy.[27]

[26] Revoltella (1864, 17).
[27] Revoltella (1864, 18).

If Revoltella's concerns are deliberately magnified in order to stimulate a response by Vienna, this competitive attitude, in which states, nations, and ports are fighting with each other for the conquest of new spaces, marks a sharp change compared to the 'peaceful' multi-nationalism of some years earlier.[28]

Genoa, Namely, Pragmatism

Since the beginning of the nineteenth century, Genoa seemed to have found a privileged outlet for its maritime activities in the route to South America. Thus, 'already in the 1830s over three thousand Ligurians [were] present in the area of the Rio de la Plata as owners, captains or sailors of vessels and boats for medium and short range navigation'. The local governments, 'just emerging from the wars of liberation from the Spanish Empire, often entrusted' to these Ligurian sailors 'the geographical and topographical reconnaissance of their territories'.[29] Genoa's greater activism compared to that of Venice as for international trade had been preempted, in the first half of the century, by some explorations carried out by local pioneers of transoceanic navigation who sailed, in addition to the transatlantic one, also the Indian Ocean route, up to China, Japan, the Philippines and Australia. Furthermore, the shipping company *Transatlantica* had already been founded in the 1850s, whose seven steamers connected Genoa to Rio de Janeiro and New York.

In line with this early familiarity with modern transport compared to other Italian contexts, the Genoese reactions to the possibility of opening up the Suez Canal were much more concrete. They were based on a careful and rational analysis of the opportunities offered by the new sea route with the Indies and paid greater attention to the starting conditions of the local merchant navy and the transalpine infrastructure system, necessary to make the port of Genoa competitive compared to Marseille or Trieste. In local debates, published in the newspaper *La Borsa*, for which important economists like Gerolamo Boccardo and Jacopo Virgilio also wrote, the Cavourian pattern of the indissoluble combination between navigation and railways was raised once again as a far-sighted strategy to bridge the gap with the most advanced European states.[30] Cavour's design to first

[28] Reill (2012).
[29] Surdich (1994, 457–8).
[30] Surdich (2003, 347).

bring the Kingdom of Sardinia and then Italy into the ranks of European powers was certainly linked to the presence of a strong state based on the French model, and included both trade relations with other states and a colonial projection into other continents such as South America, Africa, and Asia. Accordingly, the Genoese economists were aware that the challenge was represented by the ability to carve out an autonomous space, away from the French, English, or Austrian competition. This led, for example, to the acquisition by the shipping company *Rubattino* of the commercial base of Assab Bay, the first outpost of the future Italian colonial adventure in East Africa. This expansion was favoured by the orientalist Giuseppe Sapeto, with the involvement of the Italian Chambers of Commerce, on the occasion of the expedition to Ismailia in 1865.[31]

Patriotism from the Far South: Great Expectations in Syracuse

Regarding the Sicilian case, the booklet *Il Bosforo di Suez e l'avvenire di Siracusa* by Pasquale Midolo, the first president of the Chamber of Commerce of Syracuse, is particularly interesting.[32] It was also published in several instalments in the *Bollettino* in 1859: it represents a 'bottom-up' attempt to interpret the construction of an infrastructure in a patriotic, Italian way, but with potential repercussions also on a local scale.

After recalling the ancient history of the Mediterranean, the splendours of the Maritime Republics (significantly defined as 'Italian') and the geographical importance of Italy, Midolo assesses the future prospects offered by the Suez Canal to different Italian ports.

Venice 'designs basins, piers, freight yards, warehouses, docks, banks, warrants, etc. to recover its ancient greatness'. Genoa 'spends 2 million to lengthen the new pier by 150 metres; plans to enlarge its port enormously to accommodate the ship that Suez will send; prepares docks and steam tugboats and merges the *Società Vapori* into the *Lloyd Italiano*'. Turin 'charts another route to the Mediterranean for Savona [which] doubles the size of its port', while through the piercing of the Mont-Cenis Piedmont 'will bind itself with an iron chain and with an indissoluble

[31] Sapeto (1865).
[32] Midolo (1859); on Midolo, see Adorno (2011, 343–4).

commercial and industrial tie to France'. In Tuscany, Livorno—'rich in wood and a thousand industries, the port of an educated country, headed by the Athens of Italy'—'prepares itself for the new commercial movement, and from its quay large ships are launched to sail distant seas'. Even 'the flourishing Kingdom of Naples is not deaf to the new trend: the Government promotes every industry, opens communications: a railway to Apulia will exchange products from the Gulf with those from the Adriatic: another railway for the states of the Church will bind Italy; steamships cut through its clear sea waters, while the official press of the Kingdom studies the results of the great project [of Suez]'.

However, according to Midolo, the Italian region that would have benefited more than the others was Sicily: 'a bridge between Asia, Africa and Europe, with the cutting of the Suez Isthmus, it will become the banquet of all nations; in its ports the riches of Asia and Arabia will be exchanged with products just as precious from its land; from those [same ports] ships built on its shores with timber from its woods will move, operated by its intrepid seafarers'. Although Midolo was linked to Syracuse, he did not exclude other Sicilian cities from his panegyric. Palermo, defined as 'an imitable example of agricultural wisdom', increased its trade 'with the marketplaces of the Mediterranean and other seas' thanks to the export of 'flaxseeds, sumac, oil, sulphur, citrus fruits, and almonds'. Messina, instead, 'placed at the crossing of the *Faro*, with the privilege of being a depot to foreign imports [...] increases its trade year after year'. Catania, 'jewel that sits at the foot of Etna [...] rich in capital, work and science' is 'destined to become the Antwerp of Sicily'. Essentially, according to Midolo, 'the cutting of the Suez Isthmus is a powerful development opportunity for Sicily'.[33]

It is obviously Syracuse and its hinterland at the core of his pamphlet. Here the narrative becomes twofold, even contradictory. On the one hand, Midolo celebrates the natural and historical magnificence of the Syracuse area with idyllic tones, focusing on the products that naturally grow from the land (e.g. olive oil, grains, wine, almonds, locust beans, and citrus fruits) and could be exported abroad thanks to the opening of the canal and the consequent inclusion of Syracuse in the international trade circuit. On the other hand, the backward-looking commercial situation of

[33] *Bollettino*, 25 June 1859, 376, 382–383, 389.

the city, little accustomed to economic enterprise and entrepreneurial dynamism, is described with great realism.[34]

Conclusion

Starting from the 1870s–1880s onwards, both the optimistic tones towards the technological innovations, attributable to the Sansimonian progressivism *à la* Chevalier, and the patriotic enthusiasm towards the commercial destinies of Italy that had been raised at the time of the opening of the Suez Canal disappeared almost totally from pamphlets, reports, conferences, and parliamentary inquiries on maritime issues. Instead, the analyses—often confused and partial—on the needs and limits of national maritime trade with respect to foreign competition multiplied, especially with regard to the backwardness of the merchant steam fleet, the tax system levied on import/export, and the state of port infrastructure.[35] At the same time, in the economic debate between liberalists and protectionists, which also had political-ideological origins, the latter ended up prevailing, as demonstrated by the customs policy adopted by the Depretis governments.[36] The transition from international maritime trade—considered as an opportunity for general well-being and for the union of peoples—to an increase in international competition, which would then lead to colonialism, and imperialism, was accompanied by a more marked polarization of traffic in some specific ports, such as Genoa: in the absence of a clear port policy at government level, many of the aspirations of the numerous Italian port cities were frustrated.[37]

In the medium term, 'for Italy the opening of the canal did not have the same positive consequences as other countries': it 'was the beginning of an increase in the gap with competing merchant navies, precisely because Suez played an important role in the creation of a large international market, within which, due to its slow evolution, the Italian fleet was eliminated, as it was for a long time, from world competition'.[38]

[34] *Bollettino*, 25 June 1859, 411–420.
[35] Flore (1970, 295).
[36] Grange (1994, 59).
[37] Tonizzi (2018, 29).
[38] Rollandi (2002, 245).

References

Bollettino dell'Istmo di Suez. 1859.

Adorno, Salvatore. 2011. Scenari di centralità mediterranea. Retoriche e politiche delle classi dirigenti siracusane (1860–1960). In *Il Mediterraneo delle città*, ed. Enrico Iachello-Paolo Militello, 342–353. Milan: FrancoAngeli.

Bonaini, Francesco. 1856. *Livorno considerato nelle sue presenti condizioni e nel suo avvenire*. Florence: Cellini.

Cafarelli, Andrea. 2014. *Il leone ferito. Venezia, l'Adriatico e la navigazione sussidiata per le Indie e l'Estremo Oriente (1866–1914)*. Rome: Viella.

Cegani, Gaetano. 1869. *Il canale di Suez. Lezione tenuta nell'Istituto Industriale di Venezia*. Milan: Treves.

Correnti, Cesare. 1869. Sull'istmo di Suez e sul commercio orientale. In *Bollettino della Società geografica italiana (III)*. Florence: Civelli.

Curli, Barbara. 2015. Il Piemonte e il canale di Suez, 1855–1856. *Studi Piemontesi* XLIV (2): 325–338.

Flore, Vito Dante. 1970. *L'industria dei trasporti marittimi in Italia. Parte II. L'azione dello Stato tra il 1860 e il 1965*. Rome: BIM.

Frascani, Paolo. 2008. *Il mare*. Bologna: il Mulino.

Grange, Daniel J. 1994. *L'Italie et la Méditerranée (1896–1911)*, 2 voll. Rome: École Française de Rome.

Jacini, Stefano. 1869. *Sulle opere pubbliche d'Italia nel loro rapporto collo stato*. Milan: Civelli.

Levi, Giacomo Giorgio. 1865. *Lettera sui lavori dell'istmo di Suez visitati dal barone Giacomo Giorgio Levi rappresentante la Camera di commercio di Livorno e Pisa*. Florence: Cellini.

Midolo, Pasquale. 1859. *Il Bosforo di Suez e l'avvenire di Siracusa*. Syracuse: Pulejo.

Monina, Giancarlo. 2008. *La grande Italia marittima. La propaganda navalista e la Lega navale italiana (1866–1918)*. Soveria Mannelli: Rubbettino.

Reill, Dominique. 2012. *Nationalists Who Feared the Nation: Adriatic Multi-Nationalism in Habsburg Dalmatia, Trieste and Venice*. Stanford: Stanford University Press.

Revoltella, Pasquale. 1864. *La compartecipazione dell'Austria al commercio mondiale*. Trieste: Tipografia del Lloyd Adriatico.

Romano, Giovanni Antonio. 1869. *La Valigia delle Indie e l'Europa*. Venice: Cecchini.

Rollandi, Maria Stella. 2002. Lavorare sul mare. Economia e organizzazione del lavoro marittimo fra Otto e Novecento. *Atti della Società ligure di Storia patria* XLII (2): 236–471.

Sapeto, Giuseppe. 1865. *L'Italia e il canale di Suez: operetta popolare*. Genoa: F.lli Pellas.

Spadoni, Ugo. 1970. Il canale di Suez e l'inizio della crisi della marina mercantile italiana. *Nuova rivista storica* LIV (V–VI): 651–702.

Surdich, Francesco. 1994. I viaggi, i commerci, le colonie: radici locali dell'iniziativa espansionistica. In *Storia d'Italia. Le regioni dall'Unità a oggi. La Liguria*, ed. Antonio Gibelli-Paride Rugafiori, 455–509. Turin: Einaudi.

———. 2003. Una geografia per l'espansione commerciale e coloniale. *Atti della Società ligure di Storia patria* XLIII ((CXVII), II): 337–414.

Tonizzi, Maria Elisabetta. 2018. L'Italia e il mare. I porti e la politica portuale (1861–1913). *Contemporanea. Rivista di Storia dell'800 e del '900* XX (1): 27–49.

Torelli, Luigi. 1859. *Dell'avvenire del commercio europeo ed in modo speciale di quello degli Stati italiani*. Volume II. Florence: n.d.

Virgilio, Jacopo. 1869. *Il commercio indo-europeo e la marina mercantile italiana a vela ed a vapore*. Genoa: Tipografia sociale.

CHAPTER 5

The Suez Canal and the Italian Sailing Fleet: Expectations, Problems and Alternative Routes (1869–1914)

Leonardo Scavino

On 17 November 1869, on the notes of Johan Strauss, the passage of the French ship *Aigle* through the Suez Canal represented a crucial milestone in the history of navigation. The construction of this passage, which connected the Mediterranean with the Red Sea, was aimed to provide new impetus to the world maritime traffics. Although no official delegates attended the inaugural ceremony (Amedeo of Savoia-Aosta was summoned back to Italy because of the King's sudden illness), the opening of the Suez Canal had unsurprisingly become a crucial subject within the Italian public discourse.[1] In the previous years, numerous publications related to the de Lesseps's monumental infrastructure had circulated, and the news from the canal construction site was widely discussed within the

[1] Bono (2006).

L. Scavino (✉)
Università di Genova, Genoa, Italy

B. Curli (ed.), *Italy and the Suez Canal, from the Mid-nineteenth Century to the Cold War*,
https://doi.org/10.1007/978-3-030-88255-6_5

Italian society.[2] This interest stimulated the composition of various pamphlets, articles and discourses, which shared an optimistic, and positivistic, attitude about the role of the canal for the global economy and, in particular, for enhancing the Italian position among the European powers.

In this context, intellectuals, economists and engineers—together with the Italian entrepreneurial class assembled in the regional Chambers of Commerce—tried to frame the construction of the canal within more complicated commercial and infrastructural designs. Most of their propositions addressed the improvement of land transport (railways), its integration with commercial sea routes and the renovation of the harbour facilities as the state's most urgent priorities. Nevertheless, within this framework, few people were concerned about the characteristics of the national merchant marine which, in the 1860s, albeit in optimal conditions and even rising within the international scenario, lacked the structural requirements to profit from the maritime transport revolution brought by the canal. However, when the canal opened, sailing vessels were still the heart and lungs of the Italian merchant marine: the shipping factor and, in particular, the question whether Italian maritime operators were able or not to pass through the canal and, therefore, to participate to its commercial benefits deserved more attention from all the supporters of the Suez option. After 1869, the Italian shipping system and its position among maritime powers received no evident advantages from Suez. Instead, while Rubattino's steamships conducted their pivotal travels on eastern routes, several sailing shipowners headed for different directions, for example, through the Cape route, disregarded by steamers but sufficiently cost-effective and profitable.[3]

Expectations: Suez and the Italian Trade

The opening of the Suez Canal to navigation raised expectations among Italian manufactures and shipping—the economic sector most exposed to the transport revolution. According to Errera, in his report presented at the 1869 National Congress of the Italian Chambers of Commerce: 'there is no Chamber of Commerce of the Kingdom which has not felt the urge

[2] Boccardo (1869), *Bollettino dell'Istmo di Suez* (1856–1858), De Lesseps (1856), Sapeto (1865), Torelli (1867), Vimercati (1864), Virgilio (1869).

[3] Spadoni (1970).

to develop considerations and studies about the masterpiece thanks to which the greatest part of the world traffics with India, China and the other countries of the Far East will be conducted into the Mediterranean.'[4] And Giuseppe Sapeto summoned: 'Be the commercial fate of Italy put in your hands, providential Italian Chambers of Commerce which, anxious to see the cut of the Isthmus done, send your representatives to examine with their own eyes the first drops of that marine channel, which might be the spring of gorgeous blessings!'[5]

The geographical position of the Italian peninsula along the new route that connected Europe with the Far East was one of the most recurrent arguments to sustain the canal's positive effects on the national economy. The relative proximity of the Italian South to Suez was perceived as the most crucial factor to determine the long-sought reestablishment of the Italian trade and shipping within the international scenario. According to many contemporary observers, if Italy were able to reposition at the centre of Indo-European commerce by exploiting the 'Suez revolution', transit trade would have naturally increased while redirecting from the Atlantic to the Mediterranean. For its relative proximity to Egypt among the western European countries, Italy would have converted into a deposit for the incoming and outgoing merchandises from and to the Old Continent. To depict the role which expected Italy after the canal opening, sentences such as 'deposit of the world ports' spread widely within the public discourse; within these designs, the first and crucial objective consisted in the realisation of a flawless integration of land and sea transports. Western industrial products and oriental merchandises would have flown to the Italian ports, put in connection with Central Europe through railways. Indeed, the beneficial effects of the Suez Canal were part of more ambitious designs. In the positivistic climate of the period, de Lesseps' work was integrated within a broader transport framework, in which ports infrastructures, railways and Alpine tunnels were essential to revolutionising the Indo-European trade. The rhetoric which assimilated the Suez Canal with railways and, in particular, with similar large-scale projects such as the Brenner, Fréjus and Gotthard Alpine tunnels, was an everyday discourse.[6] For instance, during the discussions to prepare for the opening of the

[4] Relazione Errera, Congresso annuale delle Camere di Commercio del Regno d'Italia, ASGe, CC, b. 39, 1869.

[5] Sapeto (1865, IV–V).

[6] Spadoni (1970, 656).

canal, the Genoese Chamber of Commerce praised the 'coincidence of these two masterpieces'[7] (the Fréjus tunnel and the Suez Canal) and their expected positive influence on the national trade.[8] From an anthropological point of view, the Suez Canal epitomised the first step of a process leading to human takeover against nature constraints.

However, Suez was necessary but not sufficient to the Italian success among European nations and, therefore, needed to be complemented by further measures. In these schemes, the development of Genoa, Venice and Brindisi was critical, as these ports represented the continental gateways for Indo-European trade, in both directions. In his works, for instance, Giuseppe Sapeto attributed distinct roles to each of them: Genoa and Venice were responsible for all the merchandises in transit to Western and Central Europe, whereas Brindisi had to specialise in direct Italian commerce, imports and exports.[9] However, the centrality of ports within the design of the Italian re-appropriation over European transit trade was rooted in a false assumption, concerning railways superior cost-effectiveness compared to sea transports. For instance, Luigi Torelli developed a detailed comparison between cargoes discharged in Brindisi and Venice: as a result, his studies outlined that land transport was less competitive, over the same distance, to shipping.[10]

Even the role of Genoa was not so stable as commonly believed. In the long run, the competition with Marseille was increasingly disadvantageous, in particular as far as the transit trade to continental Europe was concerned. In the late 1860s already, before the inauguration of the Fréjus tunnel, most experts were aware that the massive construction costs would have determined higher transport fares from Genoa in comparison with the standards of Marseille, connected with the Valley of Rhone through an utterly flat track. Moreover, in the preparation of the unlikely event of a boost in transit trade, the Ligurian port required adequate infrastructures, in particular docks and deposits. Nonetheless, the port of Genoa was modernised and enlarged only in 1888, and these works were accomplished only through the private donation of Raffaele De Ferrari who, in

[7] Congresso annuale delle Camere di Commercio del Regno d'Italia, ASGe, CC, b. 39, 1869.

[8] Despite the common rhetoric, public investments towards port infrastructures were limited in a comparative perspective. See Spadoni (1970, 680).

[9] Sapeto (1865, 168–80).

[10] Cora (1870), Torelli (1867, 74).

1876, donated twenty million (Italian Lira) to the city for this specific purpose.[11]

If transit trade was at the core of the public discourses, similar euphoric tones targeted import and export trade as well. However, this issue was more problematic, since it had to confront with the Italian depressed industrial production and its reduced consumption of colonial goods. In the mid-nineteenth century, the size of the commercial movement between Italy and Egypt, India and the Far East was almost negligible; in contrast, it was the United Kingdom to handle the overwhelming majority of the Indo-European direct trade relationships. Indeed, the comparison between the whole Mediterranean Europe and the United Kingdom returned a one to eight proportion in terms of volume and one to twelve in value.[12] The Italian industrial sector was unable to sell considerable amounts of national merchandises to the Asian markets and, as a result, Italian merchants lacked the means to purchase lots of products from the Far East.

On the contrary, most of the Italian exports value consisted of luxury textiles (silk), foodstuff (rice) and raw materials (sulphur), which found little or no market in Asia. Quite the opposite, they were abundant in the East and, therefore, their afflux to Europe would have been nothing but counterproductive for the Italian economy.[13] Instead, Asian markets were eager for cotton textiles, in which British production outnumbered Italian producers under all regards. According to Gerolamo Boccardo, Professor of Economics at the University of Genoa from 1860, olive oil, Tuscan marbles and Venetian glasses represented the Italian most marketable commodities in Asia; wine could complete this list, under the condition of a sharp productive turn to higher quality products.[14]

Expectations: Suez and the Italian Shipping

If most of the discourses targeted internal production and transit trade, the optimistic climate which dominated in the Italian public debate involved another vital national industry, shipping. According to some contemporary observers, while redirecting all the world trade to the Italian peninsula, the Suez Canal's thaumaturgical influence would have diffused

[11] Giacchero (1970), Doria (1973).
[12] Boccardo (1869, 20–1).
[13] Torelli (1867, 66).
[14] Boccardo (1869, 22–3), Torelli (1867, 66).

to the national merchant marine which, by the adoption of temperate protectionist policies, would have risen to the high ranks of the world shipping business.

A few years before 1869, the Italian shipping company of Raffaele Rubattino had started regular connections with Port Said and Alexandria; from the Egyptian ports, then, mail and merchandises travelled on the railway to Suez where—in force of a private agreement with the English company *Bombay Bengal Steamships*—were loaded on steamers to India.[15] Nevertheless, apart from Rubattino's efforts, the dimension and frequency of the Italian direct services to the Far East were negligible. Well before the opening of the canal, British companies (in particular the *Peninsular and Oriental Company*) had established regular connections from and to Suez by integrating land and sea transport from the Mediterranean to the Red Sea. Thus, most of the strategic traffic between the Mediterranean countries and India had been previously monopolised by such companies.[16] Furthermore, the absence of substantial Indo-Italian commerce failed to stimulate further experiments.

To the eyes of contemporary observers, however, the opening of the Suez Canal was preparing the ground for dramatic change in this *status quo ante*, and in these designs, the role of the Italian merchant marine was critical. In particular, the dramatic cut of the distance which separated Italy from the Far East markets epitomised the most crucial revolution:

> In such a blessed position, the enterprising maritime spirit of the Italian seamen, perceiving the immense opportunity for shipping profits, will surely rush into it. How could it be otherwise? The vessels of the whole country would need just to set sail for entering the African channel and, through that, reaching the Far East and the western coasts of America; on the contrary, given the difficulty to pass through Gibraltar and the extreme length of the Cape route, only few Italians frequent those countries now; in a future when, instead of being further, they will be closer to them, who could ever retain from participating to such adventurous maritime enterprises?[17]

The importance of the spatial dimension emerged in numerous publications. The Suez option implying a diminution of more than 13,000 kilometres along the route between Genoa and Bombay was particularly

[15] Virgilio (1869, 122).
[16] Giuntini (2011, 66–73).
[17] Interdonato (1856, XXI).

emphasised. However, distance was not the most relevant issue. The Italian supply of maritime transport—the total tonnage made available by the merchant marine—was higher than what the Italian trade demanded. Therefore, various shipping operators engaged in cross-trade—namely maritime transports carried out between different countries from the one where the ship was registered. As the Genoese economist Jacopo Virgilio pointed out in 1869, Italian shipping depended largely on cross-trade, as its aggregate value more than doubled that of direct import/export trade.[18] Accordingly, the majority of the Italian vessels refrained from returning to the Mediterranean for months, even for years, and, when serving along oriental routes, departed from British or Northern European ports. Within this specific business framework, the reduction of distance granted by Suez could not be as relevant as expected to the national merchant marine.

Secondly, the idealisation of geographical dimensions (as in the count of the kilometres) was a widespread misjudgement by authors who were unexperienced on maritime matters. In the shipping business, indeed, spatial distance represented one of the least essential economic components. More than space, time and costs were the leading factors to rule the market for maritime transport. The adoption of a maritime perspective, therefore, might be fundamental to reconsider the relative importance of the Suez Canal over the coeval world sea-borne traffic.

Theoretically, as far as time saving was concerned, the most considerable advantages regarded the Mediterranean connections with the western Indian ports, such as Bombay, for which the reduction was enormous. However, differentiation between sailing vessels and steamers is fundamental to grasp the respective pros and cons. To this purpose, Gerolamo Boccardo mentioned the outcomes published by the Dutch Commission for the Suez Canal, according to which a clipper, from the Netherlands to the Sunda Strait and back, could save about eleven days with the Cape option.[19] On the opposite, the trend changed significantly as far as steamers were concerned; steamships able to travel at a steady pace of ten miles per hour could diminish the journey by 50–60% of its total duration.

Finally, as in most businesses, in shipping productive costs exerted the most relevant influence. For most of the Italian operators, the passage through Suez was not a profitable option. This lack of economic

[18] Virgilio (1869, 96).
[19] Boccardo (1869, 12).

attractivity stemmed from the combination of inner structural limits of the Italian merchant marine and the navigational conditions offered by the canal, as the ocean-going fleet owned by Italian shipowners was in no conditions to sail cost-effectively through Suez. Indeed, from a technical point of view, navigating into the canal represented a troublesome challenge for sailing vessels. Since the limited depth and the variable winds and currents hindered the passage of big sailing ships even in the waters of the Red Sea, captains were forced to hire steam tugboats for the entire course. Indeed, despite advertised as much as one hundred metres large, in width the canal measured more realistically around sixty metres and the central section—where the ships were supposed to navigate—stretched for no more than twenty-two metres.[20] Discrepancies were acknowledged between the official data and the reports of first-hand witnesses: in several cases, ships with a draught measuring about five metres incurred in collisions and groundings, despite the alleged eight metres depth.[21] These troubling conditions affected voyage safety and, as a result, could be translated into higher insurance premia (the same went for the Cape circumnavigation, which was listed among the dangerous routes and, therefore, required additional costs).

However, it was the usage of tugboats (mandatory for sailing ships over fifty tons) to epitomise the most evident comparative disadvantage for sails as opposed to steamships which, at the same time, were allowed to navigate without further costs. The impact of tugboats rentals and passage fares on voyage costs was considerable. In return, Suez offered just a slight time reduction which, however, in the economy of a sailing ship, determined nothing more than a proportional diminution of labour cost, insufficient to cover the higher expenditure. The equivalent factor (time saved), instead, retained a much more significant value for the benefit of steam voyages where, together with labour costs, fewer navigation days translated into a considerable cut in coal consumptions, a critical item in steamers operational budget.[22]

[20] Idem, 7.

[21] Idem, 6.

[22] The Ministerial Inquiry on the conditions of the merchant marine (1882) provided various calculations and comparisons of sailing vessels and steamers operational costs. According to the Cadenaccio Bros., ship-builders, for instance, coal consumption (together with oil, fat and routine maintenance of the engines) accounted for about 10–15% of the monthly expenditure. See *Inchiesta*, I, 86–7.

Concerning the sail against steam competition to control maritime business, the Suez Canal might have accelerated underlying processes in favour of the uncontested establishment of steam shipping in the world traffics. In studying the transition from sail to steam, maritime and economic historians have proposed a two-stage evolution.[23] During the first phase, sail and steam competed for the same cargoes (1840–1870), leading to the progressive fall of freights and the marginalisation of sailing vessels into peripheral market niches. In the second period (1870–1914), instead, it is possible to observe a gradual evolution of the freight market into complementary sub-sections until, after the First World War, new technological advancements granted a definitive triumph to machine-powered ships (owing to the substitution of coal with liquid fossil fuels).[24] According to this model, steamers carried high-nominal-value merchandises, in order to cover the massive operating costs. Meanwhile, sailing ships specialised in the transport of low-nominal-value goods, whose relatively low freight rates were not profitable for steamships.

Alternatives: Italian Sailing and Steam Shipping to the Far East

Despite Suez's fundamental role, a history of the technological transition of the Italian merchant marine from sail to steam would extend well beyond the objectives of this chapter.[25] Likewise, also the analytical description of how public and private decisions intermingled with the lack of suitable natural resources in determining the late Italian transition would trespass its thematic. Nevertheless, it might be worth pointing out some crucial features: from a purely quantitative perspective, in Italy, steamers surpassed sailing vessels—in terms of tonnage—in 1907.[26] Academic scholars have variously interpreted the resistance of Italian shipping to the technological substitution of sail with steam. Some explanations blame the conservative attitude of shipowners, deemed incapable of conceiving more innovative forms of shipping than those belonging to

[23] Stopford (2009, 108–18), Graham (1956), Williams and Armstrong (2012), Harley (2015).

[24] Corbino (1926, 191–2), Stopford (2009, 108–18), Fletcher (1958).

[25] A rather recent attempt to introduce the issue of technological transition in shipping in the Italian historiographic debate can be found in Frascani (2001).

[26] *Relazione*, 105.

their tradition. More convincing, instead, is the identification of specific features which may have slowed down the modernisation of Italian shipping, for example, the limits of the shipbuilding industry, because of the costs and limited availability of materials for modern constructions (e.g. iron and steel replacing wood), as well as the lack of required know-how. Among further factors affecting the adjustment of Italian shipping to the international standards, the inadequacy of the national maritime credit and financial systems is recurrently mentioned.[27]

Already in the 1880s, several contemporary observers emphasised how unfortunate had been the timing of the transition, since it coincided with a rising economic cycle for Italian sailing shipping (the so-called golden age of sail) which, in the 1860–1874 timespan, received massive investments. Heavily indebted and unable to access further credit lines, in conjunction with a global freight crisis (1874–1898),[28] sailing shipowners needed to protect their investments and, as a result, contributed to delay the transition after the end of the downward cycle. Furthermore, until the late 1880s, few experts would have predicted sailing vessels' disappearance from world traffics: despite steam increased competitiveness resulted in the gradual marginalisation of sailing shipping, its complete capitulation was far from being foreseeable. On the contrary, some argued a long-term superiority of sails on oceanic routes, in particular for low-nominal-value goods, whose freight rates were too low for steamers to compete profitably.[29] As a result, the opening of the Suez Canal exacerbated the technological gap between sail and steam and accelerated their partition in two distinct market trajectories, the first flowing through Suez, the second round the Cape.

A distinction should thus be made between the few pioneering steam operators, such as Rubattino, who rapidly embarked upon the canal business, and the vast majority of shipowners, who stuck to sailing shipping up to the early twentieth century. In 1868, Rubattino signed an agreement for state-subsidised regular steam services to Alexandria. According to the terms of the agreement, the company would provide five screw-propelled steamships of 1500 tons each, and maintain three services per month from Genoa, Livorno, Naples and Messina to Alexandria.[30] Then, as a further

[27] Doria (2001).
[28] Stopford (2009, 112–14), Kaukiainen (1992).
[29] See *Inchiesta*, I–VI.
[30] Codignola (1938, 280–1).

demonstration of its commitment to the Suez cause, in March 1869 Rubattino responded to the invitations of the Genoese Chamber of Commerce and communicated its willingness to organise a commercial cruise which was preliminary to the institution of an Italian version of P&O's 'valigia delle Indie' (Indian Mail):

> It is being said that Italy is made, but not completed; conversely, we should say that Italy is made, but not known. The commercial cruise which we propose will represent an effective mean to promote our country in these oriental ports; once gathered all the sample of the commodities, it will depart to cross the Egyptian Bosphorus and, during its voyage, it will touch all the significant ports of the Red Sea, the coasts of Eastern Arabia, English India, Java, Borneo and China.[31]

Such experimental travel aimed at the exhibition of various samples of Italian products in the markets of the Far East. Moreover, Rubattino carried out such endeavour with no public subsidies. Then, in order to complete this design, in the same year Rubattino financed Giuseppe Sapeto for the purchase of the Assab Bay and to cover his needs for coal stations along the route to Bombay.[32]

Hence, with the signature of a more profitable contract (1877) and after the fusion with Florio in the Navigazione Generale Italiana (NGI), Rubattino's steamers remained dominant in the Italian commercial and postal communications with India. In 1891, for instance, according to archival evidence, five NGI steamships (*Po, Domenico Balduino, Singapore, Raffaele Rubattino* and *Manilla*) were recorded connecting Genoa with Bombay, whereas other two (*Bisagno* and *Bormida*) covered the complementary leg from the Indian port to Hong Kong.[33] These liner services, sailing between Italy and India back and forth, transported large arrays of high-nominal-value cargoes, ranging from cotton fabrics (British) to sulphur, almonds, marmalades, wines and liquors of different kinds. In this regard, the logbook of the ship *Manilla*, dated to 1887 (one of the few logbooks of this type and period among the logbook collection kept at the Genoa State Archives), seems to constitute a trustworthy sample of this shipping typology. Indeed, the list of commodities included: wine, sul-

[31] Relazione Errera, Congresso annuale delle Camere di Commercio del Regno d'Italia, ASGe, CC, bb. 39, 1869.

[32] Codignola (1938, chap. 4).

[33] Bombay, Movimento generale della navigazione, ACS, MM, DGMM, Dpct, 61.

phur, almonds, national and foreign liquors, marmalades, marbles, silver bars, cement, coral, mirrors and jewellery. From Bombay, instead, the *Manilla* transported wheat, raw cotton (mainly), spices, china potteries, indigo and coffee to Naples, Barcelona and Genoa.[34]

However, notwithstanding Rubattino and its epigones, most of the Italian shipping did not qualify for navigating through the canal. Several shipowners persevered in wooden sailing constructions and for them the Suez route was not an option to trade with India. Nonetheless, in the second half of the nineteenth century, an increasing number of Italian ships called to Far East ports. This new movement was even more evident in comparison to the previous decades when the Italian presence in the eastern waters was almost negligible. Although Suez had gradually absorbed most of the highly profitable transports, the Cape route was still active, and numerous sailing vessels carried low-nominal-value cargoes (cereals, coal, timber) according to the 'old way'. Not only Cape Town retained its status of a crucial hub for Indo-European trade but it was also part of a broader international framework.

The Italian participation to this shipping movement was the result of the previous experience of several operators as cross-traders, in particular of Black Sea grain to the United Kingdom. Owing to the success achieved in the 1850–1860 period, they began to call more frequently to the British and Northern European ports. Data evidence shows that there, and in particular in Cardiff, the development of high demands for sea transports to Cape Town loaded with coal cargoes. In the biennium 1886–1888, for instance, approximately ninety Italian vessels left the British ports to the Cape: more than 60% of them departed from Cardiff, others from Newcastle, Newport and Swansea.[35]

The main cause of this commercial movement lied into steamers success in the Indian Ocean: in order to navigate, steamships needed significant quantities of coal, which was naturally scarce in these regions. In other words, steamships stimulated the regional demands for coal, whose transport, paradoxically, was performed on sailing vessels and passed through the Cape instead of the canal. As a result, steam navigation nurtured its sail counterpart by providing abundant and incessant coal freight. Meanwhile, the sailing vessels low productive costs constituted an indispensable factor for steam navigation to expand into peripherical markets, where its

[34] Piroscafo *Manilla*, Giornale di boccaporto, ASGe, GN, 1158-1.

[35] Capetown, Movimento nazionale nei porti esteri, ACS, DGMM, Dpct, bb. 57–61.

operational sustainability much depended on the availability of low-cost coal supplies.

Turning back to the Italian vessels arriving at the Cape, the records kept by the local consulate are fundamental to reconstruct the pattern of their activities. After calling at Table Mountain, some ships continued to the ports of the Far East to discharge coal or ballast and then reached Rangoon, Singapore, Batavia or Moulmein, from where most of them loaded teak or rice cargoes. Then, on their way back, several ships stopped in Saint Helena instead of Cape Town. Indeed, in the same period (1886–1888), out of 23 Italian vessels calling at the British Overseas Territory, four came from Moulmein, Surabaya and Rangoon each, two from Bangkok and Pasuruan and one each from Bassein and Singapore.[36]

Otherwise, shipmasters were required to steer west to America: accordingly, more than half of the Italian ships which arrived in Cape Town declared Pensacola, New York, Miramichi or Philadelphia as final destinations. Most of these vessels travelled westwards (from Africa to North America) on ballast: nevertheless, the Cape route was still more profitable than sailing on ballast for the whole voyage from Europe to America. Indeed, because of the chronic deficiency of outbound cargoes from Europe to the American ports (with the partial exception of migrants[37]), several ships accepted cargoes to Cape Town before going to North America. There, most of them retrieved pitch-pine timber cargoes, widely commercialised in Latin America and Europe.[38] In other words, the Cape was positioned at the centre of a multilateral and complex commercial framework, stretching from the Far East to America.

Conclusion

For Italian sailing operators, the permanence of the alternative routes around the Cape played a fundamental role for the resiliency of the merchant marine to transition from sail to steam. If Suez, from 1869 onwards, epitomised the mainstream maritime and commercial connection between Europe and India and, therefore, British shipping rapidly seized it, this

[36] Capetown, Movimento nazionale nei porti esteri, ACS, DGMM, Dpct, bb. 57–61.

[37] Cohn (2005).

[38] Brigantino a palo *Lealtà*, Giornale di boccaporto, ASGe, GN, 1061/1; Brigantino a palo *Stella B*, Idem, 832/1; Brigantino a palo *Martinin*, Idem, 1252/1; Brigantino a palo *Edinburgh*, Idem, 602/1.

gigantic redirection of international trade left room for steam late-comers to survive and thrive up to the First World War. In other words, the alleged marginalisation of sail shipping represented, instead, a fundamental tool of survival for those maritime economies (as the Italian) which needed more time to adapt to technological change.[39]

The chapter aimed to outline the divergence between the conventional expectations of Italian public opinion and the actual direction which national trade and shipping pursued in the aftermath of the opening of the Suez Canal. Despite the enthusiastic climate, neither the manufacturing sector nor shipping was able to benefit from the canal in the short term. In particular, the Italian merchant marine lacked the required structural characteristics to yield on the isthmus cut and the redirection of most of the Indo-European commerce to the Mediterranean. Technological backwardness and the dependence on imports for raw materials, which are the most recurrent arguments of contemporary scholars, must be interpreted within a more complicated framework in which Italian sailing shipping still retained profitable alternatives. Quite the contrary, while absorbing most of the high-value trade in the organisation of liner routes, the Suez Canal left to the Cape, and to sailing vessels, most low-value transports. Furthermore, as a result of the increased presence of steam navigation in the Indian ocean, the canal stimulated new market demands (coal), which were paradoxically satisfied by sail shipping. Within this transforming framework, Italian sail operators were able to introduce themselves into peripheral freight markets whose access was previously denied to Mediterranean merchant marines; thus, they were able to stretch to its limits the 'golden age of sail' and to adopt a resilient stance until the First World War.

References

Archival Sources and Abbreviations

ACS, Archivio Centrale dello Stato
MM, Ministero della Marina,
DGMM, Direzione generale della marina mercantile
Dpct, Direzione premi compensi e tasse

[39] See the analogies with the Scandinavian countries. Kaukiainen (1991, 148–200), Gjølberg (1980).

ASGe, Archivio di Stato di Genova
CC, Camera di commercio
GN, Giornali nautici
Inchiesta, *Inchiesta parlamentare sulla marina mercantile: 1881–1882*, Roma: Tip. Eredi Botta, 1882–1883
Relazione, *Sulle condizioni della marina mercantile italiana al 31 dicembre 1907. Relazione del direttore generale della marina mercantile a Sua Eccellenza il Ministro della Marina*, Roma: Cecchini, 1908

Bibliography

Boccardo, Gerolamo. 1869. *Il Bosforo di Suez in relazione con il commercio del mondo e segnatamente con il commercio dell'Italia. Cenni ed osservazioni*. Forlì: Febo Gherardi Editore.

Bollettino dell'Istmo di Suez. 1856–1858. Torino.

Bono, Salvatore. 2006. Il canale di Suez e l'Italia. *Mediterranea ricerche storiche* 8: 411–422.

Codignola, Arturo. 1938. *Rubattino*. Bologna: Licinio Cappelli.

Cohn, Raymond L. 2005. The Transition from Sail to Steam in Immigration to the United States. *The Journal of Economic History* 65 (2): 469–495.

Cora, Guido. 1870. *Da Brindisi a Suez. Sguardo fisico, politico, etnografico, storico, economico sulla linea di navigazione da Brindisi a Suez attraverso il canale di Suez redatto secondo le opere dei più celebri geografi e viaggiatori*. Casale: Tipografia Corrado.

Corbino, Epicarmo. 1926. *Economia dei trasporti marittimi*. Città di Castello: Società Leonardo da Vinci.

De Lesseps, Ferdinand. 1856. *Memorie e documenti sull'apertura e canalizzazione dell'Istmo di Suez*. Torino: UTET.

Doria, Giorgio. 1973. *Investimenti e sviluppo economico a Genova alla vigilia della Prima Guerra Mondiale*. Milano: Giuffrè.

Doria, Marco. 2001. La marina mercantile a vela in Liguria dalla metà dell'Ottocento alla Prima Guerra Mondiale. In *A vela e a vapore. Economie, culture e istituzioni del mare nell'Italia dell'Ottocento*, ed. Frascani Paolo, 83–107. Roma: Donzelli.

Fletcher, Max E. 1958. The Suez Canal and World Shipping (1869–1914). *Journal of Economic History* 4: 556–573.

Frascani, Paolo, ed. 2001. *A vela e a vapore. Economie, culture e istituzioni del mare nell'Italia dell'Ottocento*. Roma: Donzelli.

Giacchero, Giulio. 1970. *Genova e la Liguria nell'età contemporanea: un secolo e mezzo di vita economica 1815–1869*. Genova: Cassa di Risparmio.

Giuntini, Andrea. 2011. *Le meraviglie del mondo. Il sistema delle comunicazioni internazionali nell'Ottocento*. Prato: Istituto di studi storici postali.

Gjølberg, Ole. 1980. The Substitution of Steam for Sail in Norwegian Ocean Shipping, 1866–1914. A Study in the Economics of Diffusion. *Scandinavian Economic History Review* 28 (2): 137–146.

Graham, Gerald S. 1956. The Ascendancy of the Sailing Ship (1850–1885). *The Economic History Review* 9 (1): 74–78.

Harley, Charles K. 2015. The Shift from Sailing Ships to Steamships, 1850–1890: A Study in Technological Change and Its Diffusion. In *Essays on a Mature Economy: Britain after 1840*, ed. Donald M. McCloskey, 215–238. Princeton: Princeton University Press.

Interdonato, Giovanni. 1856. L'Istmo di Suez e l'Italia. In *Memorie e documenti sull'apertura e canalizzazione dell'Istmo di Suez*, ed. Ferdinand De Lesseps. Torino: UTET.

Kaukiainen, Yrjö. 1991. *Sailing into Twilight. Finnish Shipping in the Age of Transport Revolution, 1860–1914*. Helsinki: SHS.

———. 1992. Coal and Canvas: Aspects of the Competition between Steam and Sail, 1870–1914. *International Journal of Maritime History* 4 (2): 175–191.

Sapeto, Giuseppe. 1865. *L'Italia e il canale di Suez*. Genova: Fratelli Pellas.

Spadoni, Ugo. 1970. Il canale di Suez e l'inizio della crisi della marina mercantile italiana. *Nuova Rivista Storica* 54: 651–702.

Stopford, Martin. 2009. *Maritime Economics*. 3rd ed. New York: Routledge.

Torelli, Luigi. 1867. *L'Istmo di Suez e l'Italia*. Milano: Civelli.

Vimercati, Cesare. 1864. *Il canale dell'Istmo di Suez. Sua benefica influenza in Europa e specialmente in Italia*. Livorno: a spese dell'autore.

Virgilio, Jacopo. 1869. *Il commercio indo-europeo e la marina mercantile italiana a vela ed a vapore*. Genova: Tipografia sociale.

Williams, David M., and John Armstrong. 2012. An Appraisal of the Progress of the Steamship in the Nineteenth Century. In *The World's Key Industry. History and Economics of International Shipping*, ed. Gelina Harlaftis, Stig Tenold, and Jesùs M. Valdaliso, 43–62. Basingstoke: Palgrave Macmillan.

CHAPTER 6

A Venice of the Desert: The Successes and Failures of the Suez Canal in Revitalising Mediterranean Trade

Kevin A Tang

Introduction

Throughout the nineteenth century, the idea of establishing a direct trade route linking the Mediterranean Sea and the Indian Ocean had always aroused great expectations of the benefits such a route would bring to the maritime economies of Europe and Asia. It was expected that the Suez Canal would bring about a new era of economic prosperity for the cities and countries connected by it, but that it would be particularly beneficial for the Mediterranean countries of the European periphery.

For countries of the European periphery, easy and direct access to raw materials from Asia would support the region's continued industrialization. For example, direct access to the raw silk market of China would

K. Tang (✉)
Faculty of History, Smith School of Enterprise and the Environment, Said School of Business, University of Oxford, Oxford, UK
e-mail: kevin.tang@smithschool.ox.ac.uk

B. Curli (ed.), *Italy and the Suez Canal, from the Mid-nineteenth Century to the Cold War*,
https://doi.org/10.1007/978-3-030-88255-6_6

revive the French and Italian silk industry. Silk was a particularly strategic commodity for Italy in the nineteenth century, as it was by far the largest industrial export of the country that went primarily to Northwest Europe, and it formed the foundation of Italy's eventual 'industrial triangle' in the sub-alpine region.[1] Since the 1840s Italy had gained world leadership in the engineering of silk-reeling machinery, a position it held through the 1920s. The main issue for the continued growth of the silk industry in particular, and Italian industrialization in general, was the supply of silk cocoons, the production of which had collapsed because of disease, the *pébrine*, since the 1850s. Thus, securing a reliable and direct supply of silk cocoons via the Suez Canal was believed to form the basis of the country's continued industrial growth.[2] Separately, across the Mediterranean, the Suez Canal was anticipated to enlarge the entrepôt trade of port cities on either side of the canal from Bombay, Aden, Suez, Marseilles, Lyons, and Trieste.[3]

To the extent that the Suez Canal was expected to enlarge the entrepôt trade of several cities across the Mediterranean and the Indian Ocean, it was expected to be particularly beneficial to the port cities of the Mediterranean, as it would create new cities that would become, 'a Venice of the desert, another Alexandria, an Egyptian Calcutta where Mediterranean shipping and Indian Ocean shipping would meet at great entrepôts for the exchange of eastern and western produce'.[4] Where the Mediterranean's influence and participation in the global economy in the nineteenth century had slowly declined as European trade with Asia travelled increasingly via the Cape of Good Hope, the opening of the Suez Canal was expected to transfer the trade and industry of Northwest Europe to the Mediterranean.[5]

Extensive studies show how the Suez Canal contributed to initiating the first period of globalisation with reduced travel times and declining transportation costs.[6] In terms of fulfilling its economic expectations,

[1] Federico and Tena-Junguito (2014, 349–69.)

[2] Federico (1996, 250–74).

[3] Arles-Dufour to Dufour-Feronce, 1856, LCS, 182. LCS is a series of collected letters compiled by Lesseps relating to the construction and development of the Suez Canal. There were several series published ranging from one to five years, as well as those published in translation from the original French version.

[4] Lesseps to *The Times,* 30 October 1855, LCS, 278.

[5] Acemoglu et al. (2005, 546–79).

[6] Findlay and O'Rourke (2007, 380).

studies have shown that the Suez Canal fulfilled its primary strategic objective of facilitating Asia's role as a supplier of raw materials.[7] However although the effect of the Suez Canal on trade routes between Europe and Asia is well known, how the canal relatively contributed to and facilitated trade and development within the Mediterranean compared to the trade of Northern Europe has not been studied. By the early nineteenth century, the relative economic decline of the urban and maritime economies of the Mediterranean were widely apparent in relation to the rapidly industrializing countries of Northwest Europe.[8] Such economic distinctions within Europe are therefore a significant part of understanding European trade in general, and trade through the Suez Canal in particular. Part of the issue in assessing and understanding the relative economic contributions of the Suez Canal to facilitating trade on long-distance routes between Europe and Asia has been distinguishing between what is and is not trade through the Suez Canal, and to which part of Europe did canal trade pass. For example, France has major ports on both the Atlantic and the Mediterranean coasts. Most of the trade of Austria-Hungary connected between ports of Germany, since the Dinaric Alps largely separated the Mediterranean ports from the population centres of Austria-Hungary.

Most studies evaluating Suez Canal trade have only observed aggregate trade flows at a national level. This chapter introduces an originally compiled archival dataset taken from a variety of national statistical reports that specify the port city of origin and destination for trade through the Suez Canal. Disaggregated trade statistics from port cities in India to a destination port shipped via the Suez Canal is included in the annual reports of *Trade and Navigation for the Presidencies of Bengal, Madras, and Bombay.* The trade reports include a detailed summary of the quantity, value, and type of good shipped between the two ports. Additionally, this data can be matched to import trade statistics from national reports, including the French *Annuaire Statistique de la France* (ASF), the Italian *Annuario Statistico Italiano* (ASI), and the British *Statistical Abstracts for the United Kingdom.* Collectively the dataset represents the most accurate assessment of trade through the Suez Canal between India and Europe.

Expectations of the impact of the Suez Canal aroused imperialist imagination, political, cultural, and economic ambition in the decades prior to the canal's opening. The degree to which such expectations were met in

[7] Whiting (1983, 77–8).
[8] Broadberry and Gupta (2006, 2–31).

the period after the opening of the canal has been widely observed in bringing about the first period of globalisation, and the expansion of European colonialism across Asia. However, given the growing divergence in industrialization and development within Europe over the course of the nineteenth century, the degree to which expectations of fostering the entrepôt trade and economic growth in Mediterranean ports of the European periphery has not been evaluated. Through a collection of trade flows between several port cities where the trade route from India to the Mediterranean is known to pass through the Suez Canal, this chapter looks at the relative effect of the Suez Canal in expanding trade between port cities of the industrialising global core of Northwest Europe and the colonial Asian periphery, compared to less well-known trade between the European periphery of the Mediterranean and colonial Asia.

Trade Through the Suez Canal

The growing involvement of European colonial powers in India was always a predominant motivating factor for creating a faster sea route between Europe and Asia since the eighteenth century. Therefore, improved and expanded access to the India trade was a central incentive to the construction of the canal. As such, this chapter assesses the relative effects of the Suez Canal in contributing to expanded trade with Asia, based on trade with India.

Looking at the total export trade of India via the Suez Canal versus any other route, the canal was a clear success in redirecting trade through the Mediterranean based on the total export flows for Indian trade through the canal. Figure 6.1 shows the total value of Indian exports via the Suez Canal versus any other route, demonstrating a discrete growth in trade travelling through the canal. Additionally, trade via any other route does not diminish over the period either, suggesting that Indian exports were not simply rerouted through the canal, but that trade via all other routes remained stable, while trade to destinations connected by the canal rapidly expanded over the period.

From the figure, trade flows through the canal rapidly expanded to become the primary route for Indian exports by 1883, from which point the majority of Indian exports continues to pass through the canal. On this basis, the role of the Suez Canal in redirecting the Asian trade through the Mediterranean appears successful, as trade only expanded over the

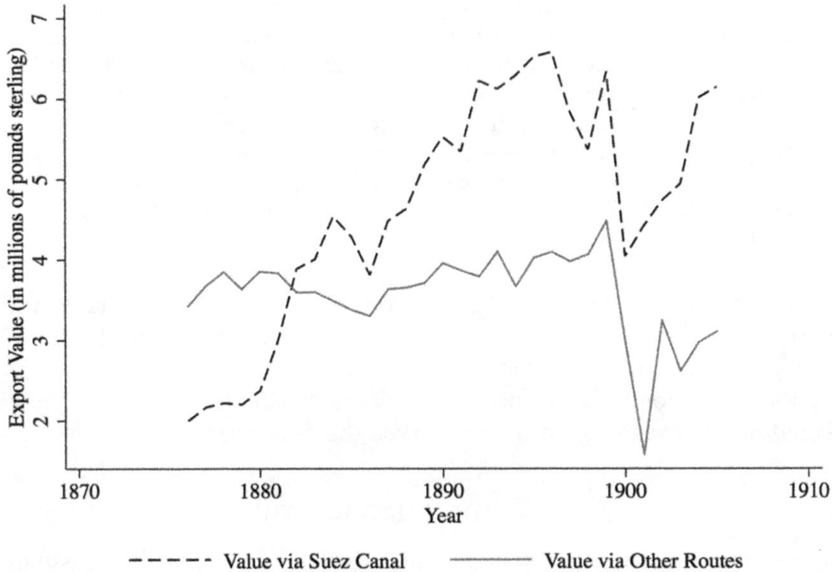

Fig. 6.1 Total Indian exports via the Suez Canal versus any other route. (SA BI, 1876–1914)

period, except for a collapse in trade at the very beginning of the twentieth century, which was most likely due to the Indian famine of 1899–1900.

The Suez Canal was intended to benefit the economies and trade of the Mediterranean, which had seen their share of trade with the East shrink compared to the rapid growth of trade for countries of Northern Europe. From the first half of the nineteenth century, Mediterranean powers such as France, Italy, and the Ottoman Empire witnessed a decline in their relative share of world trade that had been shifted from the Mediterranean to the Cape route that went directly to Northwest Europe. Although each decade since the 1830s had witnessed an increase in Mediterranean trade overall, the growth of trade going directly to Northwest Europe via the Cape route was even greater. The Suez Canal was therefore intended to challenge British domination of the global economy and global shipping by the Mediterranean World:

> The cities of the Mediterranean would become once more the centre of the world and regain the most profitable branch of economic activity in the

> entrepot trade with all its enriching ancillary traffic in unloading, pilfering, warehousing, processing, distributing and financing and its associated freight, insurance, depot, consignment and transhipment charges. England would lose in succession its commerce, its merchant marine, its maritime supremacy and finally its Indian possessions. Thus Lesseps appears as the avenger of a hundred seaboard cities ruined by the avarice of England and was accordingly decorated on the successful completion of the canal by the rulers of the Mediterranean states of Russia, Greece, Austria, Italy, France and Belgium.[9]

Based on the expectation of the Suez Canal to reroute global trade, and particularly the India trade through the Suez Canal, the trend illustrated in Fig. 6.1 demonstrates that the Suez Canal successfully created several opportunities for Mediterranean countries to expand trade ties with India, based on the expansion in trade flows via the canal route.

Mediterranean Trade Through the Suez Canal

The previous section has demonstrated that the Suez Canal successfully contributed to creating new trading opportunities for Mediterranean countries with India. This suggests that there should be a similar increase in Indian trade with those countries of the Mediterranean. While the previous section has looked at the total Indian trade, this section comparatively looks at Indian trade based on trade to countries of the Mediterranean compared to countries in Northern Europe to determine which region gained the most from the rerouting of global trade through the Suez Canal.

Since trade statistics for each country in Europe have been collected at a disaggregate level by port city, this study compares export flows from the Indian ports of Madras, Bombay, and Calcutta to port cities of the Mediterranean versus ports of Northern Europe. Figure 6.2 illustrates the major destination port cities for Indian trade, but only including those in which the port city was a regular trading destination appearing in at least three consecutive years in the Indian trade statistics.[10] Destination ports are blue diamonds for those that are located in Northern Europe, and red

[9] *The Spectator*, 17 February 1872, LCS, 198.

[10] Some European port cities appear in only one year in the Indian trade records, but not over time. Since this study creates a time series over the period 1870–1914, only those cities that are repeated and regular destinations of Indian exports are able to provide a picture of the growth in Mediterranean trade relative to trade to Northern Europe.

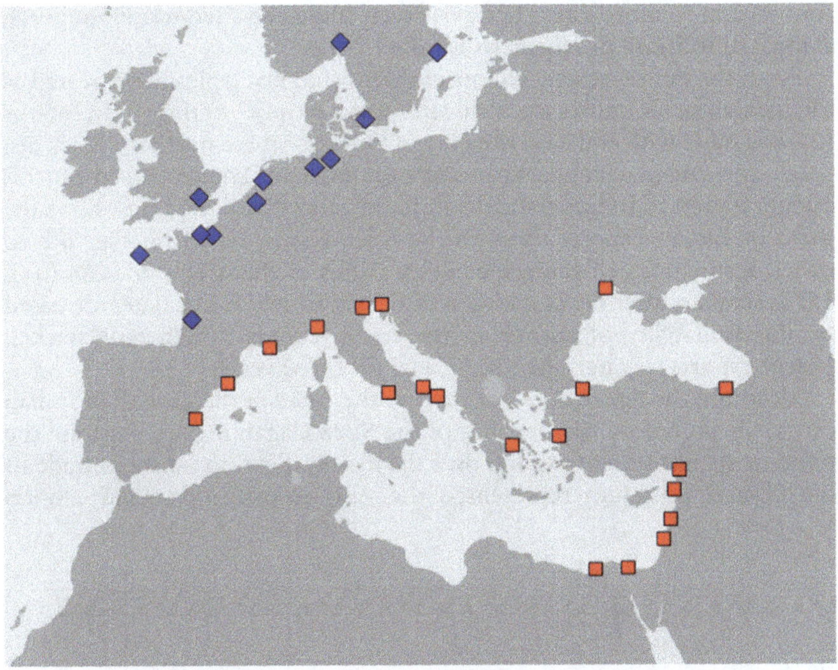

Fig. 6.2 Map of destination ports of Indian export as either Mediterranean or Northern Europe. (ASTN BE, ASTN BO, ASTN MA, 1875–1915)

squares for those that are identified as Mediterranean. Ports of Northern Europe are Brest, Bordeaux, Rouen, Le Havre, Amsterdam, Antwerp, London, Southampton, Hamburg, Bremen, Oslo, Stockholm, and Copenhagen. The port cities of the Mediterranean are Marseille, Barcelona, Valencia, Genoa, Naples, Brindisi, Bari, Venice, Trieste, Athens, Istanbul, Izmir, Alexandretta, Latakia, Beirut, Haifa, Alexandria, Port Said, Trabzon, and Odessa.

From Fig. 6.2, of all destination ports, there is a greater representation of the various ports of the Mediterranean compared to Northern Europe. In total, there are 20 ports of the Mediterranean that regularly traded with India through the Suez Canal. In comparison to Northern Europe, there were only 13 ports that traded regularly with India via any route. Based on Fig. 6.2, representation of various European ports in the Indian archival trade records demonstrates that the opening of the Suez Canal contributed to more direct trade between India and Mediterranean ports, compared to those of Northern Europe.

From the figure, several countries that would not typically be treated as Mediterranean countries are included for the purpose of this study, such as Russia's trade with India via the port of Odessa, on the northern Black Sea coast, and Austria-Hungary through the port of Trieste. Based on this categorization of Indian trade through the Suez Canal, Fig. 6.3 shows the share of Indian exports through the Suez Canal, based on Fig. 6.2 to either destinations of Europe or destinations of the Mediterranean from 1875, six years after the opening of the canal, to 1904. The dates are based on the availability of data from the trade records, which only tracked Indian exports through the Suez Canal for those years.

Although the figure does not capture the trend in the growth of Indian exports from before the opening of the Suez Canal, it does illustrate the trend in distribution of exports over the period. First, the share of trade to the United Kingdom represented the vast majority of Indian exports

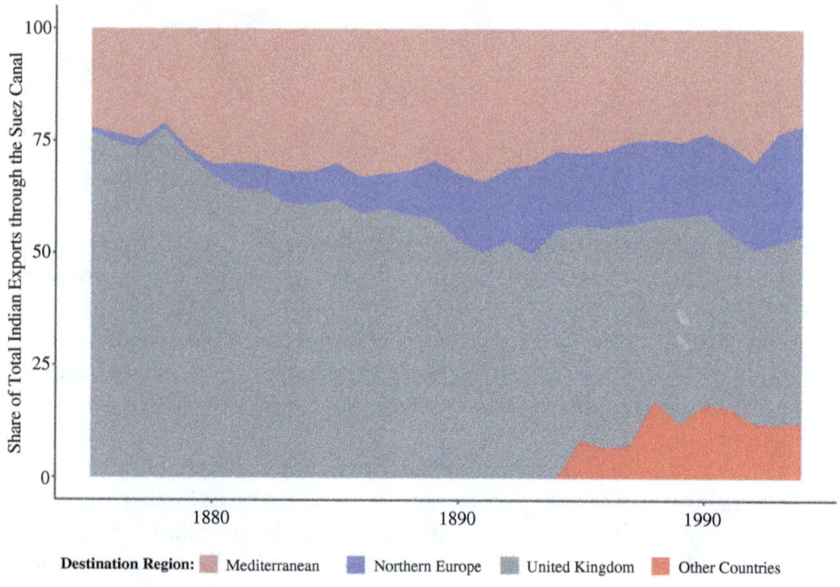

Fig. 6.3 Share of Indian export trade through the Suez Canal by destination region. (ASTN BE, ASTN BO, ASTN MA, 1875–1915; SA UK, 1876–1914; SA BI, 1876–1914)

through the canal. Second, trade to all countries with Mediterranean ports represents the second-largest share of trade, beginning with 22 percent of trade, and ending with 21 percent. Third, the region that witnessed the greatest growth in Indian trade over the period was Northern Europe, not including the United Kingdom, which represented a little over 1 percent of trade in 1875, but became the second-largest regional destination after the United Kingdom by 1904, with 25 percent of trade.

While destinations outside of the UK show rapid growth over the period, the rate of growth is greater for trade to Northern Europe compared to the Mediterranean. In order to compare the rate of growth in Indian exports, a simple linear regression is estimated on the trend levels in Fig. 6.3. Taking the value of Indian trade to the three destinations of the United Kingdom, the Mediterranean or Northern Europe, a simple ordinary least squares regression is calculated for the relationship of trade value to the linear time trend. The resulting coefficient for each of the three destinations is the mean rate of growth in Indian exports to each region for each year. The larger the value of the coefficient represents the faster the rate of growth in Indian trade. Although the absolute level of trade value was largest to Britain, a better measure for the relative success of the Suez Canal for facilitating trade is the rate of growth.

While Britain maintains the largest share of Indian exports through the canal, it actually did not significantly benefit from the opening of the canal in terms of an expansion of trade with India. The time coefficient for Indian trade to the UK is -0.001 and not significant with a standard error of (0.002). The coefficient indicates that trade to Britain slightly shrank, but that trade flows were not significantly trending positive or negative, thus the level of trade remains more or less the same. Hence despite the traditional view that the Suez Canal primarily benefitted Britain's access India, while it may have been strategic, the canal did not significantly contribute to expanded trade ties between Britain and India.

While the share of Indian trade to all Mediterranean ports was still smaller than Indian trade to the UK, the canal contributed to the rapid expansion of trade to ports of the Mediterranean, with a time coefficient of 0.033 and significant with a standard error of (0.004). The positive and significant coefficient for Mediterranean trade indicates that Indian exports to the Mediterranean grew by approximately 3 percent on average per year. Hence the extent to which the Suez Canal was expected to redirect Asian trade through the Mediterranean was successful, as trade to the region rapidly grew through the canal.

From Fig. 6.3, Indian trade to countries in Northern Europe was a similar size, and demonstrates a similar trend. Taking the linear time trend of trade to Northern Europe, the coefficient is 0.121 with a standard error of (0.007), thus confirming a significant growth rate in trade. The time coefficient for the value of trade to Northern Europe indicates that Indian trade grew on average by approximately 13 percent per year, a faster rate of growth compared to trade to Britain or all countries of the Mediterranean.

While historical treatment on the economic impact of the Suez Canal has focused on how it contributed to expanding British trade ties with India, analysis of an original trade dataset of Indian exports has shown that the canal actually did relatively little to increase trading opportunities with Britain. Instead, the canal was intended to revitalize trade for countries of the Mediterranean, and in this respect the canal was successful, with Indian exports to the Mediterranean expanding by 3 percent on average per year. However comparatively, trade to ports of the Mediterranean did not grow by as much as trade to ports of Northern Europe outside Britain, which witnessed the greatest growth in Indian trade of all regions at 13 percent annual growth.

Bilateral Trade Growth Through the Suez Canal

Part of the objective of the Suez Canal was to reroute the Asian trade that had entirely bypassed the Mediterranean, in favour of travelling via the Cape route, back to the Mediterranean. While evidence has shown that the Suez Canal did reroute Asian shipping back through the Mediterranean, it only led to a temporary growth in the share of Indian trade that came via the canal. Britain remained the main beneficiary of the canal trade, however the region that witnessed the greatest increase in trading opportunities with India was actually Northern Europe outside of Britain.

This section further evaluates the relative growth of the Indian canal trade by looking at the level of trade and growth by major trading countries. For all countries trading with India through the canal, Britain remained the largest destination for Indian exports. Figure 6.4 shows the trend of the five destinations for Indian exports in Europe by the end of the period in 1913.

From the figure, Britain is distinctly the largest destination for Indian exports, far greater than the second major destination for most of the period, which is France. The sixth-largest destination, and the smallest in the dataset is Italy, which exhibits a similar trend and level to the growth

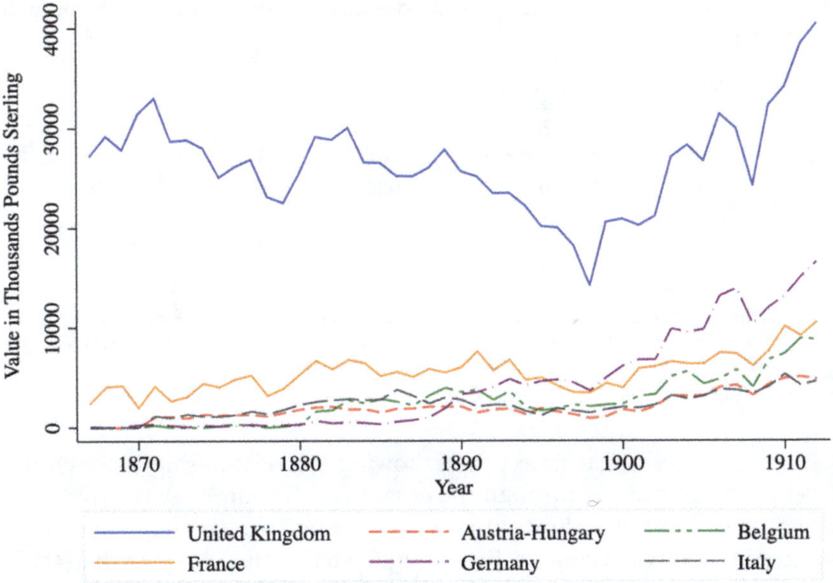

Fig. 6.4 Total value of trade flows from India to top destinations through the Suez Canal. (ASTN BE, ASTN BO, ASTN MA, 1875–1915; SA UK, 1876–1914; SA BI, 1876–1914)

in the value of Indian exports is Austria-Hungary, another country whose primary port was on the Mediterranean at Trieste.

Surprisingly, countries that exhibit the greatest growth in exports from India are actually both Northern European destinations: Belgium and Germany. At the beginning of the period, there was no record of Indian trade to Belgium until 1870, the first full year in which trade through the Suez Canal was made possible. Similarly, Germany exhibits the second lowest level of trade with India above Belgium in 1871, which was the first year in which all six countries appear on the Indian trade record. However, by the end of the period Germany was the second-largest destination for Indian exports after Britain, and Belgium was the fourth-largest after France.

Table 6.1 shows the top six destination countries for Indian exports through the Suez Canal with the time trend to illustrate which country's trade expanded the most from the opening of the canal until 1913.

Table 6.1 Trade flow trends of top six destinations for Indian exports through the Suez Canal

Rank	*Destination*	*(1) Time Trend*	*(2) P-Value*	*(3) Constant*	*(4) Std. Error*
1	United Kingdom	-0.001	0.002	10.351***	0.056
2	Germany	0.131***	0.005	4.150***	0.147
3	France	0.018***	0.003	8.136***	0.078
4	Belgium	0.149***	0.025	3.147***	0.696
5	Austria-Hungary	0.029***	0.004	6.828***	0.111
6	Italy	0.024***	0.004	7.106***	0.117

Note: Coefficients represent the logarithm of the total export value of Indian trade to each country; hence, the coefficients in column 1 represent the percentage increase in trade value. Constant represents the *y*-intercept

***, **, * indicates statistical significance at the 1 percent, 5 percent, and 10 percent level

Column 1 is the linear time trend, showing the relationship between the level of Indian exports through the canal to each country. A positive coefficient in column 1 indicates that trade expanded over the period, thus suggesting how each country benefitted from the canal via growth in trade with India. From the table, the coefficients for Belgium and Germany are the highest, thus indicating that trade grew by the greatest percentage to these countries. France, the primary financier of the Suez Canal, actually benefitted the least, as the Indian trade grew to France by the smallest percentage. The country that witnessed the second smallest growth was Italy, then Austria-Hungary, thus the Mediterranean countries witnessed the lowest level of growth in trade through the Suez Canal compared to the three primary destinations of Northern Europe.

Shipping Through the Suez Canal

Part of the reason for the more rapid growth in Suez Canal trade to countries of Northwest Europe compared to the Mediterranean can be understood from the changing shipping lines and routes used by companies to connect ports of Europe with those across the Indian Ocean. Prior to and in the years immediately following the opening of the Suez Canal, shipping lines of the Mediterranean and the Indian Ocean remained separate services. Steamships would operate on regularly scheduled services between ports of Northwest Europe, to ports of the Mediterranean, and subsequently to the Indian Ocean, with merchandise being trans-shipped via ports of the Eastern Mediterranean of Egypt or the Levant. European countries would operate regular services to a string of ports along the

Mediterranean, ending at a port along the Eastern Mediterranean, which would subsequently be connected to cities of the Indian Ocean.

For most of the nineteenth century, the ports of Egypt functioned as the entrepôt for trade between Europe and the Indian Ocean. There were five British services that ran through Egypt: two lines from Southampton to Alexandria, two between Suez to Calcutta, and one between Suez and Bombay. Similarly, five French lines served from Marseille to Bombay and Hong Kong by way of connecting between Alexandria, then overland to Cairo and Port Said. Austrian, Italian, Russian, and Ottoman lines also operated along a string of ports in the Mediterranean before connecting at Alexandria, then travelling overland to the cities of Suez or Port Said and onwards to Bombay, Basra, or Hong Kong.[11] Due to its control of the Isthmus of Suez, Egypt also controlled the coastal trade of both the Red Sea and the Mediterranean, making it the natural entrepôt for all east and west trade.

Ports of the Levant also specialized in trans-shipment via the desert route that connected ports of the Mediterranean with Baghdad, and subsequently via the river trade of the Tigris and Euphrates to Basra on the Persian Gulf coast. From Baghdad, the string of ports of the Levant from Jaffa to Aleppo were connected via the overland caravan trade, 'the caravans between these places vary from ten to 100 mules or camels. There are muleteers who are continually employed on these roads, so that there is always abundance of opportunities for the transmission of goods.'[12] Aleppo, Beirut, and Damascus were all well connected via the caravan trade, 'the importations in transit, forwarded to Damascus and Beirut, not being for the consumption of those cities only, go by the caravans to Baghdad, whence they are distributed in the adjacent countries. Those, again, destined for Aleppo find their way into Asia Minor, especially to the towns of Orfa, Diarbekir, Malatiah, Sivas, etc.'[13] From the Levant ports, there were several European companies that were, 'long established in Beirut and Aleppo, and who are largely, if not wholly, engaged in the import and export trade with the east'.[14] The Austrian Lloyd ran regular service from Trieste to Beirut and Aleppo. The *Messageries Maritimes* ran regular and frequent shipping between Marseilles and Beirut, and the various steamship companies which form the Antwerp Combine travelled

[11] Issawi (1982, 219–43).
[12] CS MacGregor (1847, 140–50).
[13] MacGregor to Pacha, CS MacGregor (1847, 140–50).
[14] BAP Weakley (1911, 81–5).

directly from Hamburg and Antwerp to the ports of Alexandretta and Beirut.

While most shipping between Europe and Asia before the opening of the Suez Canal operated through port cities of the Mediterranean, particularly Egypt or the Levant, with the opening of the canal, port cities across the Mediterranean were entirely bypassed for direct trade between primary ports of Northwest Europe and Asia, with much of the Mediterranean trade operating direct lines through the canal, rather than stopping at a series of ports along the coast of the Mediterranean. From 1870 the Austrian Lloyd switched service between Trieste to Alexandria, and instead operated lines directly between Trieste and Bombay. Marseilles had previously run services across the Mediterranean to Port Said in Egypt en route to Bombay, but instead began service that travelled directly from Marseilles through the canal to Bombay.[15] Turkey began operating direct services between Constantinople and Basra in 1870, where it had previously used the desert route through Syria to trade with Basra. Gray, Paul & Company and several other shipping services out of London also initiated a direct line between London and Bombay from 1870 onwards. In addition, Barcelona, Genoa, and Odessa all established national shipping lines that travelled directly to Bombay or Calcutta, where they had previously contracted shipping services with the Peninsular & Oriental steam navigation company (P&O) that connected to the east via Alexandria, Suez, or Port Said.[16] As most European steam services went to India, the growth in Indian trade westwards through the canal in Fig. 6.1 also represents the proportionate loss of the trans-shipment trade to ports of the Mediterranean. While all European countries created national lines that directly connected major Mediterranean and Northwest European ports with those of India, the size of the trade and the growth of the ports was ultimately tied to the size and growth of each country's economy and pre-existing trade ties, rather than based on new trade opportunities as a result of access to Asian markets provided by the Suez Canal.

[15] 'La Compagnie Fraissinet a été la première compagnie française adoptant Port Said comme escale régulière. C'est un steamer de cette Compagnie, l'*Asie*, qui, passant par le canal, a effectué le premier voyage direct sur Bombay.' (The Frassinet Company has been the first French company to use Port Said as the regular stopover. It is with the steamer l'*Asie* which travels through the Canal to make the first direct journey to Bombay.) Lesseps, 1 January 1870, LCS: 4.

[16] Farnie (1969, 140–5).

Evidence from the way steamship services and shipping companies were reorganized to take advantage of the Suez Canal by operating directly between ports of Europe and the Indian Ocean give an indication of how the advantages offered by the canal were disproportionately distributed between regions. While the Suez Canal was widely promoted as a boost to the Mediterranean economy, shipping records indicate that this was mostly true for ports of Northwest Europe, and less so for ports of the Mediterranean. Trade growth via the Suez Canal in the first period of globalisation was therefore more of a phenomenon between the global core of Northwest Europe and the global periphery of colonial Asia, rather than a general process occurring between minor ports and regions within the global periphery including the Mediterranean.

Conclusion

The promise of the Suez Canal was not only the long-held desire to bring colonial Asia closer to Europe, but also to advance the economic growth and industrialization of the port cities and countries across the Mediterranean. It was strategically intended to remake the Mediterranean Sea as the centre of global trade, revitalising the entrepôt economies of several port cities of the region. While several studies have found that there was an effect of the Suez Canal in expanding Indian trade to Europe in general, how the Suez Canal contributed to trade within Europe, comparing the industrializing economic core of Northwest Europe to the European periphery of the Mediterranean had not been similarly analysed.[17]

Findings from this chapter suggest that the gains from trade via the Suez Canal were significantly determined by destination region, depending on whether trade was between the global core of Northwest Europe, or within the European periphery of the Mediterranean. What the Suez Canal had promised in terms of advancing the trade economies of the Mediterranean and further integrating the region with colonial Asia proved to be illusory, as the greatest opportunities for trade continued to go to Northwest Europe in general, and to Germany and Belgium in particular.

Evidence of the disproportionate benefits to canal trade is similarly reflected in the history of shipping companies that traded through it. While Mediterranean port cities were formerly central to linking trade

[17] Steele in Wilson (1983, 1–25).

between Europe and Asia, acting as intermediaries and stations for shipping lines that operated between the two regions, the opening of the canal allowed companies to trade directly with Asia without having to stop at Mediterranean ports along the way. The increasing frequency of direct shipping lines contributed to expanding trade between those two regions, but it did not similarly contribute to expanding trade for Mediterranean ports, where there were fewer companies operating, and less demand for merchandise from Asia due to their smaller economic size.

In the case of Italy, although there were several ports that directly traded with India, the total trade value at each port was small, serving a relatively smaller market around each port, rather than fostering the entrepôt activities of a few Italian ports as a hub of Asian trade, as had been anticipated with the opening of the Suez Canal. Although Italian ports are overly represented in the Indian trade records for Mediterranean trade, the small size of trade compared to the larger entrepôt trade of Northern European cities is likely the result of the legacy of the relatively autonomous urban subsystems across the peninsula that predated unification and the opening of the canal.[18] The case for Italy is similar for other ports of the Mediterranean, where relatively low levels of trade within and between Italian port cities meant that there were not opportunities for economic agglomeration of a primary entrepôt.[19] Similarly across the Mediterranean, as there were more opportunities for direct trade with India for smaller cities and smaller shipping lines, Mediterranean cities traded directly with Asia, rather than through a single predominant Mediterranean entrepôt city. Consequently, despite more opportunities for trade between Italy and India, the agglomeration effects of entrepôt trade that came with the opening of the canal did not lead to rapid trade growth for Italy overall, nor for any Italian port city in particular, compared to the more rapid trade growth and urbanization at one primary port city for each country of Northern Europe.[20] Instead, while the Suez Canal was intended to remake Italy and other Mediterranean countries as centres for global trade, the direct route to Asia actually made the region more peripheral, port activity more localized, shipping lines more direct, and consequently less able to benefit from the trans-shipment business that had historically made the Mediterranean Sea the centre of global trade.

[18] Bosker et al. (2008, 97–131).

[19] Federico and Tena-Junguito (2014, 349–69).

[20] Fujita and Mori (1996, 93–120).

References

Abbreviations

ASF, Annuaire Statistique de la France
ASI, Annuario Statistico Italiano
ASTN, Annual Statement of Trade and Navigation
  BE, Bengal Presidency
  MA, Madras Presidency
  BO, Bombay Presidency
BAP, British Accounts and Papers
BOT, Board of Trade
CS, Commercial Statistics
LCS, *Lettres du Canal de Suez*
SA, Statistical Abstracts
  UK, United Kingdom
  BI, British India
TGCE, Tableau General du Commerce Extérieur

Bibliography

Acemoglu, D., S. Johnson, and J. Robinson. 2005. The Rise of Europe: Atlantic Trade, Institutional Change, and Economic Growth. *American Economic Review* 95 (3): 546–579.

Annuaire Statistique de la France. 1870–1915. Service de la statistique générale de France Annual Series. Institut national de la statistique et des études économiques, Paris: Ministère de l'agriculture et du commerce.

Annual Statement of the trade and navigation of the Bengal Presidency. 1875–1915. Report General on External Commerce, issuing body 17–23. Government at the Education Society's Press, London: India Office.

Annual Statement of the Trade and Navigation of the Madras Presidency. 1875–1915. Report General on External Commerce. Government at the Education Society's Press, London: India Office.

Annual Statement of the Trade and Navigation of the Presidency of Bombay. 1875–1915. Report General on External Commerce. Government at the Education Society's Press, London: India Office.

Annuario Statistico Italiano. 1880–1914. Statistical Yearbooks Annual Series. Rome: Ministero di Agricoltura industria e commercio.

Bosker, M., S. Brakman, H. Garretsen, H. De John, and M. Schramm. 2008. Ports, Plagues, and Politics: Explaining Italian City Growth, 1300–1861. *European Review of Economic History* 12 (1): 91–131.

British Accounts and Papers. 1829. Letter from Lord Ellenborough to A. H. Imlah, September 18, 49–50.

———. From E. Weakley, 1911, Report on the conditions and prospects of British trade in Syria, No. LXXXVII, 81–85.

Broadberry, Stephen, and Bishnupriya Gupta. 2006. The Early Modern Great Divergence: Wages, Prices, and Economic Development in Europe and Asia, 1500–1800. *Economic History Review* 59 (1): 2–31.

Commercial Statistics. 1847. British Consular Reports. Report of Ibrahim Pacha, from John MacGregor, London.

Conrad, F.W. 1969. Reizen naar de Landengte van Suez. In *East and West of Suez: The Suez Canal in History*, ed. D.A. Farnie. Oxford: Clarendon Press.

Federico, Giovanni. 1996. An Econometric Model of World Silk Production, 1870–1914. *Explorations in Economic History* 33 (11): 250–274.

Federico, Giovanni, and Antonio Tena-Junguito. 2014. The Ripples of the Industrial Revolution: Exports, Economic Growth, and Regional Integration in Italy in the Early Nineteenth Century. *European Review of Economic History* 18 (3): 349–369.

Findlay, Ronald, and Kevin H. O'Rourke. 2007. *Power and Plenty: Trade, War, and the World Economy in the Second Millennium*. Princeton: Princeton University Press.

Fujita, M., and T. Mori. 1996. The Role of Ports in the Making of Major Cities: Self-agglomeration and Hub-effect. *Journal of Development Economics* 49: 93–120.

Issawi, Charles P. 1982. *The Economic History of the Middle East and North Africa*. New York: Columbia University Press.

Lettres, journal et documents pour servir à l'histoire du Canal de Suez, 1854–60; 1861–64; 1864–69. Paris: Didier.

Statistical Abstract for the United Kingdom. 1860–1940. Annual Series. Statistical Department, Board of Trade.

Statistical Abstract relating to British India. 1860–1915. Annual Series. Office of the Economic Adviser, Dominion of India: House of Commons Parliamentary Papers, Commercial intelligence department, Calcutta Government.

Steele, David. 1983. Britain and Egypt, 1882–1914: The Containment of Islamic Nationalism. In *Imperialism and Nationalism in the Middle East: The Anglo-Egyptian Experience, 1882–1982*, ed. Keith M. Wilson, 1–25. London: Mansell Publishing.

Tableau général du commerce extérieur: Commerce de la France avec ses colonies et les puissances étrangères. 1880–1914. Statistical Yearbooks Annual Series. Paris: Administration des douanes.

Whiting, Richard C. 1983. The Suez Canal and the British Economy, 1918–1960. In *Imperialism and Nationalism in the Middle East: The Anglo-Egyptian Experience, 1882–1982*, ed. Keith M. Wilson, 76–95. London: Mansell.

PART II

Italian Cultures and the Canal Project

CHAPTER 7

Luigi Negrelli (1799–1858): A Tyrolean Engineer at the Heart of the Suez Canal Project

Andrea Leonardi and Alice Riegler

Luigi (or Alois) Negrelli's key role in the planning of the Suez Canal, which the *Compagnie universelle du canal maritime de Suez* built under the guidance of Ferdinand de Lesseps, was extremely clear at the time of the structure's realisation 150 years ago. However, over the following years, his name slowly slipped into an oblivion from which it only re-emerged during the first half of the twentieth century, when a small number of methodologically rigorous studies allowed an all-round recovery of his life story.[1] At the same time, abundant literature marred by ideological approaches of a markedly nationalist nature distorted his image and role. During the interwar period, scientific publications and government

[1] Negrelli's most accurate and well-documented biography is a fundamental read: Birk (1915, 1925). See also: Birk and Müller (1925).

A. Leonardi (✉) • A. Riegler
Università di Trento, Trento, Italy
e-mail: andrea.leonardi@unitn.it

B. Curli (ed.), *Italy and the Suez Canal, from the Mid-nineteenth Century to the Cold War*,
https://doi.org/10.1007/978-3-030-88255-6_7

propaganda alike credited Negrelli with being an 'Italian genius'. Even studies conducted with strict methodology succumbed to this bias, forcing the concept of nationality on him with no regard for the historical and cultural context in which this concept had gradually taken shape and thus painting a completely misleading portrait. The same was true for some Austrian literature with similar nationalist intent, which claimed that the engineer had a distinctly German heritage.[2]

This attitude persisted well into the second half of the twentieth century, when the disastrous outcome of exasperated nationalism should have allowed biographers to carry out their research in a climate free from ideological preconceptions. Yet, many post-war studies on Negrelli are not based on the ample documentation that he directly produced,[3] but instead based on misleading works that failed to take first-hand sources into account and thus continued to amplify the distortion of his image.[4] Very few works have succeeded in framing Negrelli's role within the European context of the Habsburg monarchy in which he operated.[5] Even some very recent studies, produced and published in different European countries, are approximate, imprecise and totally inadequate for the purposes of profiling him.[6]

This chapter is based on 1597 documents that constitute the main body of the Negrelli *Nachlaß*, a collection consisting of 2435 documents of various nature housed in the *Technisches Museum* in Vienna, that a research group based at the Department of Economics and Management at the University of Trento has made its object of study.[7] The aim of this project is to bring Luigi Negrelli's role in the planning of the Suez Canal

[2] Leonardi (2020) has extensive references to works that have improperly described Negrelli as either Italian or German.

[3] The following works are at least partly based on primary sources: Scaglione (1971/1972), Leonardi (1990, 2006), Deihsen (1993), Bussjäger et al. (1997), Marzani Prosser (1999), Gatani (2000, 2005).

[4] Leonardi (2020).

[5] Deihsen (1993), Gatani (2000), Helps (2005), Leonardi (1990, 2020).

[6] Piquet (2008, 256, 389), Piquet (2018, 219) are imprecise and erroneous on Negrelli. Other recent works (Moghira 2002; Bonin 2010) completely ignore his role.

[7] The Nachlaß Negrelli in the Archive of the *Technisches Museum* in Vienna contains 1597 documents that include letters by and to Negrelli, a series of reports, service dispatches and notes of various kind dated 1815–1858 and a further 838 documents retrieved in a search coordinated by A. Leonardi. All documents are available online https://negrelli.primiero.tn.it/it/archives/index/.

to the fore and to refute the anachronistic and fallacious traits ascribed to him, including in modern mass media, once and for all.[8]

In order to create a truthful portrait of the man and the eminent scientist and to make him, his work and the context within which he lived and operated known in the most appropriate and extensive way possible, it is necessary to take the documentation he directly produced as a starting point. The archival research and document analysis carried out up until this point have allowed the scrutiny of various parts of his life and work, lifting the veil of privacy in which the figure of Negrelli was for a long time shrouded, and have revealed a personality that was particularly rich, dynamic and open to new ideas.

Negrelli's Education, First Projects and the Interest for Railways

Luigi Negrelli was born on 23 January 1799 in Primiero, which had been part of the Habsburg empire since the fourteenth century, in a family that was profoundly Catholic and unwaveringly loyal to the monarchy.[9] This conservative atmosphere did not prevent Negrelli from displaying, from a young age, a clear receptiveness for supranational ideals, albeit in a spirit that always aimed at being of service to his Austrian fatherland. In his private correspondence, Negrelli professed his allegiance to the emperor while at the same time stating his desire to support elements of contact and cooperation between different peoples.[10] It was against this backdrop that Negrelli's interest in the realisation of the Suez Canal took shape.[11]

In October 1820, after completing an internship and undergoing a theoretical assessment, Negrelli passed the state exam in Innsbruck, becoming a qualified *Zivilingenieur* and joining the Imperial Austrian Civil Service.[12] Between 1821 and 1824 he worked in several Tyrolean valleys, following road and waterway regulation works.[13] His planning and management skills, together with the effective nature of his projects, soon led his reputation to exceed the boundaries of the Austrian empire. In

[8] The 'Italian' Negrelli of most websites is a prime example of this.

[9] Francesconi to Negrelli, 7 January 1843, TMW, NN, 05_0198_000.

[10] Negrelli to Caroline Negrelli, 11 December 1854, TMW, NN, 11_1112_000.

[11] Negrelli to Escher, 30 December 1854, TMW, NN, 11_1118_000. His declared intention was for the canal to create new relations between peoples.

[12] Francesconi to Negrelli, 7 January 1843, TMW, NN, 05_0198_000.

[13] Deihsen (1993, 21).

1832, Negrelli was appointed Inspector of Road and Waterways in Switzerland, where he remained until 1840, overseeing the construction of numerous roads, bridges and hydraulic structures.[14]

It was during his Swiss period, following an assignment for the study of the layout of a railway line between Zürich and Basel, that Negrelli started to develop an interest in railway technology.[15] In 1836, Negrelli undertook a study trip to France, Great Britain and Belgium, where railways were already a tangible innovation, to observe all the implications that accompanied their construction.[16] The engineer drew the conclusion that: 'For industrial nations, railways are indispensable, just as factories are indispensable'.[17] Negrelli therefore believed that the Alpine area should become a transport hub for trade and travellers, if it was to continue to play a mediating role between the various parts of Europe. He advocated a reorganisation of Alpine communication routes that would allow them to function as a link between different countries and peoples and to smoothly connect the ports of the Mediterranean with Central Europe.[18]

Two years after returning to Austria to work for a private railway company, in 1842 Negrelli entered the state service as Inspector, at a time when the Austrian railway network was going through a phase of rapid expansion.[19]

From the Railways to the Suez Canal

Negrelli followed several projects during this period, starting with the *Nordbahn* line Lundenburg–Olmütz–Prague in the Czech provinces, as well as the preliminary studies for the Galician railway and for the Lemberg–Czernowitz section in modern-day Ukraine.[20] The design, construction and management of these railway lines were not just proof of his undeniable technical skills; they highlighted that Negrelli was a *civil* engineer, in the broadest sense of the term. He analysed transport networks in

[14] Birk (1915, 72–81), Bussjäger et al. (1997, 75–84), Gatani (2000, 87–127).

[15] Brugger (1909, 27–28).

[16] Negrelli (1838, 6–7).

[17] Ibidem, 192.

[18] Leonardi (2016).

[19] Francesconi to Negrelli, 23 March 1843, TMW, NN, 05_0175_000. See also: Birk (1915, 169–170). Sandgruber (1987).

[20] Joseph to Negrelli, 22 March 1842, TMW, NN, 05_0174_000. The long viaduct next to Prague is still known today as the 'Negrelli viaduct'. Leonardi (2016).

their entirety, paying particular attention to the economic and functional characteristics of railway lines. Looking at both German and French engineering cultures, Negrelli successfully combined 'the ability of an engineer and the perspicacity of a transport technician and economist'.[21]

In the midst of this phase of intense planning, supervision and consultancy in the railway sector, Negrelli started to draft his first projects for cutting the Isthmus of Suez, into which he had first started to enquire between 1837 and 1838 during his stay in Switzerland.[22] He forged contacts with some of the leading exponents of Saint-Simonianism, who were particularly active in advocating the construction of the canal, and began corresponding with them.[23] Amongst the movement's followers, the railway engineer Barthélemy Prosper Enfantin, who had organised a scientific expedition to Egypt to study the feasibility of cutting the Isthmus of Suez in 1833, became a prominent backer of the project. The geognostic and topographical surveys that the expedition produced—which hypothesised a difference of about ten metres between the levels of the two seas—provided the bases for various preliminary canal projects.[24] The most significant was that of the French engineer in the service of the Egyptian Viceroy Louis Maurice Linant de Bellefonds (Linant Bey), who envisaged a direct connection, with locks at each end, between the two seas.[25]

Karl Ludwig von Bruck, the Austrian minister of commerce and finance, and chancellor Metternich himself endorsed the idea of the canal, as their numerous letters preserved in the *Nachlaß* testify.[26] This led Negrelli to intensify his international efforts to realise the project.[27] Thus, when Enfantin presided over the foundation of the *Société d'Études du Canal de Suez* in Paris on 27 November 1846, Luigi Negrelli was a co-founder. The *Société* was composed of three groups: the 'German group', and not 'Austro-German-Italian', as most Italian historians wrongly labelled it,[28]

[21] Contro (1990).

[22] Negrelli (1856).

[23] See: Jouve (2001), Coilly and Régnier (2006).

[24] Bono (2006, 412). On the age-old quest for a connection between the Red Sea and the Mediterranean, see: Bono (1990).

[25] Mazuel (1937).

[26] The correspondence between Negrelli and von Bruck in the Nachlaß consists of over 80 documents, dating from 3 October 1846 (TMW, NN, 06_0259_007) to 26 March 1858 (TMW, NN, 12_1457_000).

[27] Agstner (1995).

[28] This 'nationalistic' misrepresentation has affected also prominent scholars like Angelo Sammarco, and Armando Levi Cases.

which included, in addition to Negrelli, the Leipzig-based bankers Albert Dufour-Féronce and Louis Sellier; the 'French group', which, besides Enfantin, included François Barthélemy Arlès-Dufour and the three brothers Jules, Léon and Paulin Talabot, while the 'English group' was composed of Robert Stephenson and Edward Starbuck.[29]

As a result of the establishment of the *Société d'Études* Negrelli became more committed than ever to the project, especially since the three heads of each group, Paulin Talabot, Stephenson and Negrelli, each had specific tasks to carry out. The German group's assignment was to survey the Mediterranean coast in order to find the ideal location for the outlet of the canal. The French group was to study the terrain where the canal layout was to be excavated, and the English group was responsible for examining the canal's outlet in the Red Sea, in the proximity of the port of Suez.[30] Many documents from the *Nachlaß* give evidence of Negrelli's care and enthusiasm in preparing for the mission in Egypt. Asking his close friend, the Swiss architect Felix Kubli, to be his right-hand man during the trip, Negrelli wrote: 'I cannot offer you a reward, because we will sacrifice our services on the altar of Europe and Europe will be grateful to us only if the canal will be built'.[31]

A series of obstacles, which included the revolutions of 1848 as well as the divergences with Stephenson, who kept on postponing the date of departure, prevented the three head engineers from travelling to Egypt. Some historians have interpreted Stephenson's behaviour as an attempt to boycott the canal project, given that, like his country's government, the English engineer feared a potential French dominance over the project but backed the construction of a railway link between Alexandria and Suez, which was completed in 1857.[32] Negrelli nonetheless arranged for a group of technicians, the so-called German brigade, to travel to the Mediterranean shores. In July 1847, the brigade had finished the survey of 7000 *Klafter* of coast around the point identified as the ideal location for the canal outlet and was about to bring the survey and cross-section of the area's seabed to an end.[33]

[29] Charles-Roux (1901, vol. 1, 208–209).

[30] Ibidem.

[31] Negrelli to Kubli, 14 November 1847, TMW, NN, 07_0469_000. Felix Kubli (1802–1872) accompanied Negrelli to Egypt in 1855.

[32] Algardi (1988, 33–34), Gatani (2000, 248–257).

[33] Jaßnüger to Negrelli, 1 June 1847, TMW, NN, 08_0554_000. Jaßnüger to Negrelli, 16 June 1847, TMW, NN, 08_0555_000. The Klafter is a Central European unit of length that corresponds to 1.89 metres.

Negrelli concerned himself also with the financial side of the project, considering how the work could be funded without appearing like a predominantly French enterprise. This possibility was causing concern in Britain as well as in Egypt, where the Khedive or Viceroy Mohammed (or Méhmét) Ali was growing increasingly sceptical of the project.[34] Dufour-Féronce, founding member of the German group, recounts this in the first letter regarding Suez found in the *Nachlaß*, dated 13 November 1845: '(...) everyone who is interested in the project believes that the best thing would be for Germany to take the lead. The whole affair would thus appear to be purely commercial, given that the two great German powers [Prussia and Austria] do not have notable navies and cannot give rise to concerns anywhere'.[35]

Following this advice, Negrelli started to press for the involvement of the Habsburg monarchy and of big Central European financial groups in the canal project. Within the German group of the *Société d'Études*, Negrelli was able to secure five out of ten places for Austria and to find as many institutions willing to invest in the project.[36] In regard to the financing of the work, Negrelli corresponded assiduously with both Dufour-Féronce and the Triestine banker Pasquale Revoltella, who would go on to play a key role in the Austrian funding of the project and whom traditional Italian historiography has wrongly labelled the 'Italian' promoter of the Suez Canal.[37]

The Experience in the Lombardo-Venetian Kingdom and the Evolution of the Suez Canal Project

The turbulent events of 1848 that wreaked havoc all over Europe impacted on the planning of the Suez Canal in two ways: they caused the trip to Egypt of the three head engineers of the *Société d'Études* to be indefinitely postponed, as well as indirectly leading to a slowdown of the whole

[34] Treue (1970), Lo Giudice (1981, 38–65).

[35] Dufour-Féronce to Negrelli, 13 November 1845, TMW, NN, 06_0259_002.

[36] Negrelli to Thurn, 21 January 1847, TMW, NN, 07_0496_000. The institutions were: the municipality and the Chamber of Commerce of Trieste, the Austrian Lloyd, the Viennese Union of Industries and the Chamber of Commerce of Venice (which was part of Austria).

[37] The correspondence with Dufour-Féronce lasted from 1845 to 1855. The relationship between Negrelli and Revoltella is documented in the Nachlaß by more than 200 documents. After Negrelli's death, his daughter accused Revoltella of being co-responsible of the theft of her father's project. See: Scaglione (1971/1972), Introduction: XIX–XXVI.

project. In the aftermath of the war of that year, the Austrian government appointed Negrelli to oversee the reconstruction of the heavily damaged infrastructure of the Lombardo-Venetian Kingdom, the Italian 'province' of the empire.[38] The move to Verona enabled Negrelli to establish relations with all the technicians working on the realisation of a railway network in North and Central Italy. Amongst them was Pietro Paleocapa, an engineer and minister of the Kingdom of Sardinia, who was following the progress of the Suez Canal.[39] In 1849, while in Verona, the emperor elevated Negrelli to the nobility on account of his railway projects and he took on the title of Ritter von Moldelbe.

The correspondence of this period shows that Negrelli's attitude towards the protagonists of the Risorgimento was not in the least sympathetic to the cause of Italian unification, like many Italian historians affirmed in the interwar period and still do today.[40] In the light of several statements decrying the Italian quest for independence, it is clear that Negrelli's 1855 removal from his position had nothing to do with his alleged pro-Italian stance, as some authors have uncritically maintained, but was related to professional antagonisms.[41] In other words, the Suez Canal project was about to turn into reality and, as the author of the plan that was tipped to be the definitive one, Negrelli was swiftly gaining a reputation as the most famous Austrian engineer. His rivals aimed at bringing him into disrepute and at the same time hindering the whole project.[42] Less than five months after his destitution, he was reinstated in the state service and made Inspector General of the State Railways.

The Expedition to Egypt and the Project for the Suez Canal

Following Negrelli's removal from Verona and his return to Vienna, the Suez Canal project gained new momentum at the onset of 1855. The previous year, international developments had finally led to a breakthrough in the long-standing issue of the canal. Mohammed Ali, the Egyptian

[38] Birk (1925, 12–38), Bernardello (1997), Leonardi (2006).

[39] Birk (1925, 238), on the role of Paleocapa Curli (2015).

[40] Negrelli to Escher, 1 February 1848, TMW, NN, 08_0589_000. Sammarco (1939, 66–68). Algardi (1988, 145–158).

[41] Deihsen (1993, 219–233). The author has reconstructed the whole affair using documents from Vienna's Verkehrsarchiv.

[42] Deihsen (1993, 219–233), Birk (1925, 165–184).

viceroy who had taken an interest in the enterprise had died in 1849. His successors, his son Ibrahim and his grandson Abbas, were not particularly keen on the project and were more concerned with the railway line between Alexandria and Suez.[43] When Abbas was murdered in 1854, his next-in-line was Mohammed Said, who counted a very entrepreneurial man amongst his acquaintances: Ferdinand de Lesseps, an extremely well-connected French diplomat who had taken the canal matter into his own hands when the *Société d'Études* appeared to be going through a period of inactivity. De Lesseps, who had previously served as the French general consul in Egypt, succeeded in piquing the new viceroy's interest in the project and in getting him to issue a 'firman', that is a mandate for the realisation of a waterway between the Red Sea and the Mediterranean, in 1854.[44]

Before moving on to the next phase of the project, it was necessary to overcome the Egyptians' distrust, who feared that the canal might turn into a Trojan horse for a French colonial conquest of the country. The viceroy demanded that the executive company tasked to build the canal would not rely exclusively on French money, but instead on a truly international capital, thus giving strategic importance to the participation of Austria and Prussia.[45] Mohammed Said also asked for an International Commission of experts to examine all the projects that had hitherto been developed. The Commission consisted of the most prominent engineers from seven countries, including Negrelli as the Austrian representative.[46] Their first action was to send a sub-commission, of which Negrelli was also part, to Egypt, to settle unresolved issues through appropriate field investigations. Negrelli thus undertook his first and only trip to Egypt from 10 November 1855 to 11 January of the following year, eight years after having first started to plan this journey.[47]

On 5 January 1856, juggling multiple priorities with extreme confidence, de Lesseps was able to attain a new concession agreement. This one contained a clause that required de Lesseps to secure the consent of the Ottoman sultan and the approval of the statutes of the *Compagnie universelle du canal maritime de Suez*, which he went on to found on 15

[43] Sammarco (1935).
[44] Edgar Bonnet (1951, 132–144), Pudney (1969, 34–36), Algardi (1991).
[45] Cervani (1990).
[46] On the Commission see Chap. 2.
[47] Birk (1925, 238–250).

December 1858 to carry out the building works.[48] After the trip to Egypt, the Commission was due to give its opinion on the project to be adopted. Ahead of their meeting, in May 1856, Negrelli published a book in support of his project. The final part of this work described the environmental and geological characteristics of the land on which the excavation was to be carried out, as well as the characteristics of the future canal. The volume is entitled: *Egypt's present means of transport and communication in regard to the proposed piercing of the Isthmus of Suez* and was published in both German and French.[49] In Vienna there was widespread optimism about the realisation of the project, as transpires from a lecture that the geologist Franz Foetterle coordinated in 1856 at the k.k. Geographical Society on behalf of a highly regarded committee that also included Negrelli.[50] The talk detailed all the steps of the planning process, identified the various protagonists of the different phases of study and survey and recognised Negrelli's authorship of the project of the lock-free canal.

In June 1856, in Paris, the Commission chose Negrelli's project as the most outstanding and worthy of being realised. The day after the final meeting, 26 June 1856, Negrelli summarised the debate and its outcome in a letter to Revoltella: 'As you know my project was always about an open canal, free from hindrances, without locks and supplied with seawater. The English brought their own project. It was a raised canal, supplied with Nile water and only accessible through six locks. The French wanted a canal like mine, but with two locks, one at each end. Holland was also inclined towards the locks, but wanted to erect them as a precautionary measure only and otherwise leave them free. The discussion was long and animated. The representatives of the English and French navies and the Sardinian minister Mister Paleocapa supported my idea, which was finally given preference and adopted by the commission. The programme is now over and we are preparing the details. Afterwards, we will publish the general proceedings and the Viceroy of Egypt will dispatch his Foreign Minister to Constantinople to obtain the exequatur of the Sublime Porte with Austria's and France's backing'.[51]

[48] Maltese (1978, 63–69).

[49] The volume was published as an appendix to the weekly *Austria*, the French version appeared in the *Journal des Deux Mers*.

[50] Foetterle (1857).

[51] Negrelli to Revoltella, 26 June 1856, TMW, NN, 12_1336_000.

This letter sheds light on a series of elements that have until now been open to conflicting interpretations: the French and British navies' support for Negrelli's plan, despite the British government's aversion for the canal; the support of the Sardinian minister Paleocapa; and Austria's key financial role in the project's success.[52] The letter depicts a situation in which the die was cast and Istanbul's ratification of Austria's and France's adhesion to the project was the only thing missing. A letter written to his wife on the same day also offers evidence of how friendly Negrelli was with the other main participants in the enterprise, whom he describes as treating him like a 'brother'.[53] The engineer's long-standing efforts appeared to be crowned with full success.

Negrelli's project had prevailed over all the others and the International Commission had embraced it. 'We believe' record the Commission's proceedings 'that the direct route, which we will briefly outline, meets all requirements as fully as possible and in every respect. We will justify this opinion by going into the project's details. We have to add that Mister de Negrelli had devised the direct route in 1847 and that his studies of that time have found a solution to the problem that is very similar to the one brought forward by the gentlemen Linant Bey and Mougel Bey'.[54]

Yet, 1857 and 1858, the last two years in the short life of the Tyrolean engineer, were full of upheavals, given that in England there was manifest opposition to the realisation of the canal. Robert Stephenson, the former member of the *Société d'Études*, launched a virulent attack on Negrelli's project and Lord Palmerston followed suit in the political arena.[55] Their objections were not technically motivated, as Negrelli believed that the tidal flow would prevent the canal from silting up. Instead, the politically driven criticism originated largely from the English governmental and financial milieu, which was little inclined to give space or means to the enterprise.[56] Paleocapa intervened in the debate to defend Negrelli and the latter refuted Stephenson's observations, in a courteous but firm manner, via two articles published in Vienna's *Österreichische Zeitung*.[57]

[52] The Nachlaß contains one letter from Paleocapa to Negrelli, 6 June 1857, TMW, NN, 13_1407_000.

[53] Negrelli to Caroline Negrelli, 26 June 1856, TMW, NN, 12_1337_000.

[54] The Commission's proceedings are in: Dupin (1858, 21–22).

[55] Birk (1925, 279–284). Gatani (2000, 274–277).

[56] Algardi (1988, 317–334).

[57] Paleocapa (1857), Negrelli (1858a, 1858b). Birk (1925, 264–293).

Luigi Negrelli never saw the work he had designed, dying on 1 October 1858. Two months after his death, de Lesseps founded the *Compagnie universelle du canal maritime de Suez*, which started excavating the canal in 1859 and completed it in 1869. De Lesseps turned Negrelli's most ambitious project into reality and much literature undeservedly credits the former with technical merits that are entirely due to the Tyrolean engineer.[58] Luigi Negrelli fell into oblivion until 30 years after his death, when his daughter Maria Grois sued the *Compagnie universelle* and de Lesseps's heirs, accusing the latter of having passed her father's project off as his own.[59] The trial went on for years and, despite acknowledging Negrelli's authorship of the design, it did not award anything to his heirs. The court case however triggered a political-ideological competition between Italy and Austria, which both claimed the figure of Negrelli as their own, giving rise to a series of misrepresentations that persist until today.[60]

Conclusion

The eventful life of Luigi Negrelli, his studies, his *Weltanschauung* led him to seek a way of uniting different peoples, firstly the multinational Austrian state, then Europe and, last but not least, different continents. The fact that this happened at a time when the prevalent trend was to highlight elements of division rather than reasons for cooperation between people of different cultures and languages, makes his contribution even more remarkable.

Negrelli's work was not limited to providing high-level technical support for railway, road and hydraulic works; it transcended the boundaries of pure engineering and manifested itself in a marked curiosity for the context in which his projects were to be integrated. This curiosity, experienced in a pragmatic way, prompted Negrelli to study the economic and organisational aspects of many projects he was responsible for, in the belief that they largely determined the effective utility and social importance of the work.

[58] Kirsch (1971), Algardi (1988, 334–350).

[59] Scaglione (1971/1972: XIV–XXVII), Algardi (1988, 334–350).

[60] In 2018 the exhibition *L'épopée du canal de Suez. Des pharaons au XXI siècle* at the Institute du Monde Arabe in Paris did not mention Negrelli at all. The articles on this exhibition in the Italian press rightly pointed out Negrelli's role but still labelled him 'Italian'. Fuganti (2018).

Negrelli successfully combined the roles of talented professional and upstanding official, despite the obscure plots hatched against him while he was in Verona and from which, although cleared of all charges, he emerged profoundly disappointed. He was always careful to operate observing a strict distinction between private and public interests. The *Weltanschauung* that motivated him was a combination of the Christian values to which he had been exposed since childhood and the wealth of knowledge he had drawn from the scientific and epistemological debate of his time. His own first-hand experiences led him to identify with the mechanisms of progress that the society he lived in was creating and to offer a scientific contribution to the advancement of peoples in Europe and worldwide. Against this backdrop, Negrelli's greatest work, the study and project of the Suez Canal, is the peak of a decade-long engineering career.

References

Abbreviations

TMW Technisches Museum Wien
NN Nachlaß Negrelli

Bibliography

Agstner, Rudolf. 1995. *125 Jahre Suezkanal. Österreich (-Ungarn) und seine Präsenz am Isthmus von Suez. Zur Geschichte der Konsulate in Suez, Ismaila und Port Said 1844–1856.* Kairo: Österreichisches Kulturinstitut Kairo.

Algardi, Zara Olivia. 1988. *Luigi Negrelli, l'Europa, il canale di Suez.* Firenze: Le Monnier.

———. 1991. La intramontabile compagnia di Suez: magia dei nomi. *Belfagor* 46 (4): 441–448.

Bernardello, Adolfo. 1997. Il commissario Negrelli e le comunicazioni stradali e ferroviarie nel Lombardo-Veneto (1848–1849). *Il Risorgimento,* (1–2), 201–216.

Birk, Alfred. 1915. *Alois von Negrelli. Die Lebensgeschichte eines Ingenieurs, vol. I: In der Heimat, in der Schweiz, in Österreich.* Wien-Leipzig: Braumüller.

———. 1925. *Alois von Negrelli. Die Lebensgeschichte eines Ingenieurs, vol. II: Der Suez Kanal. Letzte Kämpfe.* Wien-Leipzig: Braumüller.

Birk, Alfred, and Karl Hermann Müller. 1925. *Der Suezkanal: seine Geschichte und seine wirtschaftliche Bedeutung für Europa, Indien und Ägypten.* Hamburg: Boysen & Maasch.

Bonin, Hubert. 2010. *History of the Suez Canal Company, 1858–2008. Between Controversy and Utility*. Genève: Droz.
Bono, Salvatore. 1990. Precedenti storici del canale di Suez: idee e progetti dal secolo XVI al XIX. In *Luigi Negrelli ingegnere*, ed. A. Leonardi, 141–159. Trento: Società di studi trentini di scienze storiche.
———. 2006. Il canale di Suez e l'Italia. *Mediterranea* III (8): 411–422.
Brugger, Max. 1909. *Zürcherische Eisenbahnpolitik, Inaugural-Dissertation zur Erlangung der Würde eines Doktors der Staatswissenschaften der H. Juristischen Fakultät der Universität Freiburg (Schweiz)*. Zürich: Coradi-Maag.
Bussjäger, Peter, Josef Concin, and Karl Gerstgrasser. 1997. *Alois Negrelli und seine Spuren in Vorarlberg (1822–1832)*. Eine regionalhistorische und verwaltungsgeschichtliche Untersuchung, Bludenz, Geschichtsverein Region Bludenz.
Cervani, Giulio. 1990. Gli ambienti economici triestini, Pasquale Revoltella ed il progetto della realizzazione del canale di Suez. In *Luigi Negrelli ingegnere*, ed. A. Leonardi, 187–250. Trento: Società di studi trentini di scienze storiche.
Charles-Roux, Jules. 1901. *L'Isthme et le canal de Suez*, 2 vols. Paris: Hachette.
Coilly, Nathalie, and Philippe Régnier, eds. 2006. *Le Siècle des Saint-Simoniens: Du Nouveau christianisme au canal de Suez*. Paris: BNF.
Contro, Roberto. 1990. I progetti di Luigi Negrelli in campo ferroviario. In *Luigi Negrelli ingegnere*, ed. A. Leonardi, 380–381. Trento: Società di studi trentini di scienze storiche.
Curli, Barbara. 2015. Il Piemonte e il canale di Suez, 1855–1856. *Studi Piemontesi* XLIV (2): 325–338.
Deihsen, Christian. 1993. *Alois Negrelli Ritter von Moldelbe, ein Österreichischer Eisenbahnpionier des 19. Jahrhunderts. Seine Leistungen und Verdienste um die Entwicklung des Eisenbahnwesens im Österreichischen Kaisertum und in der Schweiz bis 1858*, Dissertation, Wien.
Dupin, Charles. 1858. *Canal maritime de Suez*. Rapport à l'Académie des sciences par M. le Baron Charles Dupin. Paris: Institut Impérial de France.
Edgar Bonnet, Georges. 1951. *Ferdinand de Lesseps: Le Diplomate, le Créateur de Suez*. Paris: Librairie Plon.
Foetterle, Franz. 1857. *Bericht über die Durchstechung der Landenge von Suez an die k.k. Geographische Gesellschaft von der hierzu gewählten Commission* […]. Wien: Auer.
Fuganti, Daniela. 2018. Un'epopea lunga 164 chilometri. *Archeo* 400: 90–91.
Gatani, Tindaro. 2000. *Luigi Negrelli ingegnere trentino ed europeo*. Messina: Edas.
———. 2005. Luigi (Alois) Negrelli ingegnere ferroviario. *Civis* 29 (87): 139–184.
Helps, Hanni. 2005. Luigi Negrelli, Engineer 1799–1858: Planner of the Suez canal. *Transactions Newcomen Society* 75 (2): 317–339.
Jouve, Bernard. 2001. *L'épopée saint-simonienne. Saint-Simon, Enfantin et leur disciple Alexis Petit de Suez au pays de George Sand*. Paris: Guénégaud.

Kirsch, Walter Paul. 1971. *Luigi Negrelli. Der Schöpfer des Suez Kanals. Ein Genie, seine Zeit, sein Leben und sein Wirken*. Wien: Verlag Jugend und Volk.

Leonardi, Andrea, ed. 1990. *Luigi Negrelli ingegnere e il canale di Suez*. Atti del convegno internazionale, Primiero 15–17 settembre 1988, Trento: Società di studi trentini di scienze storiche.

———. 2006. Dall'Europa a Suez; il ruolo di Luigi Negrelli (Primiero 1799–Vienna 1858) nella promozione delle comunicazioni internazionali. In *Giovanni Miani e il contributo Veneto alla conoscenza dell'Africa. Esploratori, missionari, imprenditori, scienziati, avventurieri, giornalisti*, ed. Romanato Gianpaolo, 251–265. Rovigo: Minelliana.

———. 2016. Luigi Negrelli: un protagonista del take off ferroviario in area mitteleuropea. *Histoire des Alpes—Storia delle Alpi—Geschichte der Alpen* 21: 195–214.

———. 2020. A Prominent Figure in the Creation of the Suez Canal: Luigi Negrelli. *The Journal of European Economic History* 49 (2): 27–70.

Lo Giudice, Giuseppe. 1981. *L'Austria, Trieste ed il canale di Suez*. Catania: Università degli studi.

Maltese, Paolo. 1978. *Storia del canale di Suez. L'Egitto e il canale 1833–1956*. Milano: il Formichiere.

Marzani Prosser, Maria Beatrice. 1999. *Luigi Negrelli, un anno di vita. Tagebuch auf das Jahr 1831—Diario dell'anno 1831*. Trento: Fondazione Museo Storico.

Mazuel, Jean. 1937. *L'Oeuvre géographique de Linant de Bellefonds: étude de géographie historique*. Le Caire: Schindler.

Moghira, Mohamed Anouar. 2002. *L'isthme de Suez: passage millénaire, 604–2000*. Paris: L'Harmattan.

Negrelli, Luigi/ Alois/ Louis. 1837. Brief Negrellis vom 26 Dezember 1836. *Appenzellisches Monatsblatt* 2: 33–36.

———. 1838. *Ausflug nach Frankreich, England und Belgien zur Beobachtung der dortigen Eisenbahnen, mit einem Anhange über Anwendung von Eisenbahnen in Gebirgsländern*. Frauenfeld: Beyel.

———. 1856. *Die gegenwärtigen Transport—und Kommunikationsmittel Egyptens mit Beziehung auf die beantragte Durchstechung der Landenge von Suez*. Abgedruckt aus der Wochenschrift Austria Jahrgang 1856, Heft XVII, Wien.

———. 1858a. Negrelli contra Stephenson. *Österreichische Zeitung*, 136, 18 June.

———. 1858b. Negrelli und Stephenson. *Österreichische Zeitung*, 220, 26 September.

Paleocapa, Pietro. 1857. *Observations sur le discours prononcé par M. Stephenson, ingénieur, dans la Chambre des Communes (Séance du 17 Juillet 1857)*. Paris: Henri Plon.

Piquet, Caroline. 2008. *La Compagnie du canal de Suez. Une concession française en Egypte (1888–1956)*. Paris: Presses de l'Université Paris-Sorbonne.

———. 2018. *Le canal de Suez. Une voie maritime pour l'Egypte et le monde.* Paris: Bonnier.

Pudney, John. 1969. *The Lesseps' Canal.* London: Praeger.

Sammarco, Angelo. 1935. *Les règnes de Abbas de Said et d'Ismail avec une aperçu de l'histoire du canal de Suez, vol. IV of: Précis d'Histoire d'Égypte.* Roma: Istituto Poligrafico dello Stato.

———. 1939. *Luigi de Negrelli, La mirabile vita del creatore del canale di Suez. Con un'appendice sugli Italiani in Egitto.* Roma: Istituto fascista dell'Africa italiana.

Sandgruber, Roman. 1987. Eisenbhanzeit: Das Jahrhundert der Bahn. In *Zug um Zug. 1x Marchfeld und Retour*, ed. Reinhard Linke and Hannes Schopf, 132–141. St. Pölten and Wien: Niederösterreichisches Pressehaus.

Scaglione, Francesco Attilio, ed. 1971/1972. *Luigi Negrelli e il canale di Suez nelle carte del Fondo Maria Grois Negrelli*, 2 vols. Roma: Ministero degli Affari Esteri.

Treue, Wilhelm. 1970. Das österreichische—mitteldeutsche und das norddeutsch staatliche und private Interesse am Bau des Suezkanals. *Vierteljahresschrift für Sozial- und Wirtschaftsgeschichte* 57: 534–555.

CHAPTER 8

'This Concerns the Future of Italy': The Suez Canal as an Opportunity—Giuseppe Sapeto and His *L'Italia e il canale di Suez*

Filippo Gattai Tacchi

In the newly created Kingdom of Italy, within the narrow political elite and in the circles involved in manufacturing and trade, the opening of the Suez Canal gave rise to lively debate on the prospects and plans for a commercial expansion that would increase the country's economic importance and diplomatic prestige. In the contemporary discussion, Italy's particular geographical fortune was often highlighted. With the cutting of the Isthmus of Suez, it was thought that Italy would be transformed into a transit station; the canal would unite the nations and markets of Northern Europe with the East, functioning as a launching pad for the Red Sea, the Indian Ocean, and the Far East, where the European powers had established their first colonial bases. Debates, studies, and publications on the opening of the canal addressed different themes: from the prospects for expansion and consequent economic growth to the need for

F. Gattai Tacchi (✉)
Scuola Normale Superiore, Pisa, Italy
e-mail: filippo.gattaitacchi@sns.it

B. Curli (ed.), *Italy and the Suez Canal, from the Mid-nineteenth Century to the Cold War*,
https://doi.org/10.1007/978-3-030-88255-6_8

governmental and infrastructural measures that would prepare Italy for the great event; from the quest for a respected position within the Concert of Europe for the young nation to its potential colonial future in Africa and the East.[1]

This discussion was led by some eminent figures in Italian public life, including Luigi Torelli—who was Minister for Agriculture, Industry, and Commerce in the La Marmora government (1864–1865) and represented Italy on the board of Ferdinand de Lesseps' *Compagnie universelle du canal maritime de Suez* (Suez Canal Company), and Nino Bixio—who had been a key figure in the Italian Risorgimento, subsequently a parliamentary deputy, and then a senator. These two were part of a dynamic group of merchants, intellectuals, bankers, investors, and shipowners who had interests in Italy's naval and commercial future and were open to novel initiatives. This small but powerful pro-Suez lobby saw commercial expansion in the Mediterranean Sea and beyond as an opportunity for growth that could not be ignored, and which would be dependent on a network of consulates and commercial bases, national and private steam shipping companies, and active collaboration with religious missions in Africa, the Ottoman Empire, and the Far East. Essentially, they wanted Italy to pursue the path of Gladstonian imperialism that had been establishing Great Britain's global supremacy: a liberal imperialism based on 'the three great Cs' of Civilization, Commerce, and Christianity. The opening of the Suez Canal seemed to be the perfect moment to take action.[2] The pro-Suez lobby mainly conducted its campaign through public debates and the flowering of an 'isthmic literature' that had different rhetorical registers and themes, from patriotic enthusiasm about the potential impact of the canal to pessimistic realism about the state of Italy's infrastructure and merchant navy.[3] In 1865, in the midst of this debate, Giuseppe Sapeto published *L'Italia e il canale di Suez*, a volume intended for both the well-informed and the general public. This merits consideration as one of the most important texts on Italian colonialism, partly because of its themes and the issues raised, but especially because of the part that its author subsequently played. It is also an exemplary instance of how evangelism, nation-building, and colonialism were closely interconnected during that period. In addition, the book demonstrates how

[1] Natili (2008, 7–50).
[2] James (1994, 184–200), Morris (1968, 175–96).
[3] Correnti (1869, 497).

Sapeto, speaking for Italy's cultural and political establishment of the mid-nineteenth century in general, represented and reinterpreted Africa: it presents an intriguing mixture of fascinating orientalist imagery, narratives of the revival of the Roman Empire and a new age for the *Mare Nostrum* and the Maritime Republics in Italy's hoped-for political and commercial role in the Mediterranean, and the country's potential as a great power on the international and colonial stage.

Giuseppe Sapeto was born in 1811 in Carcare, Liguria. In 1829, he enrolled at the seminary for the Congregation of the Mission of St. Vincent de Paul in Turin, to pursue his theological and philosophical studies. He departed for Syria in 1834, where he taught young children from Syria, Egypt, and Cyprus at the mission's school in Antura. Sapeto's activity was not limited to teaching: from the beginning, he was an enthusiastic student of Arabic and the ancient texts of Syrian and Oriental Christianity, and was responsible for compiling an Arabic-Italian dictionary for the missionaries. His interests also encompassed art, the archaeology of Roman, Hellenic, and Crusader ruins, and numismatics, all of which served his attempts to shed light on ancient Syrian culture. In 1838, intellectually restless and insufficiently stimulated within the Syrian community, Sapeto asked his mission for permission to leave for Africa: this represented a decisive turning point in his life.[4]

Sapeto decided to go to Egypt, where he joined an expedition to Northern Ethiopia led by the d'Abbadie brothers, two well-known French explorers, with whom he arrived at the port of Massawa, in the Horn of Africa, in the autumn of 1837. Apart from a few brief European interludes, Sapeto continued his missionary work, studied, and conducted explorations in this region for almost the next twenty years, during which time he collected an enormous amount of information on all aspects of Abyssinian culture. He compiled dictionaries of local languages for missionaries and traders, and wrote studies of the liturgy of Coptic Christianity, essays on the history of the Kingdom of Abyssinia, and breviaries for priests. Finally, in 1857, he published his magnum opus, *Viaggio e missione cattolica fra i Mensá, i Bogos e gli Habab, con un cenno geografico e storico dell'Abissinia*, a three-volume compendium of everything he had seen, learnt, and discovered in the region. In the years that followed, this work came to enjoy considerable respect in both Italy and the rest of Europe, where very little was known about African geography.

[4] Surdich (2005, 5–13).

Sapeto's scholarship, thirst for discovery, and evangelizing zeal were complemented by a political and economic vision that made him one of the central figures of early Italian colonialism. At the beginning of the 1850s, he urged Napoleon III of France to intervene in the conflict between the Ethiopian *ras* (princes) in order to extend French influence into the Horn of Africa, a region that would acquire increased strategic importance with the opening of the Suez Canal, especially in regard to trade routes to India and the East. In his proposal, Sapeto first drew attention to Ethiopia's crucial location, the fertility of its land, and the wealth that could be generated by trade with central Africa. French involvement might have pre-empted an expansionist campaign by the British, which was highly likely given that the latter could already boast control of the Aden peninsula. He concluded by arguing that a French presence would protect Christian communities, perennially threatened in the Horn of Africa by Muslim persecution, and would combat the slave trade. Sapeto's attempts at moral suasion were not limited to France: much the same tone, rhetoric, and rationale were employed in a letter of 1851 to the Kingdom of Sardinia and Piedmont's consul in Alexandria, Egypt, in which he pressed for the establishment of a consulate in Massawa in order to promote and defend Italian commercial involvement in the region, implicitly offering himself as the incumbent. He tenaciously continued his campaign in the early years of the Kingdom of Italy, repeatedly asking ministers to intervene and urging the involvement of shipping companies and traders, and in 1863 Italy's Minghetti government asked him to negotiate with the British governor in Aden for the acquisition of land for a commercial base on the coast of the Red Sea. However, the reluctant involvement of the various Italian ministries, more than any British opposition, sealed the fate of this attempt.[5] By this point, Sapeto's diplomatic and campaigning activity had evidently far exceeded the reach of his evangelism and he was in open conflict with his missionary superiors; he finally left the priesthood in 1862 and went to Genoa, where he continued his persistent and impassioned lobbying on Africa while teaching Arabic and oriental languages in the local naval training schools.[6]

The city of Genoa was at that time a centre of animated debate regarding the opportunities presented by the imminent opening of the Suez Canal and the reforms that would be necessary in order to take advantage

[5] Ibid., 14–50.
[6] Ibid., 68–82.

of the new trade, with revival of both the city and the entire country in mind. Although Genoa was then Italy's main port and had a proud naval history that dated back to the Maritime Republics of the medieval era, its local elites had not kept pace with the rapid developments in international shipping: its commercial fleet, in the new steam age, still mostly consisted of sailing ships. However, Sapeto found the city to be fertile ground for his campaign, and threw himself into an intense series of lectures, debates, and public meetings. In 1865, he crystallized his efforts to sway public opinion in publication of *L'Italia e il canale di Suez* (Italy and the Suez Canal), which provided an early and fundamental text for Italian colonialism. In his brief introduction, Sapeto dedicated the volume to the Italian Chambers of Commerce, urging them to take action:

> Worthy gentlemen, you have the power to arouse in the hearts of the municipalities, manufacturers, farmers, and wealthy citizens the spirit of the great associations that will be needed if we are to increase a hundredfold the factories and workers weaving textiles of every kind, to produce our wines and oils in a better way, to reorganize and equip our ports, all in order to be ready to compete in the trade that, thanks to the canal through the Isthmus, will be unimaginably vast [...].
>
> For this exalted enterprise of generosity, industriousness, and encouragement towards the great transformation, you will have the gratitude of today's Italians and the blessing of those to come. (V–VI)

Sapeto wanted to ignite and then sustain the interest of the Italian establishment in the Suez Canal; just as Bixio and Torelli had also done, he issued a clarion call for the transformation of words into deeds. His book represented a sort of compendium of the views expressed in the lively series of debates about the importance of the canal for Italy's future. He divided it into two parts: its first eight chapters, largely descriptive, covered the history and geography of the Isthmus of Suez, including a wealth of information on the peoples that had inhabited the region at different times, its geological and hydrological features, flora and fauna, and potential for mineral extraction; the seven chapters of the book's second half dealt with the purely political and economic considerations and made suggestions regarding the reforms necessary in order to take full advantage of the canal.

In the first half of his book, Sapeto brought together the historical and scientific literature of the Greco-Roman era and Islamic period with the latest work by British and French scientists. He then retraced the long history of attempts to cut a channel across the Isthmus, from the age of the Pharaohs to the modern era, complementing his extensive knowledge of classical and medieval historiography with descriptions of the ruins he himself had observed while travelling through the territories of the Isthmus in the 1840s, employing a somewhat romantic and Orientalizing narrative style. With the forthcoming opening of the Suez Canal, what had only been a dream for past civilizations and cultures was to become reality:

> Finally, as last century came to a close, that miracle of a man, Bonaparte, turned his plans to these regions; not being able to implement them, he left their creative concept, like all the aspects of his thinking, as a legacy to his descendants, who after many centuries will see the awesome cutting of the Isthmus.
>
> It seems that humanity, to bind the East to the West, needed to remove that obstacle, and revealed this idea to its great sons [...] Modesty will not permit Lesseps' soul to hear the heroic language, but humanity will acknowledge and extol him, for having brought into being perhaps the best idea of the great minds of the past. (50)

Presenting the Isthmus as an ancient meeting point of civilizations and cultures, and the creation of the canal as a great historical moment that would unite peoples and continents, Sapeto detailed the political and planning phases between the studies by Napoleon's engineers and the final establishment of de Lesseps' Suez Canal Company in 1858. Much of the Ligurian scholar's attack focused on the strong opposition mounted by successive British governments, which had feared that the opening of the Suez Canal would not only undermine the established routes to India and the East, but also damage the stability of the Ottoman Empire, whose existence had helped to maintain the strength of the British Empire:

> Britain, which no longer had the Muscovite incubus to worry about after victory in the Crimea, began, like a gang, to scheme against the enterprise, whose opposition wanted to see it as a future trap aimed at the British colonies and Indian trade, much more worrying than Russian incursions in the Asian interior. It remained silent between 1833 and 1857, believing that it had relegated to obscurity the spectre of the canal, which risked bringing Europe closer to the Eldorado of the Indies. However, seeing it reappear in flesh and blood, the journalism of all the opposing countries opened Pandora's box and let its evil spirits loose on the four winds. (83)

The stubborn opposition by the British government, voiced in particular by prime minister Lord Palmerston, provoked heated debates in that country. De Lesseps organized a tour of the main British commercial and industrial cities in order to present his initiative and garner financial support for the Suez Canal Company; ample investments were made both by private individuals and by shipping and trading companies, placing Britain, ironically, among the countries giving the most support to de Lesseps' venture. Sapeto saw the British government's behaviour as dishonourable:

> Europe censured this opposition of injustice and selfishness, and if the personal honour of Lord Palmerston and the greater honour of his country had been susceptible, they would have lost value, as [Member of Parliament] Griffith said, in the eyes of all nations.
>
> [...]
>
> As one of its great admirers, I can therefore conclude, alongside Mr. Griffith, that Britain would have been better and more laudably advised to leave the canal at the mercy of the predicted technical difficulties, rather than put itself in jeopardy of a defeat that would be merited by its arrogance, but would scarcely be consistent with the honour of a great nation. (85–7)

Having provided an extensive description of the Isthmus, covering its geography, geology, hydrology, history, and politics, in the last chapter of the first half of his book Sapeto offered an effusive celebration of the Suez Canal as a vehicle for civilization and brotherhood between peoples and cultures united 'in one civilization: one single universality of mutual interests and trade' (109). This was almost a hymn to liberalism:

> The new canal of Suez is not a project of governments; it is instead a product of the human spirit, which wants to embrace humanity with loving kisses, protecting it, after its division for many centuries. (98)

Having portrayed the Suez Canal as a major resource for the international community, a gift from Providence 'when it ordains that great events take shape' (100), Sapeto dedicated the second half of his volume to the commercial and political opportunities that the canal offered to the European powers, especially Italy with its favourable geographical position and impressive naval tradition. He gave substantial attention to the civilizing vocation of these countries, extolling the three central features of liberal imperialism mentioned earlier: civilization, commerce, and Christianity. In Sapeto's opinion, it was evident that the great civilizations had emerged only when they were accompanied by commercial and trading activity: peoples that lived far from trade routes could never develop. Enhanced trade was necessarily beneficial:

> Trade lives on ease and freedom, products arrive in relation to the demand for them, industry proliferates with the sale of its production, the quantity matches the needs of the customers, these needs grow in relation to civilization, and civilization spreads with frequent trade and negotiation.
>
> My God! What unlimited horizons lie before us! (125)

The Suez Canal, connecting Europe to the East, was thus to aid regeneration of the West and lead the East towards development:

> The Isthmus canal will perhaps bring its best fortunes to the East. With the eastern and western sea waters wedded together, the marriage of the two humanities will soon occur; ideas will spread and contact with the vitality of the West will reinvigorate those half-exhausted spirits. [...] The brotherhood of all peoples in the union of civilization can no longer be delayed; science, trade, and progress must return to those sources that they abandoned with the passage of time: and for this, for the most part, we will be in debt to the Suez canal. (129)

Sapeto held that no other country could profit more from the advantages that the canal offered than Italy, which had the potential to become 'the most convenient transit point and richest centre of world trade' (130). It did not yet have a steam-powered commercial fleet like those of the British and French, but Sapeto reassured his readers that it would be guaranteed many commercial benefits by the canal. He presented the African and Arab coasts of the Red Sea with great enthusiasm, as gateways to the inexhaustible riches of Arabia, Abyssinia, and Central Africa. The trade in timber, food products, oil, wine, marble, glassware, and ceramics would lead Italian traders into a sort of earthly Eden. However, Italy's geographical location was no automatic guarantee of its economic success. If it was to move towards a stable future for its economy, government, and international relations, its commercial, industrial, and political elites, which Sapeto alleged had led the country into a longstanding, conservative, and debilitating lethargy, had to play their part:

> Until Italy holds a suitable position in industry and commercial activity, when it engages in trade through the Isthmus it can only reach a level that

> will be secondary, dependent, vulnerable, scarcely lucrative, and possibly disastrous for the country's general wealth, for the economy and finances of the state and individual families, for its reputation in Europe, for its politics, and perhaps for its existence as a united nation. [...] While efforts are not made to properly complete the railway network; while Italians do not use prizes and every manner of encouragement to press for the formation of huge companies with unlimited capital; while the nation does not throw itself as one into work—the operation of all the different industries, improvements in textiles, wine, oils, and all the goods to be sent abroad—we will be unable to draw much advantage from the cutting of the Isthmus. (158–9)

Sapeto therefore turned his attention to the reforms and measures needed in regard to logistics, transport, and the quality of goods for Italy to enjoy the full advantages offered by the Suez Canal, enabling it to pursue a policy of rapid commercial expansion that would improve not only its economic strength but also its international prestige. An essential element was modernization of the national railway network connecting the main political and industrial centres, such as Rome, Florence, Milan, and Turin, to the major ports of Genoa, Livorno, Naples, and Brindisi, which Sapeto put forward as Europe's terminus for connections with the East. If Italy wished to become the centre of trade between Europe and the East, it therefore also needed to develop routes through the Alps (especially in relation to the Gotthard and Lucomagno passes) that could connect the country's railway network and markets to the commercial and industrial worlds of Switzerland, the Habsburg Empire, and Southern Germany. Sapeto believed that the Northern European markets would prefer these new lines of communication with the South (towards the ports of Genoa and Brindisi) to the traditional commercial networks and major ports of Hamburg, Antwerp, and Rotterdam. Development of the railway network was to be complemented by modernization of the commercial ports of Genoa, Naples, Brindisi, and 'Habsburg' Venice (which Sapeto hoped would soon become Italian, as in fact happened in 1866). New piers, dry docks, and shipbuilding sites had to be constructed; most especially, merchant shipping legislation needed to be liberalized, involving tax relief, the abolition of duties, and more incentives for foreign shipping companies. Sapeto argued that Italy should not be just a profitable transit centre for European trade; instead, the newly unified kingdom ought to be a protagonist on the new trade routes made possible by the Suez Canal, fostering companies with large steam-powered fleets and plenty of investment capital, and with a commercial sector capable of competing with that of any other European power.

> Italy needs to have a large company, similar to the largest in Europe, which has available huge amounts of capital and the biggest ships, with all those conveniences that the other companies offer their passengers, and which can transport goods and travellers with speed, security, and the least cost possible. Such things are impossible for any company without large grants of several millions and concessions from the Italian government: these millions and concessions would not be wasted, but would in fact benefit both the nation and the selfsame government, [...] and would make Italy the rival of the most commercial nations, and prominent in world trade. (162)

In eastern ports, this sort of shipping company would need 'officials, with Italian consuls, who would give it priority and consideration, and by their work could help it develop, finding it credit, goods, and passengers' (162). This assistance would increase the Kingdom of Italy's commercial and international influence both in the Red Sea and further east. In Sapeto's view, only one piece was missing for Italy to match Great Britain, the model of liberalism and informal imperialism par excellence: the creation of an Italian commercial centre in Africa, and in particular in the Horn of Africa, where he had lived, studied, and conducted explorations over many years. Britain had had control since 1839 of the key Aden Settlement, by the southern entrance to the Red Sea, and after occupying the small but strategically important island of Perim, in 1856, was keeping a watchful eye on developments in Abyssinia; France, in anticipation of the opening of the Suez Canal, had obtained control of Obock, in Djibouti, through the Messageries Impériales shipping company; and the Habsburg Empire was negotiating the purchase of a bay on the Eritrean coast. Sapeto demanded that the Italian government become fully involved in the competition for African territory, as there was no time to waste if the nation was not to be relegated to a secondary role:

> Our government cannot stand back from the acquisition of a port near the strait as a rest point for the cargo boats, a station for some steamers, and a depot for coal and merchandise, unless it is to rely on the British, French, or Ottomans, who will overburden Italian vessels with anchorage fees and customs charges. A port, as anchorage for Italians, would also be an aid to finding replacement cargo on both the outward passage and return journey from the Indies.

> [...]
>
> Let the government, Chambers of Commerce, trade associations, municipalities, and capitalists consider this carefully: without a landing place belonging to the nation as a stopping point for our goods—both for import and export—and coal, and as a station for Italian vessels to have safe harbour and repairs, with some tugboats, to assist our increased trade with the Far East, in no way could we compete in trade: we will not last long in the struggle with our powerful rivals, and we will not have the advantages that make companies prosper and render nations thriving and respected. (161–6)

In his final chapters, Sapeto emphasized the urgency of the situation and the consequent importance of dynamism and action:

> This concerns the future of Italy: whether or not it will be the queen of Mediterranean trade, whether or not great in wealth and politics. Any delay or failure to take these measures, which give the country this hope, is therefore a crime that wounds the nation, and would be the premature death of our future, from which will follow our political oblivion, difficulties for our finances, and limited horizons for our families. (189)
>
> [...]
>
> We have to lay the ground, organize the routes, and make ready the means of transport, so that the Alpine crossings are open, the ports fit, Italy rich with industry and its finished products, its capital funds reckoned, and, above all, its great companies strengthened, invigorated by the miraculous spirit of association [...] We will have achieved the most beautiful, useful and honoured of revolutions: we will have forged the true destiny of Italy, which, surpassing its former greatness and strength, is its guarantee of enhanced strength and greatness in the future. (209–10)

Sapeto's volume thus employed a highly charged rhetoric as it drew towards its close: it made a final appeal to Italy's government and, in particular, to its industrial and commercial classes, which stood accused of a conservative immobility damaging not only to their own interests, but to those of the entire nation.

The book exaggerated Italy's real importance in contemporary trade and international relations and expressed unwarranted optimism about its future position on the colonial map. In the period that followed, a discouraging divergence between the high expectations of the pro-Suez lobby and the actual results and consequences of the canal's opening in 1869 would become apparent.

As discussed in other chapters in this collection, the euphoria based on Italian commercial ventures of the past and the peninsula's geographical location at the centre of the Mediterranean came up against the chronic underdevelopment of Italy's infrastructure and merchant navy. Successive governments of the newly unified kingdom engaged in a significant programme of infrastructural modernization during the first twenty years of state-building, channelling substantial investment into the railways, roads, Alpine passes, and tunnels, but Italy could not compete with industrial and commercial powers such as Great Britain, France, and Germany. The Italian merchant fleet was among the least prepared for grasping the opportunities presented by the Suez Canal: it largely consisted of sailing ships, and the transition to steam power (completed in the early years of the twentieth century) was a slow process due to the lack of expertise, modern docks, and shipyards, and the resources necessary for building new ships, including iron, steel, and coal. The Italian merchant fleet therefore made little impact on subsequent trade through the Suez Canal, accounting for just 1 per cent of the total tonnage over the next forty years. Competition with shipping from the great powers such as Great Britain and France was an uphill struggle; due to its slow development, Italy's merchant navy was to play a secondary role for a long time, and not only in the Suez Canal.

In the period immediately after the publication of Sapeto's text, however, there seemed to be some movement within the ministries in Rome and among the country's entrepreneurs, and not just because of Sapeto's ongoing public campaign, pursued through conferences and open debates. In February 1865, in response to an invitation from de Lesseps, Genoa's Chamber of Commerce selected one of its prominent members, the merchant Giacomo Millo, as its representative to observe the progress of work in the Isthmus of Suez. In his final report, Millo restated Sapeto's interventionist position.

In 1868, the Navy Minister, Admiral Augusto Riboty, made an appeal to Italian shipowners encouraging them to form commercial steam companies, so that they would be ready to enter the new markets created by the imminent opening of the canal. Among those who responded was Raffaele Rubattino, whose ships had previously only operated in the Mediterranean, but who had recently opened an office in Alexandria in the hope of using government assistance to establish new sea routes between Genoa and Bombay. These were also the years in which lobbying activity by the Italian Geographical Society, founded in Florence in 1867 and modelled on its British and French counterparts, became increasingly insistent and made an urgent appeal for greater Italian involvement in Africa, on both the scientific and commercial fronts.[7] In the momentous year of 1869, the Menabrea government considered turning to Giuseppe Sapeto—teacher of Arabic in Genoa, scholar of ancient cultures, campaigner, former missionary who had spent twenty years in the Horn of Africa—in order 'to buy on the coasts of Asia or Africa those lands, beaches, bays, ports or inlets' suitable for establishing a trading base.[8] Fearful of negative reactions from the other European powers, however, the government instead asked the Rubattino company to establish a station to serve as a depot for its coal reserves. With this aim, Sapeto left Brindisi in October 1869 bound once again for Africa, where on 15 November 1869, just two days before the opening of the Suez Canal, he signed an agreement with a Sultan of Eritrea for the purchase of the Bay of Assab by the Rubattino Shipping Company, thus initiating, albeit indirectly, Italian colonialism. Having completed this acquisition, Sapeto resumed his teaching career in Genoa and continued his lobbying activity by means of essays, pamphlets, and conferences.

Sapeto went back to Assab briefly in the 1870s, as a Rubattino representative, published a further volume advocating the importance of an Italian colonial presence in the Red Sea area (*Assab e i suoi critici*) in 1879, and finally returned to Italy in 1880. Two years later the Italian government formally acquired the bay, which thus became Italy's first African colony. Sapeto died in Genoa in 1895, a year before the humiliating Italian defeat at Adwa. For a long time, his contribution was largely forgotten; he then found renewed fame and respect under the Fascist regime, which revived and honoured the memory of those figures who had laid the foundations of early Italian colonialism and, ultimately, the Italian empire declared in 1936, contrasting them with the weakness and ineffectiveness of nineteenth-century liberal governments. Only long after the Second World War, when imperialistic rhetoric had faded, was the figure of Giuseppe Sapeto subjected to further study and better located in its historical context.[9]

[7] Natili (2008, 51–123).
[8] Surdich (2005, 129).
[9] Castelnovi (1996).

References

Castelnovi, Michele. 1996. In margine ad un anniversario "dimenticato": il centenario della morte di Giuseppe Sapeto (25 agosto 1895). *Miscellanea di Storia delle esplorazioni* 21: 293–321.

Correnti, Cesare. 1869. Sull'istmo di Suez e sul commercio orientale. *Bollettino della Società Geografica Italiana* 2 (3): 489–498.

James, Lawrence. 1994. *The Rise and Fall of the British Empire*. London: Little, Brown.

Morris, Jan. 1968. *Pax Britannica: The Climax of an Empire*. London: Faber and Faber.

Natili, Daniele. 2008. *Un programma coloniale. La Società Geografica Italiana e le origini dell'espansione in Etiopia (1867–1884)*. Rome: Gangemi.

Sapeto, Giuseppe. 1857. *Viaggio e missione cattolica fra i Mensâ, i Bogos e gli Habab, con un cenno geografico e storico dell'Abissinia*. Rome: Congregazione di Propaganda Fide.

———. 1865. *L'Italia e il canale di Suez. Operetta popolare*. Genoa: Pellas.

———. 1879. *Assab e i suoi critici*. Genoa: Pellas.

Surdich, Francesco. 2005. *L'attività missionaria, politico-diplomatica e scientifica di Giuseppe Sapeto. Dalla evangelizzazione dell'Abissinia all'acquisto della Baia di Assab*. Millesimo: Comunità montana Alta Val Bormida.

CHAPTER 9

The Trieste-Suez Connection: How Businessmen and Explorers Reshaped the Mediterranean in the Eighteenth and Nineteenth Centuries

Giovanni Modaffari

Introduction

Throughout the nineteenth century, the port city of Trieste, later destined to become part of Italy, played a pivotal role in the design and implementation of the project of cutting through the Isthmus of Suez. At the beginning of the 1860s, Trieste was already the major outlet to the sea of the

The author wishes to thank Dr. Frederica Daniele (University of Rome 'Tor Vergata') and Judy Moss for reviewing the text. Heartfelt thanks are extended to the Director of Trieste State Archive, Dr. Antonietta Colombatti, for making part of this research possible. All quotations from works in languages other than English are translated by the author.

G. Modaffari (✉)
Università di Milano-Bicocca, Milan, Italy
e-mail: giovanni.modaffari@unimib.it

B. Curli (ed.), *Italy and the Suez Canal, from the Mid-nineteenth Century to the Cold War*,
https://doi.org/10.1007/978-3-030-88255-6_9

Austrian Empire and thus a foreign port for the new-born Kingdom of Italy. Subsequently, the latter would inherit the extraordinary network of cultural, economic, and trading exchanges which had Trieste as its centre. The town on the Adriatic Sea, with its port infrastructure and industry, was considered essential in connecting the East with the Mediterranean Sea. A few dates: the year 2019, which marks the 150th anniversary of the Suez Canal, is also the 300th anniversary of the creation of the modern port of Trieste, which has recently been chosen as one of the European docking points in the Chinese global development strategy named the *Belt and Road Initiative*.

Framing the planning and realization phases of the canal as a crucial geographical event enables us to view it as a key step in the configuration of a new Mediterranean Sea. Since the fifteenth century, the Mediterranean had been the busiest laboratory of experimentation for Modernity, representing a major revolution in thinking about the world. For Saint-Simonians, the opening of the Suez Canal represented the tracing on the Earth's surface of 'a sign of peace and, in truth, a *trait d'union* between the two parts of the Old World, between the East and the West; not a pathway, nor a canal, but rather a new sea'.[1] As geographers, we view the series of *Mediterraneans* which exist around the world—such as the Mexican Gulf and the Caribbean Sea, the South China Sea, the Yellow Sea, the Philippine and Indonesian Seas—as all being areas of mediation, of 'immense *space-movement*, a whole circulation system in which we cannot distinguish land-routes from waterways'.[2]

In the second half of the nineteenth century, Modernity found ever-increasing expression as a cultural paradigm. The project of cutting through the Isthmus of Suez was rooted in the technicist socialism of Saint-Simonians, with their absolute faith in the ability of science and industry to make the actions of humans in the world and society more effective. Indeed, thanks to the extraordinary progress in infrastructure and transport as well as in communication technologies, it would soon be possible to literally open up a sea which was fast becoming too small a container for rapidly growing commercial powers. In the preceding decades, Egypt had been ruled by Mehmet Ali (died 1849), a reformist who carried out a far-reaching programme of European-inspired

[1] Enfantin, 28 June 1845, in Georgi and Dufour-Féronce (1913, 12).

[2] Farinelli (2003, 98–99). See also: Braudel (1985, 76–80).

modernization. Mehmet Ali succeeded in making Egypt one of the main interlocutors in the Southern Mediterranean economic area.

Both Modernity and the Suez Canal as a Mediterranean endeavour are hence the models for a project, for an era, and also for an idea of progress, developing alongside the emergence of an urban feature rarely seen before to this extent in the Mediterranean: the high degree of cosmopolitanism in the cluster of cities located on its two great shores. Philip Mansel has highlighted how, at the time, the cities of the Levant were 'a third way between states and individuals, laboratories of new worlds, economic, cultural and political'.[3] Among cities on the European shore, Trieste was eager to present itself as the main gateway to the Orient, through the enhancement of its port facilities and the brokerage of a dynamic entrepreneurial class.

Trieste, Austria, and Suez Towards 1869

The Austrian involvement in the Suez Canal soon revealed the friction which comes to the surface when a naturally continental empire begins to contemplate a more active and effective Mediterranean policy. In 1846, the *Société d'Études du Canal de Suez* set up by Barthélemy-Prosper Enfantin involved an Austro-German group headed by the engineer Alois Negrelli on behalf of the Austrian government (Albert Dufour-Féronce and Ludwig Sellier); a French group (François Barthélemy Arlès-Dufour, Jules, Léon and Paulin Talabot); and an English one (Robert Stephenson and Edward Starbuck).[4] The Austrians' interest had increased after the resolution of the crisis between the Ottoman Empire and Egypt (1839–1841) and the stipulation of the London Straits Convention (1841)[5] as a result of which Austria, Prussia, and above all Great Britain supplanted Russia in ensuring the Sultan's power and controlling access to the Bosporus and the Dardanelles. The *Société*'s initial capital was 150,000 francs; together, the *Austrian Lloyd* and the Municipality and Stock Exchange of Trieste invested 5000 francs.[6] The declared purpose of the *Société* was to construct 'a free way of moving between the Red Sea and

[3] Mansel (2010, 6).
[4] *Statut de la Société d'Études du Canal de Suez*, 30 November 1846, in Georgi and Dufour-Féronce (1913, 32).
[5] Cervani (1962, 20).
[6] Babudieri (1980, 59–60).

the Mediterranean, […] a sort of Bosphorus in the Suez desert'.[7] The intention of the Austrian Empire, on the other hand, was to focus its ambitions and strategies on Trieste, as Metternich declared in a report dated 14 March 1843:

> Trade with the East Indies is to us a political and commercial concern of primary importance. Trieste and Fiume may become as important for Southern Germany and Switzerland as Liverpool is to England and Hamburg and Bremen to Northern Germany.[8]

The report further emphasized that the canal should be the route of choice, as opposed to the already tested routes which required goods to pass through London and Amsterdam before reaching Southern and South-Western Europe. In 1843, the Stock Exchange Delegation of the Trieste Chamber of Commerce sent a group of experts headed by Pietro Erichsen and Alessandro Giovanni Conighi to the East, with the task of 'exploring the convenience in transport for the Isthmus of Suez compared to the *all sea route* around the Cape of Good Hope; commerce in the Red Sea and East Indies and the potential means to improve commercial relations between Austria and these countries'.[9]

These are the aims stated in the letter sent by the Stock Exchange Delegation to Pietro Jussuff as a request for a presentation to his brother Boghos Jussuff—who was then secretary to Mehmet Ali Pasha—in the hope of paving the way for the group of experts itself. The Jussuff family was of Armenian descent and Boghos had been the Pasha's translator and Minister of Commerce and Foreign Affairs; he had also worked to host and encourage the expeditions to Egypt led by prominent archaeologists such as Johann Ludwig Burckhardt and Giovanni Battista Belzoni, and to maintain contacts with Bernardino Drovetti. His brother Pietro founded a wholesale company in 1814, and then transformed it into a financial and commercial intermediation group acting in the port of Trieste. His name is best remembered for the gift of the Sekhmet Goddess to the Kunsthistorisches Museum in Vienna, while Boghos donated a sarcophagus to the Mekhitarists of San Lazzaro in Venice. Around the middle of

[7] *Statut de la Société d'Études du Canal de Suez*, 30 November 1846, in Georgi and Dufour-Féronce (1913, 28).

[8] Babudieri (1980, 61).

[9] Deputazione di Borsa to Pietro Jussuff, 19 September 1843, AST CC, VII, box 19, folder 489.

the century, Pietro's heirs Giuseppe and Abram Abro were able to grant loans to the Khedive and to the Austrian consulate in Alexandria, and to provide transport from Trieste to Egypt of the timber to be used by the *Company of Carinthia* to build the Nile Delta Barrage.[10]

Another document from the Erichsen-Conighi mission to the East, dated 14 November 1842, shows their keen global awareness when contemplating the implications of the upcoming project of building the canal:

> [I]n this gigantic conception, in this superhuman enterprise, there lies a profound transformation in the commercial relationships that have existed up to now, and a great future is being prepared for the trading importance of the Austrian ports, which are all links in this vast chain.[11]

In his report, Erichsen stressed the need to create an Austrian trading facility in Cairo in order to promote the Empire's industries in Egypt and open a direct line of commerce between Trieste and the Orient.[12] It is worth noting that in 1842 the Treaty of Nanjing between Great Britain and the Qing dynasty had created expectations, for the Austrian Empire as well, as regards the opening of new routes to the Red Sea.[13]

When the *Compagnie universelle du canal maritime de Suez* was founded at the end of the 1850s by Ferdinand de Lesseps, the Austrian government officials were still committed to not irritating Great Britain, as they needed British cooperation in their Balkan policy. Nevertheless, of the initial capital of the Company of 300 million francs, a share of 20 million was to be subscribed in Austria, although as time went by, the Imperial Court would adopt an increasingly cautious attitude.[14] The minister of public works of the Kingdom of Sardinia, Pietro Paleocapa, wrote to Lesseps that they could not 'count on the backing of Austria, which is always ready to sacrifice the interests of its trade for those of its politics'.[15]

[10] Crevatin and Vidulli Torlo (2013, 14).

[11] *Scopo della Missione*, 14 November 1842, AST DB, box 17, 2072/1842.

[12] Erichsen (1846, 2).

[13] Babudieri (1982, 100–3).

[14] Stefani and Astori (1938, 333–34).

[15] Paleocapa to Lesseps, 21 November 1858, in Monti 1937, 274. On the use of documents published by Monti the same warning applies as in Chap. 2 of this book, note 22.

In order to obtain the concession from the Sultan, a highly influential Austrian figure was asked to intercede: Baron Karl Ludwig von Bruck,[16] a liberal statesman and disciple of List's economics, a businessman and banker whose name was indissolubly linked to the Austrian Lloyd and who at the time was internuncio in Constantinople.[17] His role was closely coordinated with that of Baron Pasquale Revoltella, a dynamic Venetian-born entrepreneur with a leading place in the Trieste establishment. Other figures involved directly (and indefatigably) in what became known as the *affaire* of Suez were Baron Kübec, president of the Hofkammer in Vienna, the Earl of Fiquelmont, president of the Hofkriegsrat, and Count Stadion, governor of Trieste. According to Giulio Cervani, these men came to interpret 'a precise relation between some great problems inside the Austrian monarchy and the general lines of European policy, increasingly conditioned by the economic boost coming from the bourgeoisie and high finance'.[18] At the same time, they also saw the *affaire* as a handy way to invest the large amount of gold available in the Austrian reserves. Moreover, with regard to the German area, Metternich was worried that the experiment of the Austro-German *Zollverein* might become a threat if the new railways envisaged in the plan were to favour the Rhineland over the Danube-Adriatic region.[19] Certainly, the customs union represented new opportunities, but these were less attractive than those being created in the Mediterranean space.

Hence the great project of Suez continued to be seen by the Austrian government as an excellent opportunity 'to get substantial gains in transport, because of the time convenience, the access to new markets and the increase in goods traded': the Suez Canal would enable Austria to have more say in the 'Eastern Question' and it would enhance Austria's status as a commercial empire.[20] These reasons kept alive the grand ambition which had first led Charles VI to declare Trieste and Fiume (today Rijeka in Croatia) free ports between 1717 (when *free* navigation along the Adriatic was proclaimed) and 1719, and guided Maria Theresa in her

[16] Arlès-Dufour to Negrelli, 20 December 1854, in Georgi and Dufour-Féronce (1913, 154–5).

[17] For a closer look at von Bruck's role, see Lo Giudice (1979, 90–98).

[18] Cervani (1962, 20).

[19] Cervani (1962, 22); Stefani and Astori (1938, 136–7).

[20] Cervani (1962, 25). See also Metternich to Kübeck, 28 April 1843, in Metternich (1883, 704–7).

Haupt-resolution, as well as Joseph II with his innovative measures throughout the eighteenth century.

After an encouraging start, however, Austria underwent a phase of insecurity. Following the collapse of the Holy Alliance system, the reactionary government led by Schwarzenberg and the short-sighted one during the Bach years from 1852 onwards, and the rise of nationalist movements, the Empire developed a closer relationship with Great Britain; from then on, the Khedive of Egypt showed ever-diminishing trust in the Austrians and would say that he had no faith in Austria except for its alliance with Great Britain; according to Cervani, Austria showed 'a genuine incapacity to adopt a modern layout'.[21]

This phase, however, had no effect on the dynamism of Trieste. The business was conducted in a peculiarly private way by Baron Pasquale Revoltella, who would become the most influential actor in the story of Trieste and Suez. A banker and a businessman linked to the *Assicurazioni Generali*, he entertained a far-sighted vision of a Mediterranean project which had to include a role for Trieste; indeed, he was well aware of the importance of the city's new entrepreneurial class, which had no time to waste dealing with imperial matters and—in Revoltella's view—needed instead to be able to take part in the wider international trade network of the time. In 1864, he wrote:

> [W]ere the hopes placed in the Suez canal not to be realised, but assuming that Austria were nevertheless to set foot in world trade, tying up connections with the Indies and with East Asia, it could very well cultivate these relations via the route of the Cape of Good Hope, and also via that of Suez by means of the Egyptian railway, thus progressing at the same pace as the other merchant nations; on the other hand, if the canal were to be built, and this happened soon, as everything leads us to hope, and Austria were to be found unprepared for this eventuality, it will be unable to derive any advantage from the new trading route, and will find itself suffering serious damage from other nations.[22]

In November 1858, Revoltella would underwrite 50,000 shares on behalf of Austria (the largest private investment after that of Muhammad Said) and was then appointed vice-president of the *Compagnie*, which marked the beginning of an Austrian involvement in the project

[21] Cervani (1962, 30–2).
[22] Revoltella (1864, 16).

exclusively on a financial basis. However, the Austrian government gave him no assistance in selling the shares, and according to Giuseppe Lo Giudice, this might be one of the elements in a broader picture connecting Revoltella to Bruck's suicide and the premature death of Negrelli. Indeed, these two departures from the scene were soon followed by the mysterious disappearance of all the documents concerning the *affaire* of Suez from their respective archives.[23] At this stage, however, Revoltella became the key interlocutor for Lesseps in Trieste: '*Monsieur le chevalier* Revoltella has signed, in his capacity as private banker, a commitment for which his country will surely thank him in days to come'.[24]

Significantly, on 23 February 1859, Revoltella, Lesseps, and the Archduke Maximillian met at the inauguration ceremony of Palazzo Revoltella, today the Revoltella Museum, which would later host the sculpted group by Pietro Magni (1863)[25]: a young woman—Europe—is holding hands with an elderly figure—representing the *old* Mediterranean—and a young Egyptian *warrior*—representing the Red Sea—under the auspices of the divinity of commerce. Between 1861 and 1862, Revoltella led a mission to Egypt on behalf of Trieste, the City Council and the *Austrian Lloyd*, in order to see how work on the canal was progressing; a fascinating journal of this mission survives to this day, as well as the paintings on the subject of Suez he commissioned from Albert Rieger, and hosted in the Museum Revoltella as well. But when he was received by the Viceroy on December 17, 1861, he had already been forced to give up his shares following the failure to place them.[26]

Trieste and Egypt

In the middle of the century Trieste was experiencing a series of events which closely mirrored those occurring in France in the same years. The French determination to complete the great Suez enterprise was clearly a consequence of the substantial flows of capital occurring during a phase of increasing wealth and industrial development, but it also stemmed from the committed cosmopolitan approach of the project itself.[27]

[23] Lo Giudice (2010, 279–90).
[24] Lesseps to Bruck, 2 December 1858, in Monti (1937, 286).
[25] Crevatin and Vidulli Torlo (2013, 18).
[26] Cervani (1962, 85). On Revoltella's mission to Egypt, see also Crusvar et al. (2015).
[27] Cervani (1962, 19–20).

As we have seen with the brothers Pietro and Boghos Jussuff, there was a dual connection linking Trieste and Egypt, namely archaeology and trade. In fact, anyone delving into the history of Trieste and Suez—and Egypt—soon discovers just how long that history is. Above all else, the port of Trieste participating in the business venture of connecting the Mediterranean to the Red Sea was a Mediterranean city, in close communication with the *skalas* (ports) in the Levant and whose reach stretched all the way to the Far East. But the long history of the relationship between Trieste and Egypt can be traced all the way back to when the city was a satellite port of Roman Aquileia, representing 'a maritime emporium with vibrant Eastern interactions'.[28] Many archaeological remains testify to these ancient contacts.

Trieste's vocation as a cosmopolitan city with a decidedly Mediterranean focus dates back to Charles VI's project to make the port city the Empire's main trade outlet to the world, and thus a serious competitor to Venice (whose glory was beginning to fade) in its pivotal role for traffic between continental Europe and the Levant. Indeed, the idea of cutting a canal through the Isthmus, which 'dal mar rosso metesse a drectura in questo mare de qua',[29] had already occurred to the Venetians between the fifteenth and the sixteenth century, as they were eager to undermine the role of the Portuguese in the lucrative trade with the East Indies.[30] But since the early nineteenth century Trieste had taken upon itself the role of replacing Venice, as it enjoyed a crucial advantage over its former rival, as Karl Marx pointed out in 1857:

> Venice was a town of reminiscences; Trieste shared the same privilege of the United States of having no past at all. Formed by a motley crew of Italian, German, English, French, Greek, Armenian and Jewish merchant-adventurers, it was not fettered by traditions like the City of the Lagunes.[31]

An important step in the strengthening of commercial relationships between Trieste and Egypt was the founding in 1770 of a *Society for the Promotion of Austro-Hungarian Commercial and Industrial Interests in Egypt*, which managed to obtain the privilege of gaining entry for European

[28] Crevatin and Vidulli Torlo (2013, 9).

[29] 'From the Red Sea flows straight into this sea here' (i.e. the Mediterranean), ASV CD, *Misti*, filza 16, n. 60.

[30] Babudieri (1980, 29).

[31] Marx (1857).

ships into the Red Sea.[32] A few years later, an *Egyptian and Syrian Company* was established, in which several merchants, expatriates, and Levantine capitalists were present, including the Grand Douanière of Egypt, Antonio Cassis Faraone.[33] The company was founded between 1781 and 1782 by Domenico Belletti, the Armenian Teodoro Zaccar, the Milanese Baldassarre Rossetti, and an anonymous Venetian associate, with the aim of connecting Trieste, Izmir and Cairo, and via Suez, establishing a strong commercial link for the trade of imperial manufacturing goods with the Indies.[34] As Daniele Andreozzi has recounted, this initiative soon met with the opposition of the British authorities, as well as the conflicting interests of the Ottomans and Venetians, thus triggering an international intrigue which ended with the flight of Giuseppe Zaccar (Teodoro's brother and Douanière in Damietta) in 1782, and of Cassis Faraone who took refuge in Trieste in 1784, after all his goods had been seized.[35] Even Giacomo Casanova was present in Trieste at the time, having been given the task of monitoring the progress of Rossetti and his associates.[36] The importance of relations between Trieste and Egypt was given official recognition in 1781 with the opening of the Austrian consulate in Alexandria, and with the trade agreement according to which goods exported to Egypt from Austrian ports would only be subject to a 3% duty.[37]

Lastly, the archaeological link between Trieste and Egypt was strengthened by further prominent individuals such as Giuseppe Nizzoli, Joseph Passalacqua, and Consul Anton Laurin; the latter helped Archduke Maximilian to create his Egyptian collection, including the sphynx which can still be seen in the Miramare Castle in Trieste, whereas the rest is now part of the Kunsthistorisches Museum collections in Vienna.[38]

The *Austrian Lloyd*

The *Austrian Lloyd* steam-navigation department was established in 1836 as a section of the *Lloyd* shipping company and a part of the Hapsburg Monarchy's grand designs for Mediterranean expansion. By the middle of

[32] Babudieri (1980, 32).
[33] Panjek (2011, 44); On Cassis Faraone, see Degli Ivanissevich (1996b).
[34] Andreozzi (2017, 77–8).
[35] Andreozzi (2017, 78–9).
[36] Preto (1994, 511–12, 528).
[37] Panjek (2011, 43–4).
[38] Crevatin and Vidulli Torlo (2013, 15–8).

the century, Great Britain and France had been pushing for a way to reach their territories in the Indian Ocean which would avoid the circumnavigation of Africa. In 1835, the overland route was opened: a new route with a transfer from Alexandria to Suez and consequently including the crossing of the desert. This solution reduced the journey time from London to Bombay from 45 to 35 days, but significant inconveniences remained.[39]

The route from London to Alexandria had Marseille as its transit point, and the *Austrian Lloyd* attempted to propose Trieste as the alternative. A trial journey was made to test the two routes: in 1842, Captain Bloomfield travelled from London to Trieste in less than 6 days and from London to Alexandria in 14 overall (the same journey via Marseille required 19 days). This was one of several episodes which were part of the challenge for the acquisition of the Indian Mail, first limited to letters and parcels, later extended to valuables such as gold bars and precious stones, silk, water and coal supplies for the city of Suez and for steamboats in the Red Sea and the Indian Ocean.[40]

In 1845, the *Austrian Lloyd* recruited the well-known pioneer Thomas Fletcher Waghorn to make the journey from Suez to London, providing him with the *Imperatore*, one of the fastest steamboats in its fleet. The overland route he followed included the segment from Trieste to Ostend via Kempten, Meiningen, Ulm, Stuttgart, and Karlsruhe; from there, he took another steamboat down the Rhine to Cologne and finally a train across Belgium to Ostend. This successful experiment in terms of time (the complete journey from Bombay to London in 29 days) and its possible benefits were not enough to persuade Great Britain and France to renegotiate the previous agreements, not least because of the objective uncertainty caused by the need to cross several borders and the absence of a railway between Trieste and its German hinterland. Indeed, the Austrian government seemed to have no intention of beginning a major railway project.[41]

The *Lloyd* continued its actions by setting up a direct line to Alexandria which opened in 1848, confirming its role as the main actor in the development of the Adriatic city.[42]

[39] Babudieri (1980, 45–8).

[40] Babudieri (1980, 51–3).

[41] Babudieri (1980, 53–7).

[42] On the Lloyd's sea-route lines to and from the Ottoman lands, see Degli Ivanissevich (1996a).

Sir Richard F. Burton Between Trieste and Suez

Among the actors who played a key role in this story between Trieste and Egypt, special mention must be made of British consul Sir Richard Francis Burton, envoy to Trieste in the aftermath of the opening of the Suez Canal, from 1872 to his death in 1890.[43]

Accounts of his experiences as an explorer, mostly provided by his wife Lady Isabel Arundel or recounted in his own books, give a fascinating testimony of the shift from travelling to India by circumnavigating Africa to going via Suez. In a note to his account of the 1853 expedition to Mecca and Medina, he shared his thoughts on the idea of the canal and the view expressed by Linant de Bellefonds which he claims to have overheard.[44] On 29 October 1875, he published *The Port of Trieste, Ancient and Modern* for the *Journal of the Society of Arts*, a skilful and historically accurate analysis of the port of Trieste and of plans for its development made by Paulin Talabot—who had been a member of the French group in the *Société d'Études*—in relation to the other Mediterranean ports. Furthermore, in 1876, Burton wrote a wide-ranging review of the documents written by Ferdinand de Lesseps on Suez and the development of the canal, for *The Academy*.[45] In the same year, the Burtons embarked on the *Austrian Lloyd*'s ship *Calypso*, leaving Trieste, on the first stage of a trip to India. On their arrival in Port Saïd, among other prominent figures, they met the grand old Baron de Lesseps. Here is Isabel's account of the new canal and their impression of how Suez had been transformed:

> We had the north-east monsoon dead against us the whole way after going out of the canal, which made the ship pretty lively [...] From being a town of importance, Suez was ruined by the opening of the canal. She has become a big village of three thousand natives, and about seventy-five Europeans, employed in telegraph, post-office, steamers, and railways. She sits solitary under the sky in the sand on the borders of the sea, far from all civilization or progress. She has had a past, and Richard says she *will have a future*.[46]

A longstanding relationship with Ferdinand de Lesseps is evident from other experiences recounted by Lady Isabel and the fact that, after Burton

[43] Modaffari and Zilli (2019).
[44] Burton R. F. (1893b, 113).
[45] Burton R. F. (1876).
[46] Burton I. (1893a, 53–54).

had returned from his archaeological mission in Midian, Lesseps offered to help him have the rock samples he had collected assayed in France.[47] On 28 October 1885, when Lord Salisbury asked him what policy should be adopted towards Egypt, Burton's response was crystal-clear about the importance of securing the Suez Canal:

> First and far away, annex Egypt and all its territory entirely; but if the Government does not decide on this bold stroke, at least have no half-measures [...]. Give them the freedom of Foreign Trade; foster National Industries; build yourselves harbours and docks and fortresses in the Mediterranean, the Suez Canal, the Red Sea, and especially lighthouses in the Red Sea. You should make roads and railways, encourage irrigation, form wells every six hours' march throughout the country, employ native labour.[48]

In fact, if we are to judge from a series of comments scattered through his biographies and in other works, combined with Burton's interest in Egypt and the political situation there after the opening of the Suez Canal, and the new political and commercial relationships which the Austrian Empire was striving to create in the Mediterranean area, it is quite possible that during his time in Trieste, Burton's role was not limited to that of a mere observer.

Conclusion

Once the canal was opened, the share of imports coming to Trieste from the countries east of Suez, which was less than 1% before 1869, rose to 19.5% in 1880.[49] Ever since then, in moments of uncertainty and weakness on the part of modern states, which often coincided with crises in the modern cities, the Suez Canal has continued to perform its original function, namely that of embodying the modern instinct to connect the globe. The cutting of the Isthmus of Suez may be considered one of the greatest engineering achievements in history, and the canal's identity as an infrastructural link is reaffirmed today, as one of the nodal passages in China's new global development strategy.

[47] Walton (2021, 59).
[48] Burton I. (1893a, 296).
[49] Panjek (2003, 307).

Trieste is thus a fitting historical metaphor for the Suez achievement: a cosmopolitan city where a crisis in imperial authority left room for an initiative fostered by a forward-looking group of enthusiasts, keenly aware of the importance of being connected through trade in order to prosper: this is the vision so beautifully depicted in Magni's allegorical marble group. The successor to Revoltella as president of the Board of *Assicurazioni Generali* was Giuseppe de Morpurgo. It fell to him to travel to Suez to attend the great ceremony for the inauguration on 17 November 1869, and he left a brief but intriguing description of the event in one of the letters to his family, dated the day before the opening:

> We met the Emperor on his ship: he greeted us politely and we went to attend a solemn religious ceremony, unique in its kind, I believe. In the middle of a wide esplanade, a huge pavilion for the sovereigns was erected with two galleries for the distinguished guests and those of convenience. On the right, a Turkish mosque, on the left, a Catholic Chapel. The Empress came along with the Emperor of Austria, followed by the Khedive with the Prince of Prussia and all the other princes who were here in attendance. What a magnificent sight!.[50]

References

Abbreviations

ASV, Archivio di Stato di Venezia
CD, Consiglio di Dieci
AST, Archivio di Stato di Trieste
CC, Camera di Commercio
DB, Deputazione di Borsa poi Camera di Commercio e d'Industria di Trieste

Bibliography

Andreozzi, Daniele. 2017. From the Black Sea to the Americas. The trading companies of Trieste and the global commercial network (18th century). In *Mediterranean Doubts. Trading Companies, Conflicts and Strategies in the*

[50] Morpurgo to his family, 16 November 1869, in Morpurgo (1998, 36).

Global Spaces (XV–XIX Centuries), ed. Daniele Andreozzi, 65–87. Palermo: New Digital Press.

Babudieri, Fulvio. 1980. L'«Overland Route» dell'Egitto dell'Ottocento e suoi riflessi adriatici. *Miscellanea*, 3, Pisa: Pacini, 29–85.

———. 1982. *La missione orientale della Deputazione di Borsa di Trieste nell'Ottocento*. Milano: Giuffrè.

Braudel, Fernand. 1985. *La Mediterranée: l'espace, l'histoire, les hommes et l'heritage*. Paris: Flammarion.

Burton, Richard F. 1875. The Port of Trieste. Ancient and Modern. *Journal of the Society of Arts*, 25, 979–86 and 996–1006.

———. 1876. Review of Ferdinand de Lesseps. Lettres, Journal, et Documents pour server à l'histoire du canal de Suez. *The Academy*, 9, 5–12 February.

Burton, Isabel. 1893a. *The Life of Captain Sir Richard F. Burton KCMG, FRGS*. Vol. 2. London: Chapman and Hall.

Burton, Richard F. 1893b. *Personal Narrative of a Pilgrimage to Al Madinah and Meccah*, Memorial Edition. London: Tylston and Edwards.

Cervani, Giulio. 1962. *Il "Voyage en Egypte" (1861–1862) di Pasquale Revoltella*. Trieste: ALUT.

Crevatin, Franco, and Marzia Vidulli Torlo, eds. 2013. *Collezione egizia del Civico Museo di Storia ed Arte di Trieste*. Trieste: Civici Musei di Storia ed Arte.

Crusvar, Luisa, et al., eds. 2015. *Trieste-Suez: Storia e modernità nel "Voyage en Égypte" di Pasquale Revoltella*. Trieste: Museo Revoltella.

Degli Ivanissevich, Sergio. 1996a. *L'apertura delle linee del Lloyd austriaco con l'Impero Ottomano*. In *Trieste e la Turchia*, ed. Gino Pavan, 58–63. Trieste: Samer & Co. Shipping.

———. 1996b. *Una caratteristica figura del '700 triestino: Antonio Cassis Faraone*. In *Trieste e la Turchia*, ed. Gino Pavan, 65–67. Trieste: Samer & Co. Shipping.

Erichsen, Pietro. 1846. *Missione alle Indie orientali della Borsa di Trieste: rapporto sommario*. Trieste: Weis.

Farinelli, Franco. 2003. *Geografia. Un'introduzione ai modelli del mondo*. Torino: Einaudi.

Georgi, Otto R., and Albert Dufour-Féronce. 1913. *Urkunden Zur Geschichte des Suezkanals*. Leipzig: Dieterich.

Lo Giudice, Giuseppe. 1979. *L'Austria, Trieste ed il canale di Suez*. Catania: Università degli Studi di Catania.

———. 2010. *Karl Ludwig von Bruck. Un ministro liberale alla corte degli Asburgo*. Udine: Del Bianco.

Mansel, Philip. 2010. *Levant: Splendour and Catastrophe in the Mediterranean*. London: John Murray.

Marx, Karl. 1857. The Maritime Commerce of Austria. *New York Daily Tribune*, 9 January.

Metternich de, Richard. 1883. *Mémoires, documents et écrits divers laissés par le Prince de Metternich*. Vol. 6. Paris: Plon.

Modaffari, Giovanni, and Sergio Zilli, eds. 2019. *Sir Richard Burton, Trieste e l'esplorazione*. Trieste: Comunicarte.

Monti, Antonio, ed. 1937. *Gli Italiani e il canale di Suez*. Roma: Vittoriano.

de Morpurgo, Giuseppe. 1998. *Diario dal canale di Suez, 1869*. Trieste: Dedolibri.

Panjek, Giovanni. 2003. Una "commercial officina" fra vie di mare e di terra. In *Storia economica e sociale di Trieste*, ed. Roberto Finzi et al., vol. 2, 235–348. Trieste: LINT.

———. 2011. I rapporti economici fra Trieste e Alessandria d'Egitto nell'Ottocento. In *Le rotte di Alexandria: convegno di studi, Trieste, 1–2 dicembre 2008*, ed. Franco Però and Patrizia Vascotto, 43–63. Trieste: Edizioni Università di Trieste.

Preto, Paolo. 1994. *I servizi segreti di Venezia*. Milano: Il Saggiatore.

Revoltella, Pasquale. 1864. *La compartecipazione dell'Austria al commercio mondiale*. Trieste: Tipografia del Lloyd Austriaco.

Stefani, Giuseppe, and Bruno Astori. 1938. *Il Lloyd Triestino 1836–1936*. Verona: A. Mondadori.

Walton, Michael. 2021. *Sir Richard Francis Burton e la sua cerchia/and His Circle*, 2nd ed., Pordenone: Edizioni L'Omino Rosso.

CHAPTER 10

'Iddio si serve mirabilmente dell'uomo per adempire i suoi altissimi fini': The Suez Canal in the Nineteenth-Century Italian Catholic Press

Carlo Bovolo

The Suez Canal and *La Civiltà Cattolica*

The challenging idea of constructing a canal between the Mediterranean and the Red Sea attracted the interest of the press across Europe, including Italy. In mid-nineteenth-century Italy, a political and cultural struggle between Catholicism and modern ideas, such as liberalism, positivism, materialism, and secularisation, started and then characterised the Italian context until the beginning of the twentieth century. Catholic newspapers and journals flourished in the middle of the century, gaining a key role in defending Catholicism's and the Pope's role, authority, and influence in Italian society, pursuing an apologetic and propagandistic discourse in every field, from politics to morals, culture to science. The clerical press

C. Bovolo (✉)
Università del Piemonte Orientale, Vercelli, Italy

B. Curli (ed.), *Italy and the Suez Canal, from the Mid-nineteenth Century to the Cold War*,
https://doi.org/10.1007/978-3-030-88255-6_10

intended to inform and educate Catholic readers and, at the same time, to propagate ideas and interpretations approved by the Church in Italian public opinion. Therefore, their aim was to present a valid alternative to the lay press and a public platform for Catholic positions and attitudes.

The Catholic press's interest in the Suez Canal project began in the mid-1850s, when Ferdinand de Lesseps (1805–1894) relaunched the challenge. After some brief references to it, the first significant article about the project appeared in June 1855 in *L'Armonia della religione colla civiltà*. This was a Catholic newspaper established in Turin in 1847 by a group of ecclesiastics and clerical nobles and politicians and directed by the priest Giacomo Margotti (1823–1887), one of the most influential champions of the intransigent press. The article, titled 'Incanalamento dell'istmo di Suez', briefly described the planned canal, underlining the advantages in trade and maritime transportation and expressing a generally positive opinion.[1]

In 1856, the topic was reviewed and deeply examined by the most influential, authoritative, and widely distributed Catholic journal *La Civiltà Cattolica*. Founded in 1850, *La Civiltà Cattolica* was the Italian Jesuit fortnightly and presented a faithful reflection of the Church's positions and a laboratory and reference point for intransigent Catholic thought. Its aim was to defend the role of the Pope and the principles of Catholicism, hence establishing a programme of apologetic and propaganda commitments.[2] The Suez Canal project was the subject of a detailed article, published in September 1856. First, *La Civiltà Cattolica* recognised the exceptional importance of the expected piercing of the Egyptian isthmus, a project fundamental not only for international trade and transportation, but also and especially for civilisation: 'It does not seem to be anything but a gigantic hydraulic work, and a high-value question of trade. However, the piercing of the Egyptian isthmus, considering its consequences, has such close relations with politics and universal civilization that overstep its material importance.'[3]

In harmony with the leading and influential role of *La Civiltà Cattolica* in the Catholic press and culture, the article pursued both educational and apologetical aims. It first summarised previous projects for the creation of a canal, from that of the Egyptian Pharaoh Sesostris to that of Napoleon

[1] 'Incanalamento dell'istmo di Suez' 1855, 495.

[2] Traniello (2007) and Forno (2012).

[3] 'Il canale di Suez' 1856, 34. All translations from Italian are by the author.

Bonaparte and the recent attempt by the French Saint-Simonian Barthélemy Prosper Enfatin (1796–1864) and the engineer Luigi Negrelli (1799–1858). After the historical review, the Jesuit journal provided an update on the development of the project relaunched by Ferdinand de Lesseps, who in 1854 had obtained from the Khedive of Egypt and Sudan, Sa'id Pasha, 'the exclusive right to set up a Company composed of capitalists from all nations, aiming to pierce the Suez isthmus and to build a canal between the Mediterranean and the Red Sea'.[4] Then, the article examined in detail the observations of the scientific committee sent to Egypt to prepare for the beginning of the work, which in 1856 appeared certain and imminent, and the related advantages:

> If the peace of Europe will last steadily, without new storms of revolts or wars returning to disturb its happy plans; then, then in a few years we will see a marvelous and universal revolution taking place in the world, which has not been seen since the century of Columbus. And who could foresee all the consequences and enumerate the intricate and distant series of effects that would result not only in material commerce and wealth, but also in political and moral civilization?[5]

According to the Jesuits, the realisation of the Suez Canal project could provide several advantages to the Italian States and in many different fields. Indeed, the article highlighted the expected outcomes in four different fields: economic, with the increase in trading and navigation; political, with the improvement of international relations and the possibility of reducing the threat of insurrections in post-1848 Europe, as a sort of relief valve for revolutionaries and rebels; scientific, with the great technological progress involved in the project; and, finally, moral and apostolic, with the opening of the canal expected to allow the easier infiltration of Catholic missions into Muslim and pagan regions that were previously hardly reachable:

> Opening a new and wide field of industriousness, that eager fever of agitations and revolts, which has been keeping Italy and other surrounding people in constant distress for a long time, would find an outlet and a remedy, and the ills of Europe would be exhausted mostly in [building] the passage towards Asia. Indeed, the discontents foment riots and rebellions because

[4] 'Il canale di Suez' 1856, 37.
[5] 'Il canale di Suez' 1856, 44.

> they hope for themselves a better political status from the new political orders; if they could obtain it through less violent and dishonest means, there is no doubt that they would take this, leaving the atrocities and the dangers of conspiracies to the few fanatics.[6]

In addition, the piercing of the isthmus of Suez could create another effect, evaluated fundamental by the Jesuit review: 'Once this barrier has fallen, it will not take long to fall [...] and to disappear, in the face of the blazing light of Christian civilization, that dark rock of barbarism which, beyond the Arab Gulf, includes many millions of mortals.'[7] The chance to penetrate apostolically new territories in the Arab peninsula and in East Africa determined the Jesuits' favourable judgement of the project because the opportunity to civilise and convert several populations to Christianity had an immeasurable value, far more important than the commercial advantages: 'The civilization of Western peoples has Christianity as its main and substantial element, and it derives from Christianity the splendour and that marvellous power, for which civilized Europe is queen of the universe. Now, Christian civilization is essentially beneficial even in purely natural orders; it conquers peoples not with impetus and violence, but with the irresistible gentleness of its attractions, [...] making them happier and more glorious every day.'[8]

Fervent praise of Catholic missionary activities followed. In harmony with its strong apologetical and propagandistic ability, *La Civiltà Cattolica* tried to insert the Suez Canal project, as it had with several other scientific and technological innovations, into the providential divine design:

> God admirably makes use of the man to fulfil his highest aims; as supreme Author of nature and grace, he leads the material progress of human intelligence and activity to the sovereign purpose of his glory and to the sanctification of his elects, wanting that grace, as queen, has in nature always a handmaid always ready to his services. Our century shows us a very bright example of this, thanks of its great achievements in nature and in industry, and especially by the discoveries of steam and telegraph, with which, by shortening the intervals of space in a portentous way, the mutual bonds of human commerce are admirably tightened.[9]

[6] 'Il canale di Suez' 1856, 46.
[7] 'Il canale di Suez' 1856, 47.
[8] 'Il canale di Suez' 1856, 48.
[9] 'Il canale di Suez' 1856, 49.

Therefore, after claiming and summarising the advantages, the Jesuit review stated its favourable opinion towards an impressive work that could be connected so perfectly and so usefully to the providential Christian perspective and to the apologetical and missionary strategy of the Catholic Church:

> Religion no less than civilization must rejoice in the rapid increase of human commerce: and, if among these, the enterprise of the Egyptian isthmus will take place, as hoped in a short time, it [religion] will bless and greet it as one of the most adventurous events of our century, because, by expanding the gates of Europe to East, it prepares to the apostles of true faith a field and an immense harvest of sweats and conquests.[10]

The Attitudes of the Catholic Press

This favourable opinion of the Suez Canal characterised not only the clerical and intransigent press, such as *La Civiltà Cattolica* and *L'Armonia della religione colla civiltà* but also the Catholic moderate press, which was more open to lay scientific mentality and political debate. The Genovese astronomer and politician Paris Maria Salvago (1831–1899) gave his support to the canal project on the pages of *Annali Cattolici*. This review was founded in 1863 in Genoa and directed by Salvago himself, with the collaboration of the young Genovese marquis Manfredo Da Passano (1846–1922), who was a key figure in the Catholic moderate press in the second half of the nineteenth century. At the beginning of 1864, Salvago asserted the usefulness of scientific and technological progress for Catholicism, at the same time alerting his readers to its dangers: 'The progress of the industry, the press, the telegraphs, the supreme ease of communications, the travels, for some can be an effective means of spreading mistakes, but they also help the Providence to propagate the truth. And we are witnessing the spectacle of ever new conquests that Catholicism is carrying out.'[11] The piercing of the Suez isthmus fitted perfectly into this idea: besides the extensive commercial and economic advantages, Salvago underlined the apostolic and missionary outcomes of the opening of the canal, which demonstrated and represented the power of 'the

[10] 'Il canale di Suez' 1856, 50.
[11] Salvago (1864, 35).

influence of the Catholicism destined for struggle, but certain of the victory'[12] of Christianity.

Catholic public opinion gave the Suez Canal affair wide and constant attention. This interest was manifested not only in the periodical press but also in the publication of descriptive and apologetical pamphlets dedicated to the piercing of the Egyptian isthmus. The most significant example was *Lo Stato pontificio e l'istmo di Suez*, written by Erasmo Fabri Scarpellini, who was assistant astronomer at the Osservatorio astronomico of Campidoglio in Rome and also devoted to historical and literary studies.[13] As a member of a group of Roman intellectuals and scientists close to the Catholic hierarchy, he was the director of the journal *Bullettino della Corrispondenza Scientifica di Roma per l'avanzamento delle scienze*, another review that presented a positive evaluation of the idea of the canal. The pamphlet was published in December 1856 by the Roman Tipografia delle Belle Arti and presented an extract from an article that appeared earlier that year on the pages of the *Giornale arcadico di scienze, lettere ed arti*, one of the most widely distributed cultural journals of the Papal States.

First, Fabri Scarpellini declared the Suez Canal 'one of the most splendid glories in the modern industry'.[14] Besides mentioning the missionary opportunities and the importance of the Pope's validation of the project, he focused mostly on the expected economic and commercial advantages for the Italian States in general, and the State of the Church in particular. Fabri Scarpellini's point of view was not original: it was based on data and observations already published and circulating in the international press. However, his opinion was very naïve, as it over-estimated the actual economic effects without considering the geopolitical context and the marginal commercial role of the Papal States, which was further reduced some years later as a consequence of Italian unification. In order to strengthen this optimistic prediction, he resorted to using statistics, including tables that showed the distance reduction between Bombay and several European and American ports and the potential tonnage of the merchant navy of the Italian States.

[12] Salvago (1864, 35).

[13] Erasmo Fabri, Feliciano Scarpellini's pupil, took the surname Scarpellini after marriage to the more famous Caterina Scarpellini (1808–1873), Feliciano's niece, a pioneering figure in astronomy and science; Favino (2020, 63–116).

[14] Fabri Scarpellini (1856, 10).

In addition, Fabri Scarpellini's pamphlet presented a tribute to technological innovations and their applications, from the railways (defined as 'application of scientific principles to locomotive art [...] summary of the progress of physical sciences'[15]) and their development, then under construction also in the Papal States with Pope Pius IX's approval, to the telegraph and steam-powered vessels, concluding with general praise for scientific and technological progress:

> The science, we will repeat it again, is the first element of the current civil progress: its high influence on the wellbeing of the peoples in the present age is the most important event which characterised the nineteenth century spirit: the science, after departing from the hovels of the philosophers, from the porticos of the schools, from the classrooms of the academies, invaded the masses of the nations, took possession of the principles of their trade, their industries, their arts; it changed into another form, and after giving them a very sublime place for social things, began and still continues to excel and master all the elements of the current human perfectibility.[16]

This apologetical approach, based on Catholic efforts to appropriate scientific and cultural advances, referred to an enduring and well-established tradition, maintained for years by Catholic scientists and apologists and by clerical periodicals, especially *La Civiltà Cattolica*. The railways together with the telegraph were the best examples of the Catholic propagandistic use of technological progress, also referred to by Fabri Scarpellini in his pamphlet and by *La Civiltà Cattolica* in 'Il canale di Suez'. The Suez Canal project flawlessly fitted this apologetical discourse; indeed, the Jesuit journal stated it clearly: 'A very important part is reserved for the Suez canal as well as for steamers and telegraphs in the great task of bringing peoples closer together and preparing them to join in one society and in one Church.'[17]

Therefore, the main attitude towards technological progress, shared by most of the ecclesiastic hierarchy and press, was a cautious welcoming, aware of the increasing centrality of science and technology in society, culture, politics and economics. However, on one hand they rejected the positivistic celebration of progress and blind faith in it; on the other hand, they considered technological innovation to be the result of the intellect

[15] Fabri Scarpellini (1856, 4).
[16] Fabri Scarpellini (1856, 3).
[17] 'Il canale di Suez' 1856, 50.

given by God, including it in the providential design. For these reasons, Catholic periodicals and publications focused on the railways and the telegraph and on the development of their networks. Railways and telegraphs became popular and emblematic topics: the Catholic press not only underlined their pragmatic usefulness to transportation and communication but also cleverly used them as part of an apologetic and propagandistic strategy. Indeed, the railways and telegraphs had the potential to become powerful instruments for spreading Catholicism. For example, *L'Armonia della religione colla civiltà*, on the occasion of the inauguration of a railway in Piedmont in 1856, reported the discourse of the bishop of Biella, Giovanni Pietro Losana (1793–1873): he praised the role of the new inventions in the diffusion of the catholic faith, 'bringing, on land and on the sea, thanks to the wings of the steam-power, the word of civil and eternal life across the globe, in order to gather the peoples of the earth under one banner, the august and sacred Catholic banner: [...] the modern industry, crowned with the brilliant diadem, with which the genius of the discoveries encircled her forehead, will bring gladly and gloriously peoples, Duces and Kings to the Vatican'.[18]

From the 1860s, many Catholic newspapers reported specific blessing formulae, devised by the Church and bishops, for trains, telegraphs, infrastructure projects (such as ports, bridges, and canals), boats, and, later, aeroplanes.[19] In an 1875 article Giacomo Margotti showed the Catholic appropriation of new technologies and theorised their apologetic and propagandistic use, declaring 'that among the glories of the reigning Pontiff it is also the sanctification all the recent discoveries of true progress in material things. [...] The Church sanctifies progress, does not fight it, on the contrary it uses it to bring peoples to God.'[20]

The Papal Commission for the Suez Canal

The apologetic discourse needed to satisfy the desire for information; to update the public, Catholic readers; and to deny the accusation that Catholicism was in opposition to progress and science. However, in the program of Catholic science, it also needed to harmonise technological progress with Catholic morals and principles. In this context, as shown by

[18] Losana (1856, 901).
[19] Lagreé (1999, 46–49).
[20] Margotti (1875, 433).

both *La Civiltà Cattolica*'s article and Fabri Scarpellini's pamphlet, widespread as models among the clerical press, Catholic discourse was able to include the Suez Canal project (together with railways, telegraph, steam power etc.) in an apologetical perspective, proposing an idea of progress led by and in harmony with Christian faith.

The article published in *La Civiltà Cattolica* fully reflected the opinion of the Papal States and of most Catholic periodicals concerning the construction of the Suez Canal. Thanks to Giuseppe Milesi Pironi Ferretti (1817–1873), interest in the canal rose also in the Church: on January 1857 Pope Pius IX appointed a specific commission to analyse the expected effects and advantages of the construction of the canal.[21] Milesi Pironi Ferretti, Minister of Public Works, Commerce, Industry, and Agriculture for the Papal States from 1854 and later appointed cardinal (1858), was the key figure in building a railway network in the State of the Church and in improving the ports of Ancona, Anzio, and Civitavecchia, making a strong effort to promote their development. He was a man interested in technological applications to improve transport and trade.[22] The commission, chaired by Milesi Pironi Ferretti himself, was composed of officials and scientists from different fields and institutions: Girolamo Petri, a jurist of the State Secretariat; Luigi Moreschi, an official of the Ministry of Internal Affairs; Stanislao Sterbini, vice director of Papal States' Customs; Nicola Cavalieri, a Roman mathematician; Giuseppe Sgariglia, an official of the Ministry of Commerce; Vincenzo Pericoli, vice director of the Primaria Camera di Commercio of Rome; Luciano Milan, a Roman diplomat and Papal consul in Odessa; Adolfo Boscaini, an attorney; and Alessandro Cialdi, an engineer, explorer, and commander of the Papal navy, who later contributed to the construction of Port Said. Giovanni Contedini, an official of the Ministry of Commerce, was appointed as secretary of the commission, and Erasmo Fabri Scarpellini, the author in 1856 of *Lo Stato pontificio e l'istmo di Suez*, was appointed as assistant secretary.

With its authoritative role in Italian Catholicism, the Jesuit journal influenced the opinion and the results of the commission. Indeed, the Papal commission assumed a favourable judgement, convinced of the advantages in increasing the trade, in developing marine transportation and in favouring the Christian apostolic mission, echoing *La Civiltà*

[21] Lefevre (1960, 182–203).
[22] Lefevre (1995, 293–299).

Cattolica's thoughts. Also in this case, as in many others during the second half of the nineteenth century (the most famous case is related to the Congregation of Index), the Jesuit journal assumed a leading role not only in the press but also in the Catholic movement, using its authority to influence and support Papal institutions.[23]

Despite this positive attitude, the Papal commission on the Suez Canal collided with Italian political history. The work of the commissioners started on March 1857 and lasted for three years, until 1860. In this time, the commission dealt with topics related to the opening of the canal, such as the government legislation, the terrestrial and marine communication routes, the desirable development of agriculture, trade and industry, although any significant measures were taken. Then, the realisation of the Risorgimento profoundly changed the Italian order. The new situation, born from the Italian unification process (in particular due to the Second Italian War of Independence in 1859 and the following annexations in Central and Southern Italy), meant that the conclusions of the Papal commission were no longer as significant for the State of the Church: reduced to Rome and the Lazio, it lost relevant ports (such as Ancona) and its political and especially commercial importance decreased.

However, the Catholic press continued informing and updating their readers about developments in the Suez project, focusing on the establishment of the *Commission internationale pour le percement de l'isthme de Suez* and later the *Compagnie universelle du canal maritime de Suez* (1858). After the beginning of work in April 1859, clerical periodicals proceed to follow constantly the progress of the construction of the canal. *La Civiltà Cattolica* published many updates about the work, often with detailed scientific and technological descriptions and explanations. The news published in the newspapers was more concise. Particularly active in informing Catholic readers in this phase was *L'Unità Cattolica*, founded by Giacomo Margotti in 1863 which, after having rejected the moderate and prudent direction of *L'Armonia della religione colla civiltà*, shortly became a new point of reference in the clerical intransigent press. Margotti's new newspaper approved the Suez Canal project as well, focusing on its development.

The inauguration of the canal took place on 17 November 1869, with an impressive opening ceremony, including a Christian blessing: a

[23] Palazzolo (2010).

symbolic and tangible sign of the favourable attitude of the Catholic Church and the clerical press.

After the Inauguration

Attention to the Suez Canal remained significant after the official inauguration. For example, *L'Unità Cattolica* in February 1870 published an important announcement: the first passage of an Italian ship, the steam-powered vessel 'Africa', 'a bold surprise that the Rubattino company makes to Italian trade'.[24] In April, an English journal was quoted in relation to the expected advantages of the canal; then, *L'Unità Cattolica* itself positively evaluated its usefulness.[25] In the following years, the general opinion remained positive when the Suez Canal was mentioned in the clerical press.

Not only the expected economical, commercial, apologetical, and missionary effects but also the more general Italian political situation helped to affect the clerical press's favourable judgement. A comparison of the Suez Canal project with two other significant infrastructures, the Mont Cenis (or Fréjus) Rail Tunnel (1857–1871) and the Panama Canal (1881–1914), shows how political events influenced Catholic approaches and how they functioned in apologetical and propagandistic strategies.[26]

The project to build a rail tunnel under Mont Cenis between Piedmont and Savoy was developed by the Sardinian government, especially Prime Minister, Camillo Benso of Cavour, and his Minister of Public Works, Pietro Paleocapa. In a context characterised by conflict created by the emanation of anti-clerical laws, the Piedmontese clerical press, supported by *La Civiltà Cattolica*, led strong criticism of the project. In this controversy, it merged many arguments: political opposition to the Piedmontese, and later Italian, liberal government; a polemic against the slow advancement of the work and its cost, which was considered exorbitant; and disapproval of blind faith in science with no recognition of the role of the providential design.[27]

[24] 'Cronaca' 1870, 145.

[25] 'Cronaca' 1870, 328; 364.

[26] Bovolo (2017, 85).

[27] 'Il traforo del Cenisio e le promesse del Conte di Cavour' 1863, 1174–75; 'Traforo delle Alpi' 1867, 277; 'Mezzo spiccio di avere un traforo delle Alpi' 1867, 1032; 'Il traforo delle Alpi' 1871, 901.

In contrast, the case of the Panama Canal represents a third approach, characterised by a degree of indifference from the Italian Catholic press.[28] When, in the 1880s, Ferdinand de Lesseps launched the idea of a canal in Panama connecting the Atlantic and Pacific Oceans, the Catholic press's attitude was colder than it had been towards the Suez Canal project and at times somewhat ironic.[29] Only *L'Unità Cattolica* made a weak effort to include the new canal project in an apologetical discourse, quoting the favourable opinion stated by Bernardo Augusto Thiel Hoffman, archbishop of Costa Rica, defined by the newspaper as 'lover of progress and friend of the canal, powerful means of civilization'.[30] However, after Lesseps' failed attempt, work on the construction of the Panama Canal only started in 1904 thanks to a US initiative, and it was inaugurated in 1914. Italian Catholic newspapers and reviews followed the development of the project in the international news section, assuming a cold attitude, without detailed analysis or explanations: in a changed geopolitical and historical context, assuming the Italian Catholicism's point of view, the missionary and religious considerations, so important in the apology of Suez Canal, could not be strongly proposed in the Panama case. Therefore, this difference of attitudes confirms how deeply and clearly the apologetical and apostolic arguments contributed in shaping the Catholic judgement of the Suez Canal.

Conclusion

Especially during the 1850s and 1860s, Catholic journals, newspapers, and pamphlets informed and updated their readers about the project and construction of the Suez Canal, as one of the most challenging and iconic infrastructure works ever realised in the world. However, besides an informational purpose, the aim of these periodicals was to offer Italian Catholic public opinion the interpretation of the project approved by the Church on economic, political, scientific, and moral grounds. The Italian Catholic press, led by the highly influential *La Civiltà Cattolica*, gave particular importance to religious and moral arguments. The press attempted to insert the realisation of the Suez Canal project, apologetically and

[28] In contrast, the Panama Canal project was strongly debated in the Central American Catholic movement and hierarchy.

[29] 'Cronaca contemporanea' 1884, 479.

[30] 'Il signor Lesseps e l'istmo di Panama' 1886.

propagandistically, into a Christian providential perspective, claiming the work as result of the divine design and the outcome of God's gift of intelligence to humankind. The attitude of the clerical press encompassed an intention to use topics such as the Suez Canal in an apologetic way, trying to simultaneously establish and reinforce Catholic public opinion in nineteenth-century Italy.

References

List of Periodicals

Annali Cattolici
Bullettino della Corrispondenza Scientifica di Roma per l'avanzamento delle scienze
L'Armonia della religione colla civiltà
L'Unità Cattolica
La Civiltà Cattolica

Bibliography

Bovolo, Carlo. 2017. *I cattolici italiani e la scienza. Il discorso apologetico sulla stampa clericale nell'età del positivismo*. Milan: Bibliografica.

Fabri Scarpellini, Erasmo. 1856. *Lo Stato pontificio e l'istmo di Suez. Parole*. Rome: Tipografia delle Belle Arti.

Favino, Federica. 2020. *Donne e scienza nella Roma dell'Ottocento*. Rome: Viella.

Forno, Mauro. 2012. *Informazione e potere. Storia del giornalismo italiano*. Rome-Bari: Laterza.

Lagrée, Michel. 1999. *La benediction de Prométhée: Religion et technologie XIXe-XXe siècle*. Paris: Fayard.

Lefevre, Renato. 1960. La Commissione pontificia relativa all'istmo di Suez, 1857–1860. *Economia e storia* 2: 182–203.

———. 1995. Ai tempi di Pio IX: mons. Milesi, Ministro dei Lavori pubblici, Belle arti, Commercio, Industria e Agricoltura. *Strenna dei Romanisti* 55: 293–299.

Losana, Giovanni Pietro. 1856. Inaugurazione della strada ferrata di Biella ed allocuzione del vescovo. *L'Armonia della religione colla civiltà* 223: 901.

Margotti, Giacomo. 1875. Pio IX e il telegrafo nel lietissimo giorno del 13 maggio. *L'Unità Cattolica* 110: 433.

Palazzolo, Maria Iolanda. 2010. *La perniciosa lettura. La Chiesa e la libertà di stampa nell'Italia liberale*. Rome: Viella.

Salvago, Paris Maria. 1864. Cronaca. *Annali Cattolici* 1: 35.

Traniello, Francesco. 2007. *Religione cattolica e Stato nazionale. Dal Risorgimento al secondo dopoguerra*. Bologna: il Mulino.

PART III

Colonial Spaces, Colonial Encounters

CHAPTER 11

Assab and the Suez Canal: Traditional Networks and Imperial Agendas in the Red Sea

Luca Castiglioni

Proposals of Empire

When the Suez Canal was opened in November 1869, Italy, like many other European powers, was ready to use this new path of intercontinental trade to find new markets and profitable exploits. This disposition was not the consequence of a sudden rush: in the years prior to the opening of the canal, a number of private or even individual initiatives tried to gather momentum for a more active involvement of the Italian government in the Red Sea and the East African region. Among these active proponents

L. Castiglioni (✉)
Università degli Studi di Milano, Milano, Italy
e-mail: luca.castiglioni@unimi.it

B. Curli (ed.), *Italy and the Suez Canal, from the Mid-nineteenth Century to the Cold War*,
https://doi.org/10.1007/978-3-030-88255-6_11

was Giuseppe Sapeto,[1] who in 1863 managed to convince Minister Torelli that the canal would open a great number of commercial opportunities for the struggling economy of Italy and that a survey of the Red Sea, in search of coal deposits, was key to support future endeavours.[2]

The mission proved to be a failure: coal mining was not possible in the area around Bab-el-Mandeb. Nonetheless, Sapeto pushed for further Italian projection in the southern Red Sea: Italy needed to finance a private trading company, at a loss, and to build an *ad hoc* port settlement,[3] functional to European trade and transport. In 1865 Sapeto was essentially proposing that Italy built something akin to the port of Suez, but under Italian control.[4]

Sapeto also suggested that Italian trade and shipping could succeed in the local Red Sea inter-trade: the notion of replacing traditional actors in the management of the 'pilgrim' trade and to compete with modern means in the small trade and transport of the region was novel.[5]

The Italian shipping entrepreneur who seemed to understand the possibilities the canal offered was Raffaele Rubattino.[6] His low-risk, low-cost connections to Egypt started in 1865 and by 1868 he launched the Marseille–Genoa–Alexandria line. He bet on the last-minute rush of the Italian government to take advantage of the canal's opportunities, his final goal being to monopolise government-financed work in the region.[7] Rubattino counted on the 'flag' going first and taking all the risks, he was just providing transportation.

[1] Born in Carcare (Savona, Italy), Giuseppe Sapeto (1811–1895) lived an adventurous life, in open contrast to his initial clerical vocation. His most remarkable success was without a doubt his involvement in the acquisition of Assab. See Surdich (2017). See also Pellegrineschi (1932) and Giacchero and Bisogni (1942) for Sapeto's fascist representation as a founding father of the Italian empire.

[2] Surdich (2005) and Del Boca (1976, 33–49).

[3] Codignola (1938, 303–321).

[4] Sapeto (1865, 118–166).

[5] Low (2008).

[6] A Genoese shipowner and entrepreneur, Rubattino became a key figure in the Italian economic landscape of the nineteenth century, although through alternating fortunes. See Giulianelli (2017).

[7] Doria (1990).

The Egyptian Empire in the Red Sea

The Red Sea was a space of borders and intertwining peripheries. It was the periphery of the Egyptian Khedival Empire, the dividing rift between Asia and Africa, the farthest reach of the Indian Ocean networks and of the British Raj's influence. Also, its African coastline was the edge between Muslim port cities and Christian Ethiopia. Finally, the Red Sea was alive with many smaller native connections: movements of people and trade between smaller coastal centres was the living infrastructure that made the Red Sea nothing like a desert of water.[8]

All along the coast line, from Massawa to Berbera, ruled a number of local chiefs, of varying wealth and enjoying different degrees of independence, often forced in overlapping suzerainties. These local Sultans made Egyptian imperial control fragmented and usually limited to port cities, where garrisons were stationed. Smaller harbours where left unchecked, becoming essential points of free contact between local polities on the two coasts. The survival of many local Sultans depended on these blind-spots for the continuation of the slave trade.[9]

Prior to the opening of the Suez Canal, Egypt had actively pursued its imperial expansion on the African coast of the Red Sea, using Massawa as a starting point for raising 'seven flags' between 1860 and 1862. This was an attempt to force the Sultans ruling over seven port settlements on the long stretch of coast between Suakim and Zeila to formally submit to Egyptian control. Significantly, the Egyptian envoy was the son of the Naib of Massawa.[10] The choice between yielding to Egypt, in exchange of leeway and support against local rivals, and resisting them deeply upset local politics.[11]

This uncertain political situation was defined as a 'blockade' by British officials.[12] The Egyptians held the key to the Highlands through defined gateways like Massawa and Zeila, where they levied taxes and duties that made trading with Ethiopia expensive. But they exerted no real control beyond these cities' boundaries, which made trade very risky. The idea of

[8] Miran (2014, 200) and Huber (2013).

[9] Clarence-Smith (1989).

[10] Erlich (2002), Moore-Harell (2010) and Talhami (1979). Also see Munzinger (1869, 214).

[11] *Memorandum on the Italian Claim to Sovereignty over Assab Bay*, FO 881 4231. Also *British Documents on Foreign Affairs* Part I Series G, Vol. 13 pp. 81–82.

[12] FO 881 4231 p. 17.

directly acquiring a port was played with by the French and British respectively long before the opening of the canal, but the difficulty of actually settling along the cost and maintaining a stable commercial activity confirmed the impossibility to break the monopoly of the existing ports. Nonetheless, their failed openings towards Obock and Amphylla in the 1840s showed the local Sultans that leasing or selling remote areas of their domains could prove profitable, counting on the fact that any attempt to create a stable European settlement would invariably fail.[13]

1869: The First Settlement of Assab

In 1869 Giuseppe Sapeto returned to the Red Sea. This time he was meant to find a suitable location for a naval station for the Rubattino Company. The political need for a private intermediary was strongly advised so to not upset British interests in the region of Aden: Sapeto had to settle for Assab, not his first choice. Sited on the western coast in the Danakil region, close to the southern exit of the Red Sea, Assab had a wide bay with many inlets and he deemed it *res nullius*. According to Sapeto, a port settlement there would have been ideal for European steamships: close to Yemeni ports like Mocha and Hodeida, Assab could provide a new gateway for Italian trade and exploration.[14] His conclusions would have proved flawed.

As a matter of fact, he was buying through unauthorised proxies a piece of land belonging to Sultan Berehan of Rahaita.[15] Sapeto was unaware of the balance of power between the local Sultans, taking possession of a buffer zone between two Sultans who were struggling to maintain their own relative independence, compressed between Egyptian suzerainty and the strong outreach of Sultan Anfari of Awsa, whose realm covered and controlled the Afar salt plains.[16] Assab was not as isolated or empty as the Italians thought, but their arrival stirred the complex intricacies of local power, which spanned over large territories all over the western Red Sea. The Sultans of Rahaita, Beilul, Obock, Tadjurra, Zeila and many others

[13] FO 881 4231 pp. 23–28. Also *Correspondence respecting Egyptian Jurisdiction over the Somali Coast*, FO 881 4099.

[14] Sapeto (1879).

[15] Report, Loch to India Office, N.12, 10 March 1880, IOR L/P&S/9/54. The Sultan's name is spelled Beheran or Burhan in British sources, while Berehan it's the most consistent traslitteration used in Italian sources.

[16] The historiography on the Afar people and their traditional institutions is quite limited. See Chailley (1980), Hassen (2011), Rahem (2001), Saint-Veran (1977) and Soule (1986).

based their wealth and power on the control over the necessary ganglia of the Indian Ocean Slave trade, that from Central Africa passed through the Ethiopian highlands, through the salt plains of the Sultan of Awsa and out of the badly controlled ports and anchorages of the coast, using small vessels and moving a small number of people at a time. This trade was excruciatingly difficult to control by the British and Egyptian authorities.[17]

The unplanned sale of Assab to the Italians in 1869 was undoubtedly going to impact the interests of the Sultans of Rahaita and Beilul, but the lack of private investment and commercial interest on the Italian part in Assab made this first Italian foray in the lands of the Danakil a very short one. The acquisition took place on 15 November 1869, formalised on 11 March 1870. The sellers were three 'Sultans', who accepted a fraction of the payment, in exchange for its immediate solvency, and gave the representatives of the Rubattino Company full possession of the site.[18] The steamship *Africa* of the Rubattino Company deployed a small squad of workers, led by Sapeto, and left for its inaugural voyage to India. The small settlement was not in such amicable surroundings as Sapeto believed: raiders and soldiers of the Sultan of Beilul—the northern neighbour and rival of the Sultan of Rahaita—encroached on the settlement and had to be threatened away by the sailors of the navy escort.[19] The impossibility of soon using Assab for big steamships was immediately clear.

But the 'Sultans' who sold Assab resettled to the vicinity of the Italian settlement in 1869: among them were the sons of the Sultan of Rahaita, who might have been following a personal political agenda of cutting out the father and master by relying on the interposition of the Italians.[20] When *Africa* returned with uninspiring profits from India, Sapeto was invited to return with them to Italy. The settlement was left abandoned, with the idea of returning with supplies and men to improve it. There would be no return for a decade. Most likely acting on information by the Sultan of Beilul, soldiers from Massawa came to the area in the summer of 1870 to dismantle what little there was, beating the 'Sultans' responsible for the sale, and remarking that Egypt owned those lands.

[17] Blight et al. (2014).

[18] *Acquisto di Assab e sue conseguenze*, doc. I, ASDMAE, *Africa Italiana*, Cartella 1–1, Fascicolo 1.

[19] Ruggiero ad Acton, Suez, 7 May 1870, IIA Vol. I Tomo I, N°16.

[20] Report, Aden to Bombay, N°16, 12 April 1879, IOR L/P&S/18/B 22.

Disrupting the Periphery

During the 1870s the Red Sea became Lord Salisbury's 'Lifeline of the Empire' and all attempts to change its balance in that decade were ostracised. Egyptian attempts to bring the region to heel were also defeated and the massacre of Werner Murzinger's expedition by the hands of the Sultan of Awsa in 1875 sent shockwaves across the region: Egypt regional credibility collapsed, in no small part because the complicity of the omnipotent Egyptian-invested Sultan of Zeila Abou-Bakr in sabotaging the expedition of his superior.[21] Mohamed Anfari once again extended his control over the Salt Plains and all the local chiefs, Berehan of Rahaita among them.

The return of Sapeto in 1879 caught most of the local actors by surprise.[22]

The negative impressions of 1869 and the total lack of interest by the Rubattino or any other company to follow through with the Assab experiment did not discourage its supporters, many of whom belonged to the Italian Geographical Society. Italian foreign policy was aimed at Tunisia and the small acquisition in Dancalia had been an afterthought at best.[23] But the interest for Ethiopia, access to its resources and supporting the efforts of Italian explorers and 'pathfinders' became relevant enough to warrant a return to Assab, still under the cover of a private endeavour. This cover was essential given Italy's delicate position in foreign policy matters.[24] The goal for Assab had changed since 1869. It was meant to achieve local dominance, in an area still deemed void of real competition Assab as a counter-point to Suez was replaced by the notion of bridgehead for Italian penetration in East Africa.[25]

Italian activities immediately caught British attention, with predictable resistances and some unexpected exceptions: Charles Gordon deemed the Italian presence an opportunity. First, it would engage the Italians away from Europe, probably entangling them in something that would bound

[21] On Lord Salisbury's policy see Cain and Hopkins (1993, 362–369), Robinson et al. (1961) original, 1981 2nd Edition. Regarding Abou-Bakr Ibrahim of Zeila see Fontrier (2003).

[22] Loch to India Office, Aden, 2 January 1880, IOR L/P&S/18/B 22.

[23] Natili (2008).

[24] Paget to Lord Salisbury, Rome, 9 January 1880, FO 881 4429 N°17. See also Loch to the Secretary of Government in Bombay, Aden, 6 February 1880, IOR Vol. 4 PS/9/54 N°43.

[25] Rapporto, De Amezaga a Cairoli, Aden, 29 April 1880, DDI 2 Vol. XII N°933.

their foreign policy for years. Secondly, he noted that Assab had very little water and therefore no real chance to grow, meaning that it would become dependent on surrounding ports, allowing the British to have the settlement's traffic easily under control. Finally, in Gordon's view, the desire of the Italians to reach the highlands and redirect all possible trade in the area to Assab was clearly unrealistic, given the presence of the Sultan of Awsa. Gordon saw no future for Assab itself, but he did foresee a function for the Italians in the Red Sea, especially if they were to move their attention to the North, closer to Massawa and the Tigré. The Italians could have proven a reliable partner in future, but Assab itself was worthless: a stable settlement there would prove fruitless and pointless, if not even dangerous.[26]

At a local level the settlement of Assab forced native local networks and polities to react and adapt to a stable foreign element, settled in a contested area. The lack of any Italian military or technological superiority in this impoverished context made the Italians containable, giving leeway to the local Sultans to balance the Italians' presence in a complex equilibrium of divided loyalties and agencies. But the Italian's scarce military prowess also meant that they were not strong enough to guarantee the safety of the Sultans in their struggle against Mohamed Anfari's stranglehold.

The Italian settlement posed a risk to the slave trade, which was the base for the wealth of the local leaders, something that sparked tension between them. The contrasts between the Sultans of Rahaita and Beilul intensified, as did the bitter disapproval of their liege. Berehan's greed and lack of control over his family and subjects allowed a foreign presence in the region. His rival, the Sultan of Beilul, took every opportunity to push the British against the Italians, by presenting even false accounts of Italian military activity. British agents attempted to buy land in order to contain Assab's acquisitions, but the officials were not willing to violate Egyptian sovereignty outright, given that it was the legal basis for all formal objections to an Italian settlement in the area.[27]

[26] Memorandum on the Operations of the Italians in the Red Sea by Colonel Gordon R.E, FO 881 4429 N°52, Inclosure 1.

[27] Statement made by Hassan Mohamed of Bailool, Agent of Sultan Mohamed Hanfali, regarding Assab Bay, Aden, 26 February 1880, FO 881 4429 N° 65, Inclosure 3. British acquisitions attempts are referred in the Memorandum by Major Ardagh, Londra, 20 February 1880, FO 881 4429 N°47, Inclosure 1.

Meanwhile, Sapeto acted as a self-invested Sultan of Assab[28]: he demanded to be addressed as Sultan Yusuf by the locals and attempted to sway all Danakil traders in the neighbourhood to stop trading with Aden and to trade only with him. British accounts report that Berehan received Sapeto's promise to defend him against the Egyptians,[29] revealing the Italians' offers and dealings to the British envoys, hoping to spark a bidding war.[30] Dealings he conducted, according to De Amezaga, in the middle of the night, so not to "spark the envy of his courtesans"[31] or not to reveal that he was still acting with an autonomy that neither Egypt nor Awsa allowed. Meanwhile, the Italian Royal Navy officers were less and less certain that they could actually turn Assab into a functional port without a massive investment. An opinion shared also by the British Navy.[32] British informers returned many accounts of the settlement and its progresses in 1880 and 1881. The officers in charge, Hunter and Berners, gathered the opinions of the locals, who agreed that no trade would ever flourish in the region.[33]

What worried the British was that the only ostensible advantage of Assab was military in nature: its bay, if properly developed, could become a remarkable naval base in the Red Sea.[34] The British account may have been overly positive, since only smaller vessels could navigate the bay effectively. Assab needed a long pier and extensive dredging before being able to serve its purpose, which was repeatedly stated not being military in nature.[35] But Amezaga's heavy-handed attitude against British inquiries brought too much embarrassment to the Italian Government: President Cairoli needed cordial relations with Britain and level headed agents on the ground. Governmental intervention in case of a contrast would be

[28] Loch to Gonne, Aden, 12 April 1879, IOR L/P&S/18/B 22.

[29] Report, Loch to Secretary of Government in Bombay, Aden, 25 March 1880, IOR Vol. 4 PS/9/54 N°14/475.

[30] Translation of Statement, Loch to India Office, 20 February 1880, IOR Vol. 4 PS/9/54 N°10/342.

[31] Rapporto, De Amezaga a Cairoli, Aden, 3 January 1880, DDI 2 Vol. XII N°524.

[32] Report, HMS Seagull to Read Admiral Gore Jones, 20 January 1880, IOR L/P&S/9/54.

[33] Report, Loch to Secretary of Government in Bombay, Aden, 25 March 1880, IOR Vol. 4 PS/9/54 N°14/475,.

[34] Heron to Read Admiral Jones, HMS Seagull, 20 January 1880, FO 881 4429 N°43, Inclosure 3.

[35] Rapporto, De Amezaga a Cairoli, Aden, 3 January 1880, DDI 2 Vol. XII N°524.

difficult.[36] Sapeto ineffective 'native policy' of dealing with the Sultans as a Sultan himself didn't help matters.[37]

Containment and Balance

Meanwhile Mohamed Anfari's patience with his vassal Berehan was growing thin.[38] The position of the Sultan of Rahaita worsened when the new Italian administrators, Commander Frigerio and Commissioner Branchi, replaced Sapeto in 1881: they were under orders to finally make contact with the interior with the least possible investment of manpower and money.[39] The road to Awsa was rarely travelled by Europeans, little accounts of it were available after Munzinger's death and only a few people knew it well enough to travel on it. The main trade that made use of it was that of the slavers, who were not willing to share it. Nonetheless, the suggestion of the renowned explorer Orazio Antinori[40] in April 1881 to send 2000 Remington rifles to Shewa from the old Vatican armoury to support king Menelik against Yohannes IV made finding a safe way across Awsa an impellent necessity.[41] Assab was becoming part of greater regional dynamics, in which the Italians were playing a small but ever-growing role.

The Italians had no alternative but to employ the services of the Sultan of Rahaita to contact Mohamed Anfari. Berehan had convinced Sapeto that visits to Anfari's court were too perilous to be attempted: gifts and communications to Awsa had to go through him instead.[42] Effectively, Berehan ran a soft containment policy on Assab, since the Italians didn't have the knowledge needed to cross the Salt Plains. Meanwhile attempts

[36] Benedetto Cairoli (1825–1889) was a key figure of the Italian Risorgimento. In 1880 when he was the President of the Council of Ministers and Foreign Minister, he was forced to deal with the Assab situation. See Cattane (2020). The need for a quiet situation in Assab is expressed in Cairoli a Boccardo, Roma, 12 February 1880, IIA Vol. I Tomo II N°94.

[37] Loch to Cranbrook, Aden, 20 April 1880, FO 633 48 N°3 Inclosure.

[38] Lt. Inglis to Loch, Perim Island, 28 January 1880, Reports of Awsa gathering forces against Assab, FO 881 4429 N°64, Inclosure 2.

[39] Dispaccio, Cairoli a Frigerio, Roma, 14 August 1880, IIA Vol. I Tomo II N°139.

[40] Orazio Antinori (1811–1882) founding member of the Italian Geographical Society, he travelled extensively along the Nile and in Eastern Africa. He saw in Assab a way to circumvent Zeila and Egyptian access to Ethiopia. See Ambrogi (1992).

[41] Relazione, Antinori alla Commissione…, Let Marefià, 6 April 1881, IIA Vol. I Tomo II N°177.

[42] Rapporto, Branchi a Cairoli, Assab, 18 February 1881, ASDMAE, *Archivio Eritrea*, Busta 6, Corrispondenza 1881–1883, N°18.

were made by Italian agents to bring the interest of the Indian traders to Assab so that their vast network could bring 'colonial' goods, but Aden proved to be too much of a safe bet for the higher echelon of traders, selling their goods to local traders, who in turn served smaller settlements, Assab among them.[43] This proved to be a significant handicap for the Italians.

In September 1880 Aden's assistant political resident Hunter, who had kept a very close eye on the Italians since their arrival, made an important anti-slave trade cruise along the Danakil and Somali coasts. He determined that the trade was rampant, with the clear support of important Sultans who doubled-duty as Egyptian administrators. Local policing was minimal, authorities complacent. Britain had little means to stop the trade on its own. The Italians were vocal about their desire to support the fight against the slave trade, but British agents turned down such offers, noting the contradiction between the Italians extending their protection to Berehan in late 1880 and them offering to help stop the main avenue of profit for his peers.[44]

During this cruise, Hunter checked on Berehan. Unbeknownst to the Italians, the Naib of Massawa, acting as an agent of the Egyptians, visited Berehan. He had informed him that a Khedival envoy was coming to arrest him and strip him of his rank as an Egyptian vassal, to pass it to his nephew, Uaziz. Uaziz had previously tried to convince the British consul at Aden that his uncle was untrustworthy and if the British would support him in overthrowing Berehan, he would deal with the Italians and turn Assab to the British.[45] Given the negative response of the Consul, he turned to the Sultan of Beilul and, in turn, to Massawa.

The Sultan of Beilul, contrary to Berehan, had gained nothing from the Italians and was to lose his share of the slave trade due to their vicinity. Moreover, after Berehan hid under the Italian flag, he was going to be even more under the watchful eye of Mohamed Anfari. By April 1881, Mohamed Anfari invited the Italians to come visit him and to have Berehan accompany them. Berehan must have known the meaning of Anfari's invitations and refused to move outright. Berehan's position had become

[43] Frigerio a De Amezaga, Assab, 10 July 1880, IIA Vol. I Tomo II N°138 Allegato 3.

[44] Report by Commander Hunter, Aden, 18 September 1880, IOR Vol. 4 L/PS/9/54. Regarding Berehan's protection see IIA Vol. I Tomo II N°159 Rapporto, Frigerio a Cairoli, Aden, 9 December 1880.

[45] Goodfellow to Bombay, Aden, 9 September 1880, IOR Vol. 4 L/PS/9/54 N°300.

unsustainable: he had grown dependent on Italian protection, even if the Italians didn't realise it. Given Berehan's resistance, commissioner Branchi resolved to more forceful action.[46]

Forcing the Blockade

The experienced adventurer Giuseppe Maria Giulietti,[47] officially the settlement's secretary imposed to Branchi by the lobby urging for reaching Shewa, requisitioned a small squad of Italian sailors and pushed into the territories of the Sultan of Beilul, towards Awsa. To reach his purpose he stole camels from the Danakil and threatened the chieftains under the Sultan of Beilul to show him the way to Awsa.[48] He and his escort were massacred in late May 1881 and this violent action brought an Egyptian investigation in Beilul, thus allowing the Egyptians to act directly with justification in a volatile environment to restate their sovereignty.

The Governor of Massawa, Fehemy Pasha, came to Beilul to lead the inquiry and Commander Hulton of *HMS Dragon* kept an accurate report of the proceedings, noting the deliberate negligence of the Egyptians. The court chose to reside on an inhabited costal area near Beilul, since the Governor didn't want to 'inconvenience' the Sultan of Beilul. This gave the suspected perpetrators any chance to avoid interrogation. Evidence was plentiful against a group of men of Beilul, but the Governor ignored the plea of the Italian Commander to take action against them.[49] The blatant disregard towards law and procedure in this affair, along with the clear inability or unwillingness of the Governor to even question the Sultan of Beilul, convinced Commander Hulton that the investigation was

[46] Branchi a Cairoli, Assab, 18 February 1881, ASDMAE, *Archivio Eritrea*, Busta 6, Corrispondenza 1881–1883, N°18; Loch to the Chief Secretary to Government Bombay, Aden, 8 February 1881, FO 633 48 N°20 Inclosure 2; Branchi a Cairoli, Assab, 14 April 1881 ASDMAE, *Archivio Eritrea*, Busta 6, Corrispondenza 1881–1883, N°38.

[47] Fascist favour for the figure of Giulietti sparked a number of regime publications in his honour: De Regny (1928), Marchese (1938) and Passamonti (1937). For the most recent biographical account on him see Volterra (2001).

[48] Memoria MAE sull'eccidio della Spedizione Giulietti, Roma, June 1881, IIA Vol. I Tomo II N°199. Hulton to the Secretary of the Admiralty, Aden, 12 June 1881, FO 633 48 N°54 Inclosure.

[49] Branchi a Mancini, Assab, 20 June 1881, ASDMAE, *Archivio Eritrea*, Busta 6, Corrispondenza 1881–1883, N°62.

a farce and the Governor did willingly favour a renowned actor in the slave trade.[50]

The arrival of Fehemy Pasha did worry the local Sultans greatly, especially Berehan. The Sultan of Beilul waved the Egyptian flag in 1869 against the Italians and had made life difficult for them, proving himself a reliable Egyptian proxy. His people had stopped the Italian explorer Giulietti and made clear to Assab that its expansion was at an end. The exact opposite to Berehan of Rahaita then, whose attempt to distance himself from the regional power of Awsa and the macro-regional Egyptian control allowed breathing room to the Italians, which were getting every day more difficult to remove.

The events of Beilul were of key significance, since they revealed a complex stratification of survival strategies implemented by local powers against overlapping Imperial agendas. In light of these events, British foreign policy began to move away from simply supporting Egyptian sovereignty to attempting to create a *modus vivendi* in the region of Assab between Egypt and Italy, so to avoid further complications.[51] Assab was a 'grey area', legally speaking, and there was a generalised lack of commitment by the central governments on the issue. Its complications became such a burden for the Italian Government, compared to the lack of any income or function, that Minister Mancini even declared that 'failure would be preferable'.[52] As often was the case in colonial matters though, the actions of the 'men on the spot' conditioned the decision of the governments and both the Italian and British governments reluctantly got pulled back into the matter of Assab because of local agents' initiatives.[53]

Diplomacy at the governmental level had been stuck since 1879 on the issue, with the Italians and the Egyptians debating in circles around the boundary of the Khedive's authority and its *de facto* limitations. The British government was constantly being dragged into the debate by both parties and Gladstone's scarce sense of urgency regarding Red Sea issues let the subject pending.[54]

[50] Inclosure 2, Substance of Commander Hulton's notes taken during the Beilul Inquiry; Inclosure 1, Hulton to the Secretary of the Admiralty, *HMS Dragon* at Aden, 24 August 1881, FO 633 48 N°96.

[51] Granville to Paget, Foreign Office, 7 July 1881, FO 633 48 N°57.

[52] Mancini a Menabrea, Roma, 2 September 1881, DDI 2 Vol. XIV N°192 Dispaccio.

[53] Schölch (1976) and Ramm (1944).

[54] Granville to Cookson, Foreign Office, 13 July 1881, FO 633 48 N°60.

Failed Gunboat Diplomacy and the Erosion of an Empire

After the conclusion of the inquiry in Beilul, in late August 1881, the Governor of Massawa expressed to Commander Frigerio his will to disembark an armed detachment in Rahaita, to secure such a dangerous area in order to shelter the Italians from further risks and to visit his subject, Berehan.[55] But Berehan was under Italian protection, although the legal bounds of such protection were opaque at best. Italy didn't agree on Egyptian sovereignty over Rahaita, using numerous British accounts to support this point.[56] This was causing a lot of embarrassment in London: an Egyptian landing would have sparked an unwanted confrontation.

In Cairo, British representatives found themselves cornered between Egyptian pleas and Italian remonstrations. The problem Gladstone and Granville faced was delicate: maintaining the foreign policy of the previous conservative government, thus alienating the sympathies of the young Mediterranean power or revaluating the support given in a strategically key location to a failing ally, which was becoming unable to keep is territories under control? The Egyptian Government was giving into the pressure of the military element, which had been constantly shrunk by the financial troubles left by the late Khedive Ismail and was finding his guide in Urabi Pasha: humiliated by the measures of the Control Committee, Cairo was considering dispatching more troops in the Danakil region, to reaffirm Egyptian prerogatives in the Red Sea.

On 1 September 1881 Egypt's Foreign Minister privately suggested to the British representative that, if a British vessel was to accompany the Egyptian landing in Rahaita, the Italians would never risk a confrontation.[57] But the implications of such a request were too great for Britain and the trust placed in Egypt was growing thin. By 9 September the officers' insurgence, led by Urabi Pasha, highlighted Khedive Tawfiq's weakness and pushed Britain to redefine its position towards Egypt and the Red Sea.

Gladstone sponsored an Italo-Egyptian convention for Assab, which Egypt rejected.[58] From that point on Britain and Italy excluded Egypt in

[55] Hulton to Rear Admiral Jones, *HMS Dragon* at Aden, 17 August 1881, reporting on the Egyptian's intentions, FO 633 48 N°91 Inclosure.

[56] Menabrea to Granville, London, 26 August 1881, FO 633 48 N°88.

[57] Cookson to Granville, Alexandria, 24 August 1881, FO 633 48 N°85 Telegraphic.

[58] Granville to Macdonnel, Foreign Office, 13 September 1881, FO 633 48 N°100.

dealings over Assab, eliding Egyptian influence south of Massawa, and the Italians became the only interlocutors of the British in the area. The local Sultans became an Italian problem, for them to deal with. The tragic end of Urabi's revolt definitely terminated Egyptian initiatives in the South Red Sea.

Between 1882 and 1885 the Italians became relevant actors in local politics: easier access to trade goods and work began attracting people from the territories of Rahaita, Beilul and further inland, impoverishing the local Sultans as consequence.[59] The presence of the Italian Navy made the area less viable for the slave trade, thus weakening the Sultans of Rahaita and Beilul.[60] Beilul would be soon incorporated into Assab's periphery, while Rahaita would survive until the border between Eritrea and French Somaliland was defined, but the Sultanate crumbled under the pressure of two neighbouring European powers. By 1882 Mohamed Anfari had to accept that his outreach towards the coast had been limited, which made him despondent to Italian requests to access his territory. His son would sign a treaty of friendship with the Italians in 1887, but Awsa domain would fall to Menelik in the late 1890s, thus ending its long period of independence.[61]

In 1885 Italy finally seized the fruit of its policy of cooperation with Great Britain, obtaining the control of Massawa. The acquisition of such a vital location shifted Italy's focus North, preparing the way for a progressive invasion of the highlands, which would find its conclusion at Adwa in 1896. Assab returned to being the periphery: a secondary port with no real background, little trade and no political value. The link to the highlands through Harar was sporadic and weak, only the invasion of Ethiopia in 1935 provided Assab with a hinterland that started to use the port as originally designed, but WWII terminated its Imperial function.

[59] Demographics and salaries, 1885, ASDMAE, *Archivio Eritrea*, Busta 11.

[60] Malet to Granville, Cairo, 28 November 1881, FO 633 48 N°128; Report of Slave Trading around Assab, 13 April 1886, ASDMAE, *Archivio Eritrea*, Busta 25.

[61] Friendship Treaty between Italy and Sultan Mohamed, son of Anfari, Sultan of Awsa and of all the Danakil. 10 May 1887, ASDMAE, *Archivio Eritrea*, Busta 37.

Conclusion

The history of the rulers of the area around Assab reacting to the Italians; the impact the little settlement had on the wider socio-political and diplomatic dynamics at a regional and international level; the deep impact of local European and native agents in directing the flow of events: these elements return us the dynamic reality that existed in the Red Sea in the nineteenth century, in a fitting example of micro-history, joining global and local. Diminished and forgotten by European accounts, lost inside new colonial denominations, the canal changed this reality forever allowing Italian designs of expansion to materialise in Assab, to create a new port town, a new link, to serve the new world of trade opened by the canal itself.

References

Archival Sources and Abbreviations

ASDMAE, Archivio Storico Diplomatico del Ministero degli Affari Esteri (IT)
DDI, Documenti Diplomatici Italiani
FO, Foreign Office Records, The National Archives (UK)
IIA, Opera dell'Italia in Africa
IOR, India Office Records, The British Library (UK)

Bibliography

Ambrogi, Silvano. 1992. *Un arabo perugino: vita e viaggi di Orazio Antinori in Egitto e nell'Etiopia di Menelik*. Torino: ERI.

Blight, David, David Freamon, and Bernard Harms. 2014. *Indian Ocean Slavery in the Age of Abolition* New Heaven: Yale University Press.

Cain, P.J., and A.G. Hopkins. 1993. *British Imperialism: Innovation and Expansion, 1688–1914*. Harlow: Longman.

Cattane, Michele. 2020. *Benedetto Cairoli. Il vessillo della sinistra storica 1825–1889*. Torino: Carocci.

Chailley, Marcel. 1980. *Notes sur les 'Afar de la region de Tadjoura*. Paris: Academie de science d'Outre-Mer.

Clarence-Smith, William Gervase (a cura di). 1989. *The Economics of the Indian Ocean Slave Trade in the 19th Century*. London: Frank Cass.

Codignola, Arturo. 1938. *Rubattino*. Bologna: Cappelli.

De Regny, P. Vinassa. 1928. *In Memoria di G.M. Giulietti*. Pavia: Istituto Pavese di Arti Grafiche.

Del Boca, Angelo. 1976. *Gli Italiani in Africa Orientale*. Roma, Bari: Laterza.

Doria, Giorgio. 1990. *Debiti e navi: la compagnia Rubattino 1839–1881*. Genova: Marietti.

Erlich, Haggai. 2002. *The Cross and the River. Ethiopia, Egypt and the Nile*. London: Lynne Rienner Publishers.

Fontrier, Marc. 2003. *Abou-bakr Ibrahim, Pacha de Zeyla—Marchand d'Escalves*. Paris: l'Harmattan.

Giacchero, Giulio, and Giuseppe Bisogni. 1942. *Vita di Giuseppe Sapeto*. Firenze: Sansoni.

Giulianelli, Roberto. 2017. RUBATTINO, Raffaele. In *Dizionario Biografico degli Italiani*. Roma: Istituto della Enciclopedia Italiana.

Hassen, Mohammed. 2011. Indigenous Governance among the Southern Afar (ca. 1815–1974). *Ethiopian Journal of the Social Sciences and Humanities* 7 (1–2): 1–26.

Huber, Valeska. 2013. *Channeling Mobilities*. Cambridge: Cambridge University Press.

Low, Michael Christopher. 2008. Empire and the Hajj: Pilgrims, Plagues, and Pan-Islam under British Surveillance, 1865–1908. *International Journal of Middle East Studies* 40 (02): 269–290. https://doi.org/10.1017/s0020743808080549.

Marchese, Aldo. 1938. *G.M. Giulietti*. Milano: Italica Editoriale.

Miran, Jonathan. 2014. Mapping Space and Mobility in the Red Sea Region, c.1500–1950. *History Compass* 12 (2): 197–216.

Moore-Harell, Alice. 2010. *Egypt's African Empire*. Brighton: Sussex Academic Press.

Munzinger, Werner. 1869. Narrative of a Journey Through the Afar Country. *Journal of the Royal Geographical Society of London* 39: 188–232.

Natili, Daniele. 2008. *Un Programma Coloniale*. In *La Società Geografica Italiana e le origini dell'espansione in Etiopia (1867–1884)*. Roma: Gangemi Editore.

Passamonti, Eugenio. 1937. *Dall'Eccidio di Beilul alla Questione di Raheita*. Roma: Vittoriano.

Pellegrineschi, Angelo Vittorio. 1932. *Giuseppe Sapeto e l'acquisto di Assab*. Roma: Istituto poligrafico dello Stato.

Rahem, Karim. 2001. Haral Mahis et les Afar, importance d'un mythe dans les positionnements sociaux et les stratégies de pouvoir. *Annales d'Ethiopie* 17: 151–174.

Ramm, Agatha. 1944. Great Britain and the Planting of Italian Power in the Red Sea, 1868–1885. *The English Historical Review* 59 (234): 211–236.

Robinson, Ronald, John Gallagher, and Alice Denny. 1961 original, 1981 2nd Edition. *Africa and the Victorians. The Official Mind of Imperialism*. London: The Macmillan Press.

Saint-Veran, Robert. 1977. *à Djibouti avec les Afar et les Issas*. Cagnes S/mer: Tholomier.

Sapeto, Giuseppe. 1865. *L'Italia e il Canale di Suez. Operetta Popolare*. Genova: Pellas.

———. 1879. *Assab e i Suoi Critici*. Genova: Pietro Pellas.

Schölch, Alexander. 1976. The 'Men on the Spot' and the English Occupation of Egypt in 1882. *The Historical Journal* 19 (3): 773–785.

Soule, Aramis Houmed. 1986. Un Example du Pouvoir Traditionnel Afar: le Sultanat de Tadjourah. *Le Pount* 17: 3–11.

Surdich, Francesco. 2005. *L'attività missionaria, politico-diplomatica e scientifica di Giuseppe Sapeto: dall'evangelizzazione dell'Abissinia all'acquisto della Baia di Assab*. Millesimo: Comunità Montana Alta Val Bormida.

———. 2017. SAPETO, Giuseppe. In *Dizionario Biografico degli Italiani*. Roma: Istituto della Enciclopedia Italiana.

Talhami, Ghada. 1979. *Suakim and Massawa under Egyptian Rule 1865–1885*. Washington: University Press of America.

Volterra, Alessandro. 2001. GIULIETTI, Giuseppe Maria. In *Dizionario Biografico degli Italiani*. Roma: Istituto della Enciclopedia Italiana.

CHAPTER 12

Credit, Debt and Power: Italian Foreign Policy in the Heavily Indebted Muslim Mediterranean Countries (1867–1914)

Giampaolo Conte

Introduction

The opening of the Suez Canal was the landmark event in the integration of the Mediterranean region into the capitalist world economy that had been underway since the mid-nineteenth century. The building of modern infrastructures throughout the eastern Mediterranean linked the entire region to the core capitalist economies of Northern Europe. The growth of foreign debts in Tunisia, Egypt and the Ottoman Empire from the mid-century onwards gave European creditors steadily more influence, opening the door to intervention by creditor countries which, with the aim of protecting the interests of private bondholders imposed and institutionalised foreign-controlled international financial commissions. Sovereign defaults also had political implications, for European powers did not pass

G. Conte (✉)
Università Roma Tre, Rome, Italy
e-mail: giampaolo.conte@uniroma3.it

B. Curli (ed.), *Italy and the Suez Canal, from the Mid-nineteenth Century to the Cold War*,
https://doi.org/10.1007/978-3-030-88255-6_12

up the chance to exploit them to advance their own political agendas. The Kingdom of Italy was among the players. The Tunisian (1867), Ottoman (1875–1876) and Egyptian (1876) defaults were viewed by the Italian authorities as a bridgehead for Italy's political and economic agenda in the eastern Mediterranean. The presence of Italian bondholders allowed Rome to claim a seat on several international financial commissions tasked with paying off foreign debt by managing local revenues.

The creation of financial debt commissions in Tunisia (1870), Egypt (1876) and the Ottoman Empire (1881) presented Italian foreign policymakers with new opportunities. Italy's relatively backward industrial capitalism did not justify a policy of overseas expansion based on investments and exports. Italian policy overseas, therefore, did not fit the definition of economic and financial imperialism. Nevertheless, successive Italian governments tried to make up for the country's economic and financial deficiencies with an aggressive policy aimed at securing overseas investment opportunities for Italian businessmen (and stimulating capitalists at home) for the sole purpose of buttressing Italy's political claims and justifying its expansionist territorial goals.

The Unexpected Creditor: How Italy Defined Itself in the Mediterranean Sovereign Defaults

The opening of the Suez Canal in 1869 brought the Mediterranean back to the forefront of foreign policy-making among the Great Powers including Italy,[1] though it occupied a second tier compared with its French and British competitors.[2] The mounting foreign debt of Tunisia, Egypt and the Ottoman Empire since the 1850s had allowed the main creditor countries, notably Great Britain and France, to progressively extend their influence there, in particular after 1869. The financial crisis of 1866 and the credit-crunch of the 1870s paved the way for several sovereign defaults. Local financial troubles were leveraged by the creditor countries to impose their indirect control in an area vital to international trade, especially after the opening of the Suez Canal.[3]

[1] Vimercati (1964), Biraghi (1869), Barzellotti (1869), Schonfield (1940) and Lo Giudice (1979).

[2] Chabod (1963).

[3] Cafarelli (2014).

Italy, struggling to establish and strengthen its Great Power status, found an opening to become a player in the aftermath of several Mediterranean defaults.[4] The sovereign defaults by Tunisia, the Ottoman Empire and Egypt paved the way for Italy to sit on the boards of ad hoc financial commissions. Italy's chief objective was to facilitate settlement of these sovereign debts, while also increasing its political influence from the inside.

The newly created Kingdom of Italy came to represent the diversified interests of its new bondholder subjects both at home and overseas, mainly in Paris but also in Constantinople, Cairo, Alexandria and Tunis. For instance, the annexation of Rome in 1870 allowed the Italian political elite to take advantage of ecclesiastical investments in Ottoman bonds.[5]

Unlike France, Britain and, later on, Germany, Italy was especially interested in exploiting bondholders' claims in defaulted counties in order to obtain a seat in prospective financial commissions, above all for political reasons. In fact, Italian political elites viewed these commissions as a springboard for promoting national political and economic interests in the Mediterranean. However, Italian governments did not pursue a straightforward diplomatic course of action to maximise their representation in these commissions; on the contrary, quite often Rome had no well-defined strategy. Italy's diplomatic and political leaders jacked a board, forward-looking perspective and were quite incapable of managing and exploiting these valuable bridgeheads in areas of direct political interest. Domestic issues, such as the troublesome Roman question, remained the central focus of Italian foreign policy until the mid-1870s; there were too many unresolved issues on the table for a regional power such as the new Italian state.[6]

The Mediterranean question, so vital for the future development and positioning of the Kingdom, acquired importance only when the Historical Left (*Sinistra Storica*) took power in 1876.[7] Under its leaders, from Agostino Depretis to Francesco Crispi, Italian foreign policy increasingly embraced the principles of 'power politics' and was more active in diplomatic bargaining on the future settlement of the Mediterranean. Italian decision-makers increasingly used financial commissions to promote not

[4] Bosworth (2005).
[5] Conte (2020).
[6] Chabod (1963).
[7] Calchi Novati (1990).

only bondholders' claims but broader Italian political and diplomatic ambitions as well.

Italian creditors and bondholders were very heterogeneous. Italian-Levantine communities in the main port cities of the Mediterranean, such as Constantinople, Smyrna, Beirut, Alexandria and Tunis, held bonds and treasury securities issued in domestic and foreign markets. Primarily merchants, bankers and capitalists, these bondholders had different interests to defend, which did not necessarily accord with the political strategies pursued by the Italian authorities. In fact, the latter exploited the Italian bondholders' claims to promote Italian political influence in the defaulted countries through participation in the above-mentioned financial commissions, with a view to strengthening Italy's case for recognition as an emerging Great Power in the Mediterranean.

Italian Actions and Foreign Policy in Relation to the Tunisian Debt Controversy

The opening of the Suez Canal and the growth of European financial and investment capital in Tunisia, Egypt and the Ottoman Empire were expressions of the gap in technological and financial know-how that benefitted the main European economic powers.[8] Generally speaking, the sovereign debt crisis in the Mediterranean highlighted a new cross-fertilisation of market relations between Europe and the Middle East. The Suez Canal brought the Mediterranean back to the forefront of international trade and economic relations. Then, the 1873 financial crisis and the ensuring depression hastened the imperialistic struggle worldwide.[9] According to Benedetto Croce, 'political and commercial expansionism is enhanced within the consciousness of European civilisation. Her powerful science and technique, her law and duty enabled her to act as an example for other people to ascend gradually to the same form of civilisation.'[10]

Tunisia's sovereign default in 1867 was the first time since mid-century that a modern Muslim Mediterranean country failed to honour its sovereign debt towards foreign creditors.

The history of Tunisian foreign debt was rooted in the reformist era under way since the 1830s. In particular, the Bey's abiding desire to create

[8] İhsanoğlu (2004) and Ágoston (2005).
[9] O'Rourke and Williamson (2001).
[10] Croce (1993).

a modern army drew him into an unceasing search for new domestic funding sources. Several problems afflicted the Regency's financial stability: the accumulation of fluctuating debt carrying a high interest rate; the predatory rates on the international loans negotiated in 1863 and 1865 through French banks; the mortgaging of the Regency's principal sources of revenue in order to meet debt payments; the financial uncertainty that came in the wake of the crisis of 1866; and outbreaks of cholera as well as drought and crop failures.[11]

The struggle among Tunisia's creditors quickly escalated into a political battle. Diplomatic action towards the Bey enabled the French government to regain a position of primacy at the bargaining table for debt restructuring.[12] This prompted an Italian countermove to avert the risk that France might use unpaid interest due to Tunisia's French creditors as a pretext for military intervention. In fact, the Tunisian question suffered from the ill-conceived Italian notion that France's overwhelming dominion over Tunisian foreign debt would not be transformed into outright political dominion in the years to come.[13]

The French government, through the Marquis de Moustier, required the Bey to establish a financial commission to pay bonds, many of which were in the hands of French creditors.[14] The creation of the commission was unacceptable to the Italian general consul, Luigi Pinna, and to the entire Italian government, which was well aware that a French economic domination expressed by unequal standing among foreign creditors, would also entail stringent political control.[15]

The Italians pursued a dual political and diplomatic strategy: on the one hand, they tried to forge a diplomatic entente with Britain in order to resist the French diplomatic siege of the Tunisian sovereign authorities; on the other hand, they supported the claims of Italian bondholders, especially those living in Tunisia, notably the wealthy Italian-Jewish community that came primarily from Genoa and Livorno, for example the powerful Moreno family.[16] The Italian political elite brought Italian

[11] Conte and Sabatini (2018).

[12] Emerit (1949) and Ganiage (1959).

[13] Shorrock (1976) and Oualdi (2020).

[14] The Italian Ambassador in Paris Costantino Nigra to the Italian Minister of Foreign Affairs Luigi Federico Menabrea, 17 April 1868, ASDMAE, AIL, Paris, box 20, no. 31.

[15] Menabrea to Taparelli d'Azeglio, 29 April 1868, ASDMAE, AIL, Turin, box 20 no. 33.

[16] This is the case of the powerful Moreno family in Tunis. Fiorentino to Moreno, 14 June 1869, ASL, CM, Tunis, box 2.

bondholders together in order to strengthen its opposition to French hegemonic aims.[17] According to Pinna, 'Italy will not accept any project which would put Italy in a second-rank position to any other European power, and in no way and in no case could Italy consent to a situation in which rights [in respect of Tunisian revenues] already obtained by Italians would be placed in doubt or negotiation'.[18]

In order to ward off the threat of full-fledged French dominion in Tunisia, the Italian authorities backed the interests of Italian bondholders from Tunis, who held approximately 25% of Tunisia's public debt.[19] Overall, Italian foreign policy-making sought to hamper French economic and financial semi-colonisation of the North African country. French diplomatic action, along with bankers' requests, was aimed at mortgaging local revenues in order to obtain guaranteed resources for debt payments as well as to strengthen France's sway over Tunisian decision-making.

The Italian government could not ensure the same inflow to Tunisia of loans and investments. The powerful group of French financiers led by the Comptoir National d'Escompte of Paris, working through the well-known bankers Pinard and Erlanger, forced the Tunisian government to give way to French requests for financial control.[20] The Italian government's ambitions in that matter, supported by local Italian merchant-bankers, were not realistically powerful enough to impose an independent course of action.

The Italian-French struggle over Tunisian finances involved a conflict between two Mediterranean powers interested in subduing local authority by using sovereign debt as an instrument of political leverage.[21] Thus, Italian authorities used the large Italian stake in the domestic debt to claim a political role in Tunisian affairs. However, when an international financial commission was created French political interests alone were heavily represented. The commission enabled the Italians to indirectly control the financial restructuring of local public finances. In fact, the Tunisian

[17] D'Agostino Orsini (1939).

[18] Menabrea to the Council of Ministers, 4 June 1869, ASDMAE, MAERI, Florence, box 232.

[19] Of an overall debt of 160 million francs, Italian local bondholders held fully 40 million.

[20] Stoskopf (1998).

[21] Broughall (1942) and Russell (1977).

financial commission was the only one that had representatives of both diplomatic and private financial interests on its board.[22]

When the Historical Left came to power in 1876, Italian foreign policy in the Mediterranean became more aggressive, especially under Francesco Crispi's governments. The handling of the Egyptian and Ottoman defaults on foreign debt provides ample evidence of this.[23]

Antonio Scialoja's Mission and the Unsuccessful Italian Action Following the Egyptian Default

Rooted in the building of the Suez Canal, the Egyptian default in 1875 is one of the best-known examples of a nineteenth-century debt crisis.[24] The unexpected failure of Lesseps' company's securities market placement forced Egypt to start accumulating a substantial floating debt in order to support the construction of a public works project by a foreign private company.[25] Given the strict law prohibiting independent issuing of government securities without Ottoman permission, Egypt issued Treasury securities as a way to circumvent the problem. With the issue of the first international loan in 1862, the Egyptian public debt followed an ascending trend until the sovereign default of 1875.[26] According to the private Egyptian council, Egypt faced financial turmoil as 'the deficit left from the previous government, the huge amount owed to the Suez company as well as the interests that overwhelmed the state's capability of being solvent forced the central government to seek credit in Europe'.[27]

Furthermore, the fall in cotton prices after the American Civil War, the increase in unproductive expenses and the state's inability to increase tax revenues sufficiently exacerbated the situation of government finances.

[22] The Tunisian financial commission was composed of two committees: an executive committee comprising two Tunisians and a French inspector of the Treasury, and a control committee of foreign creditors and consisting of two representatives of French creditors of loans made in 1863 and 1865, as well as two Italian and two British representatives of the holders of converted domestic debt; Bouchard (1951).

[23] Salvemini (1919), Maturi (1946) and Romano (1986).

[24] Fletcher (1958), Kalkas (1979), Saul (1997) and Wallerstein (2011).

[25] Penfield (1895) and Owen (1981).

[26] Landes (1959).

[27] Scialoja (1875).

The default led to direct British and French interference in Egyptian political and economic affairs.[28]

The rise of British influence in Egypt stemmed from London's determination to control and defend access to the Red Sea, the gateway to India.[29] As in Tunisia 15 years earlier, Egypt's financial troubles paved the way for foreign creditors and European states to intervene in order to compel loan repayments while enlarging their political sphere of influence.

As in Tunisia, Italian private interests in Egyptian foreign and domestic debt were held by a large number of small- and mid-sized bondholders and not directly by leading Italian national banks. No Italian bank intervened as an intermediary for the placement of Egyptian securities. The Italian bondholders came primarily from the Italian community in Egypt: merchants, financers and bankers that had been operating in Cairo and Alexandria for decades.[30]

Treasury bonds were sold on the domestic market and often acquired by middle-class and wealthy investors. Many of these had merchant and banking activities in Egypt and were quite confident in using this type of financial instrument and, of course, in the investment potential offered by public securities.[31] Greek, Italian and Jewish communities in Egypt were heavily involved in the business and represented an important stock of available wealth for the heavily indebted Egyptian state.[32]

Unlike France and Britain—whose financial interests were represented mainly by the larger intermediary banks and domestically based bondholders—Italy was interested more in maintaining its nationals' influence in Egypt's domestic economy, in such spheres as retail trade and small business, than in financial speculation.[33] Given the large Italian community based in Egypt, the Italian government had more interests to

[28] Owen (1969).

[29] Rose et al. (1940), Tignor (1966), Vatikiotis (1980) and Owen (2004).

[30] *Sulla politica italiana in Egitto; interrogazione del deputato Vollaro al Ministro degli affari esteri svolta alla Camera dei deputati nella tornata del 12 giugno 1882*, Rome: Tipografia eredi Botta.

[31] Since the end of Muhammad Ali's reign, the Egyptian economy had increasingly been dependent on cotton exports. International market fluctuations often put pressure on the state's ability to cover expenses, especially after the conclusion of the Civil War in the United States caused cotton prices to drop sharply; Beckert (2015).

[32] Kitroeff (1989), Petricioli (2007) and Lanver (2012).

[33] This was true until Banco di Roma opened a branch in Alexandria. Besides operating as a commercial bank, it was involved in speculation on grain prices; De Rosa (1982).

safeguard in Egyptian territory.[34] A change in the status quo meant tampering with the Italian position in Egypt. Consequently, at the beginning, the new British influence in Egypt forced Italy to seek an alliance with the French as a counterweight to London's predominance.[35]

Antonio Scialoja's mission to Egypt was pivotal in mediating between the various diplomatic positions in terms of financial restructuring, and the initiative first met with approval from the Khedive, as well as from London and Paris. An economist, politician and former minister of finance, Antonio Scialoja was a prominent figure in Italian and European political and economic circles. The goal of his mission was to reorganise Egyptian finance in a way that would meet the needs of French investors, thereby avoiding a unilateral intervention by Britain interested in controlling Egypt and the Suez Canal for strategic imperial purposes.[36] Scialoja's appointment as temporary President of the Egyptian Treasury Council enabled him to draw up an administrative and financial plan for the rehabilitation of Egypt's finances.[37] His objectives were technocratic and did not extend to finding a diplomatic solution between the states involved.

The implementation of the plan, accepted by the Khedive, was undermined by British diplomacy, which worked to thwart what it feared was growing Italian influence within the Egyptian government.[38] At the same time, however, the Italian government tried to exploit Scialoja's position within the Egyptian administration to hamper the growth of British influence at the Khedive's court.[39]

However, Scialoja's subsequent resignation from that office evinced a lack of trust in his actions not only on the part of France and Britain but also by the Khedive himself, who proved to be more interested in maintaining good relations with London and Paris after they withdrew their support for Scialoja's project.[40] Both France and Britain opposed a

[34] Petricioli (2007).

[35] Melegari to the Italian Embassy in Paris, 6 August 1876, ASFLEO, CAS, Rome, box 31, n. 141.

[36] 'L'Egitto e la missione del senatore Scialoja', *Il Corriere della Sera*, 9 October 1878.

[37] According to former Italian Ambassador in France Costantino Nigra, it was very difficult to devise a plan without direct French and British involvement. Nigra to Scialoja, 23 February 1876, ASFLEO, CAS, Paris, box 30.

[38] Società Generale di Credito Mobiliare Italiano to Scialoja, 5 October 1876, ASFLEO, CAS, Florence, box 31, no. 185.

[39] Ministry of Finance to Scialoja, October 1876, ASFLEO, CAS, Rome, box 31, no. 22 (encrypted telegram).

[40] De Cesare (1879).

well-structured and articulated plan, pushing instead for a quick solution that would accommodate their economic and political interests.

According to Luigi Melegari, then Italian Minister of Foreign Affairs, Scialoja's resignation was a consequence of an obviously weak Italian position vis-à-vis France and Great Britain. Moreover, it revealed the inability of Italian diplomacy to mount an adequate defence of interests of the Italian community in Egypt.[41] According to Melegari, Scialoja's resignation 'was only a consequence of a troublesome situation in which any stronger Italian intervention would have been useless. We [the Italian government] had many intentions to sustain your [Scialoja's] action, although the type of mission you had in Egypt as well as the importance of our capitalists' interests within Egyptian finances limited our means. [When] finances become entangled with foreign policy, rarely can we obtain satisfactory results.'[42]

Although the Italian community was the second-largest foreign community in Egypt,[43] the Italian bondholders' interests in the floating debt and in many local business activities were insufficient to impose the course of action taken with regard to defaulted Egypt.[44] That was simply a matter of raw power between the European states. The presence of an Italian representative at the Caisse de la Dette Publique in Egypt (1876), with the aim of paying off the country's outstanding debt, was only a British courtesy to the Italian bondholders.[45] London's overriding goal was to maintain the *equilibrium* among all Great Powers' interests in Egyptian financial affairs while retaining a dominant role in the country's politics, as the French did in Tunisia.[46] As a result, Italy was excluded from the future adjustment of Egypt's public finances and no official position was assigned to an Italian government representative from 1877 onwards.

[41] The Italian community lodged complaints against the general consul De Martino. Melegari to Scialoja, 28 September 1876, ASFLEO, CAS, Rome, box 31, no. 180.

[42] 'L'Egitto e la missione del senatore Scialoja', *Il Corriere della Sera*, 9 October 1878, cit.

[43] Scialoja to Melegari, 15 May 1877, ASFLEO, CAS, Cairo, box 30. See also: 'Modificazione della giurisdizione esercitata dai Consoli Italiani in Egitto, Relazione della Commissione della Camera dei Deputati', *Atti Parlamentari*, n. 88-a, 20 March 1875.

[44] 'L'Italia in Egitto', *La Stampa*, 3 October 1879.

[45] Kaufmann and Babled (1896).

[46] Hunter (1999).

Ottoman Default and Italian Policy Within the International Financial Commission

Italy followed a similar modus operandi with regard to the Ottoman default of 1875–1876. Unable to pay the interest on its foreign debt, the Ottoman Empire became involved in lengthy negotiations with the Great Powers on the terms of debt restructuring.[47] As had been the case in Egypt, given Italy's interests in the Ottoman foreign debt, an Italian representative was allowed to take part in the bargaining on the debt settlement. In contrast with what occurred in Tunisia and Egypt, Italian bondholders were less involved in the floating debt and held approximately 5% of the overall foreign debt.[48] This share permitted them, and consequently the Italian government, to become involved in the Ottoman Public Debt Administration (OPDA).[49]

Italian attendance in the OPDA was pivotal to defending and promoting Italian foreign policy in the Ottoman Empire as well as the bondholders' financial interests. Each European nation represented in the OPDA had direct political and economic interests to safeguard in Ottoman territory. Business and financial interests were the *longa manus* of European governments interested in wielding exclusive control over a well-defined territory. If and when the Empire collapsed, that exclusive 'concession zone' would be ready for exploitation by the European powers.[50] Unlike Tunisia and Egypt, which underwent French and British military occupation between 1881 and 1882, the Ottoman Empire did not suffer complete military occupation by one of the Great Powers until the Great War. This gave Italian foreign policy-makers more scope to play a significant role in the Ottoman Empire by planning long-term diplomatic action.

The Italian authorities' readiness to take full advantage of the OPDA's enormous economic and financial potential was mainly geared to securing political benefits in the Empire and beyond. Italian governments were eager to use the OPDA as a political tool to obtain further compensation,

[47] Anderson (1964), Clay (2000), Eldem (2005), Brida (2010), Conte and Sabatini (2014), Tunçer (2015) and Pamuk (2018).

[48] Blaisdell (1929), Pamuk (1987), Geyikdagi (2011) and Conte (2018a).

[49] This percentage did not account for the entirety of the Italian investments in Ottoman securities; it did not include the many Italian bondholders who used Paris and Constantinople to make their investments; Conte (2018a).

[50] Ergil and Rhodes (1975), Wallerstein (1979) and Keyder (1987).

mainly in relation to French and Austrian expansionism in the Mediterranean and in the Balkans.

The endemic scarcity of surplus capital in Italy's domestic market, as well as the parochialism of some Italian capitalists, prevented Italian private enterprises from taking part in the partitioning of the Ottoman market together with the leading European industrial and financial groups. The Italian authorities therefore had no grounds for demanding wider involvement in Ottoman political and economic affairs. Italy remained on the sidelines of the most profitable business dealings in the Empire, at least until the first industrial capitalist-driven growth after the 1890s. However, the Italian governmental elites were able to exploit Italy's position in the OPDA. The Italian OPDA delegate, formally under the direction of the Chamber of Commerce of Rome but virtually under the supervision of the Italian Foreign Minister, often negotiated his vote in the Council in exchange for some political compensation requested by Italy's political and diplomatic authorities. For instance, the Italian opposition to the Imperial Ottoman Bank in the 1880s was not based on financial considerations; rather, it was rooted in the Italian-French struggle following the Tunisia takeover. For Italian foreign policy, the protection of Italian bondholders was a secondary objective: Italian policy-makers were far more interested in using Italian investors as a wedge for advancing the country's political interests in the Empire and in Europe.[51]

However, lack of investment capital did not mean lack of influence. Italy was able to exercise some influence over international financial commissions thanks to private bondholders and financial experts appointed by these commissions. The role of Antonio Scialoja in Egypt has already been mentioned. In the Ottoman Empire, the first Italian delegate to the OPDA, Francesco Mancardi, who had been director of general debt in Italy from 1862 to 1871, contributed to the reorganisation of Ottoman public finances and debt-accounting together with the Imperial Ottoman Bank. In addition, he proposed that the OPDA should guarantee the issuing of international loans.[52]

In the second half of the nineteenth century, Tunisia, Egypt and the Ottoman Empire benefitted from the availability of European capital in search of profits. However, the lack of investors' confidence in local public debt solvency pushed up interest rates. At the same time, the outflows of profit-seeking capital, facilitated by a fixed-parity monetary system (the gold

[51] Conte (2018b).
[52] Mancardi (1882).

standard), allowed foreign banks to achieve reasonable profit margins in the Near East. But, foreign currency-denominated loans put pressure on local economies, especially after the deflationary crisis from 1873 onwards. The establishment of international financial commissions, which consistently reduced local sovereignty, helped indebted countries to obtain lower interest rates and bolstered foreign confidence in the sustainability of their finances.[53]

The glaring political and economic gap between Rome on one side and London and Paris on the other did not crimp Italy's ambition to claim political compensation, given its direct interests in the foreign, domestic and floating debts of Tunisia, Egypt and the Ottoman Empire. The Italian authorities exploited sovereign debt difficulties as a further opportunity to increase Italy's influence in the Mediterranean.

Conclusion

Italian investors' interests in the foreign, domestic and floating debt of Tunisia, Egypt and the Ottoman Empire were exploited as a possible bridgehead to enable Italy to participate in the economic, financial and political affairs of those bankrupt states. Unlike France and Britain, Italy lacked the economic and financial resources to impose itself in these countries without previously dealing with the other Great Powers. Although Italy was a member of the club of Great Powers, in practical terms it lacked the political and military clout to operate as such. According to Emilio Visconti-Venosta, Italy favoured a policy of maintaining the status quo, even after the opening of the Suez Canal.[54] Thus, especially after Tunisian and Egyptian defaults, Rome tried its best to restore the conditions that had prevailed before European direct interventions by supporting local governments' sovereign authority. When this was not possible, as in the case of Tunisia and Egypt, Italian diplomacy tried to exploit Italy's status as a Great Power in order to gain a seat at the bargaining table on debt settlement and to avert military occupation, which would have definitely excluded Italian influence. In the Ottoman Empire, which was not under occupation, Italy had greater scope to perform its role as a Great Power until the advent of the Great War. Keeping local authorities in charge meant greater bargaining power over the decision-making process of the indebted state. When that position was impossible to sustain, Rome was willing to alter its diplomatic action only if it could gain political and economic compensation, a course it actively pursued in the Ottoman Empire.

[53] Pamuk (2008).
[54] Perticone (1961).

Italian bondholders enabled the Italian government to obtain access to international financial commissions in order to fulfil its foreign policy ambitions in the area. Moreover, Italian political and diplomatic leaders missed no opportunity to rely on Italian emigrant communities and notables, such as bankers, merchants and entrepreneurs, to increase Italy's bargaining power vis-à-vis indebted states and the Great Powers. Although Italian authorities were quick to use them in financial and debt settlements, these non-state actors never really managed, however, to allow Italy to compete on an equal footing with France and Britain.

References

Primary Sources

AIL, Ambasciata d'Italia a Londra
ASDMAE, Archivio Storico Diplomatico Ministero Affari Esteri
ASFLEO, Archivio Storico Fondazione Luigi Einaudi Onlus (Rome, Italy)
ASL, Archivio di Stato di Livorno
CAS, Carte Antonio Scialoja
CM, Carte Moreno
MAERI, Ministero Affari Esteri Regno d'Italia 1861–1887

Newspapers

Il Corriere della Sera (Milan, Italy)
La Stampa (Turin, Italy)

Bibliography

Barzellotti, Pier Luigi. 1869. *La questione commerciale d'Oriente, l'Italia e il canale di Suez: cenni storici e considerazioni*. Florence: Tipografia Eredi Botta.

Beckert, Sven. 2015. *Empire of Cotton: A New History of Global Capitalism*. London: Penguin.

Biraghi, Emilio. 1869. *Il canale di Suez, le vie e gli interessi commerciali d'Italia*. Florence: Stabilimento G. Civelli.

Blaisdell, Donald. 1929. *European Financial Control in the Ottoman Empire. A Study of the Establishment, Activities, and Significance of the Administration of the Ottoman Public Debt*. New York: Columbia University Press.

Bosworth, J. B. Richard. 2005. *Italy The Least of the Great Powers: Italian Foreign Policy Before the First World War*. Cambridge: Cambridge University Press.

Bouchard, Edwin. 1951. *State Insolvency and Foreign Bondholders: General principles*. New Haven: Yale University Press.

Broughall, Metcalf H. 1942. The Problem of Tunisia in Franco-Italian Relations, 1835–1938. PhD dissertation, Maryland.

Cafarelli, Andrea. 2014. *Il Leone ferito. Venezia, l'Adriatico e la navigazione sussidiata per le Indie e l'Estremo Oriente (1866–1914)*. Rome: Viella.

Chabod, Federico. 1963. *Storia della politica estera italiana 1870–1897*. Bari: Laterza.

Conte, Giampaolo. 2018a. *Il Tesoro del Sultano. L'Italia, le grandi potenze e le finanze ottomane (1881–1914)*. L'Aquila: Textus Edizioni.

———. 2018b. L'Italia, Francesco Mancardi e le finanze ottomane negli anni Ottanta dell'800. *Storia Economica* 20 (1): 271–287.

———. 2020. Unholy Alliances: Disentangling the Economic Relations between Italy, the Holy See and the Ottoman Empire. *The International History Review* 2: 1–18.

Conte, Giampaolo, and Gaetano Sabatini. 2018. Debt and Imperialism in Pre-Protectorate Tunisia, 1867–1870. A Political and Economic Analysis. *The Journal of European Economic History* 1: 9–32.

Croce, Benedetto. 1993. *Storia d'Europa nel secolo decimonono*. Milan: Adelphi Edizioni.

De Cesare, Carlo. 1879. *Vita, i tempi e le opere di Antonio Scialoja*. Rome: Tipografia del Senato.

De Rosa, Luigi. 1982. *Storia del Banco di Roma*. Vol. I–III. Rome: Banco di Roma.

Emerit, M. 1949. Les crises des finances tunisiennes et les origines du protectorat. *Revue africaine* 3: 247–276.

Ergil, Dogu, and Robert Rhodes. 1975. Western Capitalism and the Disintegration of the Ottoman Empire. The Impact of the World Capitalist System on Ottoman Society. *Economy and History* 18: 41–60.

Ganiage, J. 1959. *Les origines du protectorat française en Tunisie (1861–1881)*. Paris: Press Universitaires de France.

Geyikdagi, Necla. 2011. *Foreign Investment in the Ottoman Empire, 1854–1914*. London: I.B. Tauris.

Kalkas, Barbara. 1979. Diverted Institutions: a reinterpretation of the process of industrialization in Nineteenth-Century Egypt. *Arab Studies Quarterly* 1: 28–48.

Keyder, Caglar. 1987. *State and Class in Turkey. A Study in Capitalist Development*. London: Verso.

Kitroeff, Alexander. 1989. *The Greeks in Egypt, 1919–1937: Ethnicity and Class*. Reading: Ithaca Press.

Landes, S. David. 1959. *Bankers and Pashas: International Finance and Economic Imperialism in Egypt*. Cambridge, MA: Harvard University Press.

Lanver, Mak. 2012. *British in Egypt: Community, Crime and Crises 1882–1922*. London: I.B. Tauris.

Lo Giudice, Giuseppe. 1979. *Trieste, l'Austria ed il canale di Suez*. Catania: Istituto di Storia Economica.

O'Rourke, H. Kevin, and G. Jeffrey Williamson. 2001. *Globalization and History: The Evolution of a Nineteenth-Century Atlantic Economy*. Boston, MA: MIT Press.

Orsini, Paolo d'Agostino. 1939. *Francia contro Italia in Africa: da Tunisi a Suez*. Milan: La Prora.

Oualdi, M'hamed. 2020. *A Slave Between Empires. A Transnational History of North Africa*. New York: Columbia University Press.

Owen, Roger. 1969. *Cotton and the Egyptian Economy, 1820–1914: A Study in Trade and Development*. Oxford: Clarendon Press.

———. 1981. *The Middle East in the World Economy, 1800–1914*. London: I.B. Tauris.

Pamuk, Sevket. 1987. *The Ottoman Empire and European Capitalism, 1820–1913*. Cambridge: Cambridge University Press.

———. 2008. From Bimetallism to the 'Limping Gold Standard': The Ottoman Monetary System in Nineteenth Century. In *East Meets West: Banking, Commerce and Investment in the Ottoman Empire*, ed. P.L. Cottrell, I.L. Fraser, and M.P. Fraser. Burlington: Ashgate.

———. 2018. *Uneven Centuries: Economic Development of Turkey Since 1820*. Princeton: Princeton University Press.

Penfield, C. Frederic. 1895. Contemporary Egypt. *The North America Review* 161: 13–24.

Perticone, Giacomo. 1961. *La Politica Estera Italiana dal 1861 al 1914*. Turin: Einaudi.

Petricioli, Marta. 2007. *Oltre il mito. L'Egitto degli Italiani (1917–1947)*. Milan: Mondadori.

Romano, Sergio. 1986. *Francesco Crispi*. Milan: Bompiani.

Russell, J. Alberti. 1977. *The Italian Community in Tunisia, 1861–1961: A Viable Minority*. New York: Columbia University Press.

Salvemini, Gaetano. 1919. *La politica estera di Francesco Crispi*. Rome: La Voce Edizioni.

Schonfield, H. Joseph. 1940. *Italy and Suez*. London: Hutchinson.

Scialoja, Antonio. 1875. *Legge sulla Mukabala*. Alexandria: Stamperia Francese A. Mourès.

Shorrock, William. 1976. *French Imperialism in the Middle East: The Failure of French Policy in Syria and Lebanon 1900–1914*. Madison: University of Wisconsin Press.

Stoskopf, Nicolas. 1998. Alphonse Pinard et la révolution bancaire du Second Empire. *Histoire, Économie et Société* 17 (2): 299–317.

Vimercati, Cesare. 1964. *Il canale dell'istmo di Suez sua influenza di pace per l'Europa e di rigenerazione per l'Italia*. Livorno.

Wallerstein, Immanuel. 1979. The Ottoman Empire and the Capitalist World-Economy: Some Questions for Research. *Review (Fernand Braudel Center)* 2: 389–398.

CHAPTER 13

Sneaky Neutrality and the Mediterranean Space: Weapons Smuggling in Egypt During the Italo-Ottoman War (1911–12)

Andreas Guidi

In February 1912, in the midst of the Italo-Ottoman War, the Italian consul in Port Said Romolo Tritonj reported suspect ships in the Mediterranean to Italy's diplomatic agent in Cairo. He mentioned the *Bodoss*, a 'Turkish steamboat' which sailed under a Greek flag, was rented by Greeks, and was led by a Greek ship captain. It moved between Hodeida and Smyrna and had transported a Turkish officer from Yemen during its last passage at Suez, an action considered illegal for neutral ships in wartime. Only one day later, the consul added that the *Bodoss* had returned to Port Said due to an alleged breakdown. In fact, he thought that the *Bodoss* had sneaked back into neutral Egyptian waters to escape an Italian battleship hunting for war contraband. The *Bodoss* unexpectedly remained in Port Said for days, raising the consul's suspicions. The alleged owner was

A. Guidi (✉)
University of Konstanz, Konstanz, Germany
e-mail: andreas.guidi@uni-konstanz.de

B. Curli (ed.), *Italy and the Suez Canal, from the Mid-nineteenth Century to the Cold War*,
https://doi.org/10.1007/978-3-030-88255-6_13

called Sokrat, a man from Istanbul who had been helping the Ottoman government smuggle arms to Derna, east of Benghazi, since the outbreak of the conflict in October 1911.[1] The 'supplier' and 'agent' of the *Bodoss*, a ship that was not recorded in any Greek port, was a certain Xidia, a 'notorious smuggler' who dealt in ships rented by the Ottoman government.[2] The Greek consul in Port Said attested the breakdown upon request by his Italian counterpart, who speculated that the steamship would change its route toward Piraeus, whence it would smuggle another load of weapons to Derna.[3]

This is a fragment of the intense correspondence on contraband between Italian consuls based in Egypt and the ministries in Rome during the Italo-Ottoman War. Other reports reveal smuggling connections including Greece, Tunisia, France, Lebanon, Anatolia, the Red and the Black Sea, and even the North Sea and the Atlantic. Contraband, as well as its interception, involved state and private actors such as Italian, Ottoman, British, and Egyptian governmental authorities, soldiers, consuls, ship captains, merchants, various brokers and informants, coastguards, sailors, and Bedouins. Through their interactions, smuggling fueled war just as war fueled smuggling. The result was a densification of preexisting Mediterranean connections gravitating around Egypt and the Suez Canal, as well as intricate discussions about the territorial status of those areas. These discussions anticipated Britain's establishment of the so-called veiled protectorate over Egypt two years after the Italo-Ottoman war, which changed the country's juridical configuration. Yet, circuits of illicit trade made visible by the war would persist at least until the 1920s, since the Italian occupation of Libya could not eradicate contraband flows from Egypt.[4]

The Italo-Ottoman War of 1911–12 is the focus of this chapter.[5] Italy invaded the provinces of Tripoli and Benghazi, ruled respectively as a *vilayet* and *mutasarrıflık* and later occupied some Aegean islands, including Rhodes. Seen from Rome, this was a colonial war. A rivalry with other Great Powers marked by an inferiority complex coupled with the perception of southern Italian mass emigration as a demographic problem

[1] Tritonj to Grimani, February 16, 17, and 24, 1912, ASDMAE AAPP 1891–1916, 140.

[2] Tritonj to Sangiuliano, March 27, 1912, ASDMAE AAPP 1891–1916, 141.

[3] Tritonj to Grimani, February 24, 1912, ASDMAE AAPP 1891–1916, 140.

[4] Revel to Viale, February 1, 1915, USMAR, 408; Folder 1923–4, USMAR, 1686 7.

[5] Other denominations are Libya War (*Guerra di Libia*), Italo-Turkish War, and Tripoli War *(Trablusgarp Savaşı)*.

synthetized Italy's imperialist and nationalist ambitions. For many Italian observers ranging from socialists to conservatives, the 'adventure' on the southern shore of the Mediterranean would establish Italy's rights as a leading European nation which, in the early twentieth century, implied keeping the pace with overseas empires. Seen from Istanbul, this was a test for the legitimacy of the Young Turks, a composite movement gathered in the name of opposition to Sultan Abdülhamid II. In 1908, a military putsch under their guidance reintroduced the Ottoman Constitution and led this formerly underground opposition to power. Less than one year later, the Young Turks survived a conservative countercoup and forced Abdülhamid to leave the throne. Gaining the political upper hand against the new Sultan Mehmed V, the Young Turks accelerated an imperial nation-building through increasingly authoritarian and militarist policies. The Ottoman patriotism they envisaged would result from unity and fraternity among the ethnic and confessional elements of the Empire, provided that Muslim and Turkish primacy remain unquestioned. The *vilayet* of Tripolitania was, like other territories such as Yemen and Albania, situated at the cultural and geographical margins of this project. Yet, after losing Bosnia through Habsburg annexation and Bulgaria through secession in the immediate aftermath of the 1908 revolution, the Ottoman government feared that a European aggression in their last African possession would cause a chain reaction. Contrary to the much wealthier Balkan provinces, Libya was a predominantly Muslim territory and its loss would impact the empire's legitimacy in the Arab provinces.

As Jonathan McCollum has argued, the Italo-Ottoman War engaged two powers experimenting with the concepts of empire and nation.[6] This invites us to think of the conflict as *Beziehungsgeschichte*, in relational terms beyond the historiographic boundaries of Italian colonialism and of the Ottoman Empire's twilight. Ottoman and Italian ministries produced documents on the same actors and events without communicating directly. Cross-reading these records can enhance relational history toward embedding Italy and the Ottoman Empire in a larger connected space, investigating the war as a Mediterranean phenomenon. It is acknowledged that Italian war operations in the Eastern Mediterranean, far from Libya, decisively turned the tide in favor of Rome.[7] Naval operations occurred at Preveza, on the already mentioned Aegean islands, at Beirut, and at the

[6] McCollum (2015).

[7] Vandervort (2018, 23). For an overview of the war, see Labanca (2012).

Dardanelles. Beyond Italian and Ottoman visions of nation and empire, the conflict redefined connections in a maritime space whose actors escaped rigid national or imperial categories.

The historiography of the conflict has focused on war operations and only superficially addressed the issue of contraband. Although, at sea, 'only a few boats managed to slip through', successful land smuggling through Egypt had 'great practical importance and boost[ed] the morale' of anti-Italian fighters in Libya.[8] Not only guns, but also prominent officers reached the battlefield through the Egyptian route. Eşref Kuşçubaşı, one of the leaders of the Ottoman 'special units' (*Teşkilat-ı mahsusa*), left Beirut with some companions and made his way from Port Said to Cyrenaica through 'shipping companies, unscrupulous agents, fly-blown quarantine cordons […] and arms smugglers'.[9] Although smuggling had little impact on fighting, the human and infrastructural web behind it is an example of 'Mediterranean circuits' revealing the mobility and economy of this region.[10] Egypt hosted dense exchanges and was central to frame smuggling in juridical terms. Away from the battlefields, a separate game revolved around British-Egyptian neutrality, Italian extraterritoriality, and Ottoman sovereignty. Studies on smuggling highlight the potential of this phenomenon in terms of spatial history, since the 'many worlds' around smugglers open new perspectives on territorialization beyond a focus on state actors.[11] Addressing war contraband in Egypt and the Suez Canal renders a sharper image of the environment in which the conflict took place and the 'sophisticated use of geography' by Mediterranean networks.[12]

Borderlands and Seaways

Italy's invasion began at Tripoli on October 5, 1911. Within a few days, both belligerents issued a declaration on goods considered as war contraband. A decree signed by King Victor Emmanuel III conformed to the London Declaration on War Contraband of 1909.[13] On the Ottoman side, the list of prohibited objects was published in the press and contained, among others, arms, projectiles, explosives, military clothing and

[8] Simon (1987, 39; 115). See also Childs (1990, 102).
[9] Fortna (2016, 66).
[10] Carminati (2017, 130).
[11] Schayegh (2011).
[12] Tagliacozzo (2005, 298).
[13] Royal Decree, October 13, 1911, ASDMAE AAPP 1891–1916, 141.

equipment, gold and silver, banknotes and checks.[14] Based on the 1909 protocol's Articles XXII and XXIV defining 'absolute' (i.e. for direct use in battle) and 'conditional' war contraband, the Italian diplomats acknowledged that the 'Ottoman measures absolutely conform to our own'.[15] The Ottoman Empire had not sent any representative to the London conference, and this declaration can be read as the intention to align with international warfare codes established by European powers.

Already in late October 1911, Egypt emerged as the main gateway to Cyrenaica for these prohibited goods. The Italian diplomatic agent in Cairo reported conversations with the Egyptian Prince (Khedive) Abbas II stressing the necessity to control the Egyptian/Libyan border around the town of Sollum.[16] In his account of state and non-state territorialization of this borderland, Matthew Ellis has emphasized how Italian infrastructure and capital penetrated Libya through negotiations with Britain. Italy had already envisioned an invasion at the dawn of the twentieth century, although the two governments disagreed on its delimitation. British authorities were concerned about the indefinite boundary separating the Ottoman province of Benghazi and the Egyptian territory west of Marsa Matruh. Local Bedouins of the Senussiya Sufi brotherhood were an active counterpart to Ottoman, Egyptian, and British institutions in expanding a cultural, political, and infrastructural *mise en valeur* of this space spanning between the Mediterranean and the Sahara. Before 1911, the triangle between Tobruk in the West, Siwa in the South, and Marsa Matruh in the East saw frequent clashes between Bedouin tribes on both sides of this 'border'. The Ottoman officials observed with irritation how Egyptian soldiers and gendarmes easily entered Cyrenaica when intervening to suppress these clashes on Ottoman sovereign territory. The arrival of Italian troops in 1911 revealed aspirations to territorialize this area based on criteria used for European borderlands.[17]

In the early months of the war, however, the Italian army still struggled to map possible routes linking Egypt and Cyrenaica while acknowledging an intense transit of caravans westward. In November, the Italian consulate's deputy in Alexandria Dolfini reported the transit of 'disproportionate

[14] Garbasso to Sangiuliano (annex), October 13, 1911, ASDMAE AAPP 1891–1916, 140.

[15] Declaration concerning the Laws of Naval War. London, February 26, 1909, https://ihl-databases.icrc.org/ihl/INTRO/255?OpenDocument (Accessed 11 June 2021.); De Bosdari to Sangiuliano, November 22, 1911, ASDMAE AAPP 1891–1916, 140.

[16] Grimani to Sangiuliano, October 28, 1911, USMAR, 204.

[17] Ellis (2018, 147–74).

food supplies'. His 'informants' also told him about trafficking in weapons and ammunition by 'Greek merchants' owning storerooms in Marsa Matruh and Sidi Barrani, a town east of Sollum. Dolfini's informants, also Greeks, mentioned an 'Arab trader' who had negotiated with Greek ship captains the terms of the operations on behalf of Prince Tussum (Tusun) Pasha, a member of the Khedivial family and head of an 'Egyptian Committee'. At the same time, a delivery of 'twenty wagons' containing weapons and ammo had arrived at Port Said through the Suez Canal, whence they would be transported by railway to Alexandria and continue their journey to Libya on a camel caravan. Another option was smuggling on small boats transporting coal between Port Said and Damietta, east of Alexandria, and from there on land caravans. This route had the advantage of transiting through ports with only 'subordinated custom authorities and Arab coastguards, upon which one cannot rely, no British authority'.[18]

Maritime smuggling routes were not a novelty of the Italo-Ottoman conflict. Port Said and the Suez Canal were already hotspots for the 'black drug' (opiates) and the 'white slave trade'.[19] This smuggling space covering the whole Egyptian coast became integrated in a trade resulting from a global surplus of armaments, which also made other Ottoman regions such as Albania, Macedonia, Eastern Anatolia, and Yemen destinations for smuggled weapons.[20] Some months before the outbreak of the war, the Ministry of the Interior in Istanbul and the governor of Benghazi exchanged information about arms trafficking toward Libya. The overland route through Egypt under British occupation connecting Benghazi with the Wadai region, located between Chad and Sudan, had witnessed an intensifying trafficking in both directions in the framework of anticolonial guerrilla against the French, organized by the Senussiya and—ironically—bolstered by Italian actors.[21] The Ottoman government feared that the Senussiya could use weapons smuggled back from Wadai to organize a rebellion which would open another conflict in the empire.[22] At the same

[18] Dolfini to Sangiuliano, November 21, 1911; Tritonj to Sangiuliano, March 30, 1912. ASDMAE AAPP 1891–1916, 140.

[19] Kozma (2011); Huber (2013, 295).

[20] Blumi (2003); Öztan (2017).

[21] Wright (1989, 103).

[22] The Ottoman Ministry of Foreign Affairs to the Ministry of the Interior, January 24, 1911, BOA BEO, 3849 288699.

time, the Greek port of Piraeus was the suspected main origin for weapons handed to the Senussiya around Benghazi.[23]

When the Italo-Ottoman War broke out, smuggling was already flourishing in the Libyan borderland, but roles within it changed rapidly. The Ottomans now brought troops from Yemen and Anatolia to Benghazi, the Senussi became vital Ottoman allies for anti-Italian guerrilla, and the Italian government intensified the cooperation between the Navy and the Ministry of Foreign Affairs to tackle trafficking. Until December 1911, the Red Sea was an important square in this chessboard; incessant communication between consuls in Massaua, Suez, Port Said, and the government in Rome aimed to intercept vessels transporting Ottoman troops.[24] The following round between the two belligerents, however, would be played on coastal routes and Mediterranean waters.

Sneaky Neutrality and the Smuggling Business

When considering smuggling through Egypt, Italian consuls were concerned about the attitude of the British occupiers. The highest British authority in Egypt, Consul-General Lord Kitchener, assured them about his empire's neutrality and even support for anti-smuggling surveillance.[25] Yet, the Italian deputy consul in Alexandria Dolfini feared that the Egyptians' sympathy for the Ottoman Empire would allow weapons to circulate 'with the utmost facility', while British custom and police authorities spent their afternoons playing 'tennis, polo, and at the club'.[26] British-Egyptian neutrality worried Istanbul as well. In January 1912, the Ottoman foreign minister regretted that his diplomacy had been weaker than the Italian efforts to gain Britain's support. His ambassador in London replied with assurances made by foreign secretary Grey about Britain's impartiality and stressed that it was, rather, Italy's '*maitrise de la mer*' which prevented Ottoman vessels to transport material and troops to Libya.[27]

[23] Correspondence between the Ottoman Ministry of the Interior, the Ministry of Foreign Affairs, and the Mutassaraf of Benghazi, March 14, 21, and 25, 1911, BOA BEO 3873 290408.

[24] Rombo to Leonardi Cattolica, December 18, 1912, USMAR, 204.

[25] Simon (1987, 113–16).

[26] Dolfini to Sangiuliano, December 5, 1911, ASDMAE AAPP 1891–1916, 140.

[27] Correspondence between Assim Bey and Tevfik Pasha, January 7 and 9, 1912, BOA HR SFR 3, 675 25.

Egypt's complex territorial status was discussed in Rome, Istanbul, London, and Cairo. Ottoman diplomats discovered a secret annex to the Anglo-French agreement of 1904 concerning Egypt and Morocco. Whereas the official text maintained the relationship of suzerainty between the Sublime Porte and Cairo, the secret article allowed for 'modifying' British policy toward Egypt. The Ottoman government thus feared that war smuggling toward Libya could justify a shift severing Egypt from Ottoman territory.[28] In his 1911 report on Egypt and Sudan, Kitchener reiterated the country's neutrality and praised the Egyptians' respect of this clause despite the 'close relationship', 'ties of trade and commerce', and 'sympathy' they felt for 'Mahomedan combatants'.[29] This situation shows the power of diplomacy but also its limits. Britain had a privileged position in this balance of power, Italy secured *carte blanche* for its anti-smuggling operations at sea, while the Ottoman Empire acknowledged the prohibition to transport war material. Yet, the reality on the ground changed through informal negotiations and oscillating allegiances among state and private, military and trading networks. Egypt's neutrality was, therefore, 'sneaky' regardless of Kitchener's intentions, since the interests inherent to smuggling were too plural and fragmented to be controlled.

In December 1911, Dolfini reported that Prince Tusun's committee pressured Greek shipowners to smuggle war material between Alexandria and Sollum. The Italian consul used a broker to contact the shipping agent Mazaraki, who stated that shipowners faced a dilemma: either they renounce the considerable gains of contraband or they risk being intercepted by Italian warships. Dolfini proposed him a deal: for every vessel intercepted thanks to Mazaraki's help, he would 'receive the half of the seized load, except for weapons'. From other informants including a coastguard and a docker, Dolfini heard that Egyptian coastguards organized the transport of rifles and ammunition through a Turkish officer of a company called 'Teftikh SARL'. The weapons had to be transferred on two vessels carrying the Greek flag after the freight had arrived in Alexandria from Istanbul on an Egyptian steamer. These rumors were taken with caution. When Dolfini talked to Mazaraki again, the latter denied the docker's allegations, and Major Giovannini, the Italian chief of the Egyptian secret police collaborating with Dolfini, expressed his doubts.

[28] Assim Bey to Tevfik Pasha, February 20, 1912, ibid.

[29] 'Report by His Majesty's Agent and Consul-General on the Finances, Administration, and Condition of Egypt and the Soudan in 1911', 1, ibid.

A few weeks later, in January 1912, the Italian consul in Piraeus received a letter by an anonymous 'Greek, friend of Italy' stating that a trader from Beirut named Abid had arrived in the port city to purchase rifles and ammunition from local merchants and ship them to Tripoli. The same Abid had been previously described by another informant as 'well known to the seamen in Tselepi' (a neighborhood in Piraeus) who loaded the material on several steamers heading south. From a further note by the Italian consul in Cairo Siciliani, we can assume that this 'rich smuggler' was Abid [K]Handekar, who was mostly based in Port Said.[30] Dolfini heard from a 'chief Bedouin' about a Greek steamer in Alexandria with Abid aboard. The latter traded in Egypt with his partner in crime Hüsseyin El-Giridli, from Herakleion. The British authorities, instigated by the Italian diplomats, carried out an investigation on this ring and banished the main culprits to Crete.[31] The Ottoman War Ministry also kept track of the smugglers, although information in its documents is fragmented. The Egyptian government had (re)arrested El-Giridli in March 1912 and exiled him to Cyprus with the accusation of brokering (*vesatat*) for the shipment of weapons to 'Ottoman fighters' in Benghazi. Some months later, in July, 'Beyrutlu' Abid and El-Giridli intended to transport considerable amounts of cartridges in line with their 'commitment to [Ottoman] ministries'. Yet, the Ottoman ambassador in Athens informed his government that the two had changed their business and now traded in rifles privately.[32] Only a few days later, the ambassador admitted to have lost track of El-Giridli, simply stating that he had 'left for Alexandria'.[33] At the bottom of these chains of command often stood workers who ignored, or only realized in the last minute, to be involved in smuggling. In one of his reports, Dolfini included an attestation by a Greek stoker who had arrived in Egypt from Fiume:

[30] Dolfini to Sangiuliano, December 5, 1911; Attestation by Schinaz, undated (attached to Dolfini to Sangiuliano, December 4, 1911); Anonymous letter, January 5, 1912; Anonymous attestation, December 5, 1911; Siciliani to Sangiuliano, January 15, 1912, ASDMAE AAPP 1891–1916, 140.

[31] Dolfini to Sangiuliano, December 20, 1911 (erroneously dated 1912), USMAR, 246.

[32] The Office of the Grand Vizier to Halil Bey, April 3, 1912; Muhtar Bey to Mustafa Asim Bey, July 15, 1912, BOA DH H, 7 26.

[33] Muhtar Bey to Mustafa Asim Bey, July 21, 1912; Muhtar Bey to Gabriel Efendi, August 1, 1912, BOA HR SYS 1571 5.

> Since I was jobless and homeless, my countryman Giorgio Salamanos invited me on board of the Syrian, a ship of the Syrian Prince Line. I boarded at 11 PM, but at midnight, while I was chatting on the deck with Salamanos, I saw a boat of the [Egyptian] government approach our ship. Straight afterwards, my companion told me that I would be paid if I helped with loading the trunks from the boat. I saw a coast guard on it with a long coat, who spoke English, next to him two Arab porters and two more coast guards in their uniform. On the steamship, two stokers, two sailors and a board official, apart from myself, received the goods lifted with a crane. Later on, one hundred long rectangular cases were moved from the boat to our ship. During this operation, one of them fell and crushed, and I saw that it contained 12 rifles. All cases looked the same, and apart from them two hundred two-feet-long squared cases were boarded, and I assume from their weight that they were ammunitions. The cases were all stored in the locker in the eating room, whose opening is under the table.[34]

This ephemeral encounter offers a glimpse of a kaleidoscopic Mediterranean reality reflected in Egypt. Smuggling thus challenges a dual reading of Italian versus Ottoman interests. Whereas Italian consuls relied on British authorities, these could not prevent Egyptian officials from supporting smuggling on behalf of the Ottomans. However, even the latter could not fully trust smugglers active between Greece, Beirut, and Egypt who could pursue their business autonomously to maximize gains and minimize the risks of being caught. The same ambivalent relationships held true for the Italian agents. They stroke deals to obtain information from Bedouins, who smuggled material between Egypt and Libya and most likely maintained their bonds with the anti-Italian resistance. Whereas battles and diplomatic relationships determined the course of the war, the more we narrow down the scale of observation, the more interdependencies and ambivalences in these relationships prevail over clear hierarchies. This was also due to the difficulties that state actors encountered in identifying with enough precision actors, routes, and vehicles of war contraband.

[34] Telemacos Defteros's attestation, December 16, 1911, ASDMAE AAPP 1891–1916, 140.

The Camouflage Feature of Smuggling

Smuggled war material was often traded and moved by private actors facing little state interference. In March 1912, Dolfini obtained letters that the Egyptian secret police had intercepted between Athens and Alexandria. The letter head contained the name Stefanos Zagoraios and the envelope was directed to a certain Emanuel, although one letter was signed by a Ioannis. The author informed his addressee that the vessel *Cleopatra* had been spotted in Sitia, Crete, with a certain Ali Sarsour on board. The load included weapons directed toward a locality called 'Arap Kule' and hashish destined to Port Said. Sarsour emerges as a valuable broker for ship captains sailing between Crete, Athens, Egypt, and Libya. In this case, Italian warships stationing off the coast at Sitia and adverse weather delayed the departure of the *Cleopatra*. Sarsour was quoted saying that the 'Prince [*Prigkips*] who had made the shipment was not satisfied because his order has not been carried out as he wished', asking for a sample of rifles and ammo of those shipped to Tripoli.[35] Is the 'Prince' in question Tusun Pasha? Without other sources available, the convoluted web of names in this correspondence only allows for speculations.

These concealment strategies also applied to smuggling routes. The *Cleopatra* had transported material between Piraeus, Syros, the Aydin Vilayet, and Beirut in the previous years.[36] The steamboat had already caught the attention of the Italian consul in Athens in January 1912. It officially headed for Crete, although the captain would allegedly change the route and head for Chios, Patmos, Rhodes, Cyprus, and Beirut. From Lebanon, its load would leave for Tripolitania on board of smaller sailing ships, deemed safer to escape the Italian navy. Italian agents contacted the captain and promised him to return the ship topped with a reward of up to 50% of the value if he facilitated the seizure. Despite these negotiations, the *Cleopatra* became a Moby Dick for the Italians. As late as April 20, the consul in Athens repeated that it had 'already escaped our surveillance twice' and that it would again depart from Piraeus to Santorini and continue toward Cyrenaica loaded with war material. A few days earlier, the

[35] Unsigned letters, February 24 and 27, 1912; Letter signed by Ioannis (surname unreadable), February 27, 1912, ASDMAE AAPP 1891–1916, 141.

[36] Refik Bey to the Sultan's Palace, November 7, 1905, BOA Y PRK ASK, 234 30; The Ottoman Embassy in Athens to the Ministry of Foreign Affairs May 5, 1908, BOA HR İD, 2125 85; The Ministry of the Interior to the vali of Aydin, January 25, 1911, BOA DH ŞFR, 665 84.

Ministry of the Navy had complained with the Ministry of Foreign Affairs about the 'evidently erroneous data' concerning the *Cleopatra*, which had 'shaken the trust in received information' with 'negative consequences on the war ships crews' forced to be constantly on guard.[37]

In the contraband game, not only traders and routes were camouflaged. The smuggling ships could also change their identity in terms of their owners, name, and nationality. In June 1912, the Italian consul in Port Said Tritonj informed Rome about Arif Bey, the Ottoman commissioner in Egypt, a key player in facilitating the passage of Ottoman ships through the Suez Canal. Arif allegedly used local Greek traders as front shipping agents to register Ottoman state owned or rented ships as private vessels. The already mentioned *Bodoss*, for instance, was called *Massawat* until 1911 and used to transport Ottoman troops to Yemen before smuggling weapons destined to Libya as a Greek private ship. Tritonj added that the Ottoman government rented another ship named *Kayseri* from Mihran Efendi, an Istanbul based Armenian publisher. Already in April 1911, the Ottoman embassy in London mentioned Mihran and the 'arrangement made with our War Ministry' that one of his ships be brought to 'Tripoli within ten days to transport old weapons and ill soldiers'.[38] The *Kayseri* had arrived from Beirut loaded with soldiers on September 29, 1911, just hours after the Italo-Ottoman War had broken out. Arif had demanded that the soldiers be allowed to disembark from the *Kayseri* and proceed overland through Alexandria to Cyrenaica. Due to the cholera outbreak in Lebanon, though, the British authorities in Egypt enforced a 14 days' quarantine. When the ship was eventually allowed to leave Port Said, the soldiers disembarked on the east side of the Suez Canal at El-Qantara but moved back to Syria, while the *Kayseri* was repainted to conceal its identity. According to the 1888 convention that proclaimed the canal's neutrality, ships could transit in peace and wartime, although they were not allowed to delay their passage. When the Italian consul insisted that the *Kayseri*, as an Ottoman vessel, leave El-Qantara, the Ottoman commissioner claimed in vain that the Sultan's suzerainty in Egypt had the priority and required special considerations. The period of British occupation actually marked a cleavage between *de jure* international conventions and

[37] Carlotti to Sangiuliano, January 10 and April 20, 1912; Newsletter by Leonardi Cattolica, April 17, 1912, ASDMAE AAPP 1891–1916, 141.

[38] Tevfik Pasha to Assim Bey, April 13, 1911, BOA HR SFR 3, 643 18.

de facto agreements embedded in negotiations on the ground.[39] The Italo-Ottoman conflict renewed discussion about Egypt's territorial status, accelerating the Ottoman loss of authority vis-à-vis the British.

The *Kayseri* subsequently left for El-Tor at the Sinai Peninsula, it took on board hundreds of convalescent Ottoman soldiers, and returned to Port Said now considered by the Ottomans as a 'hospital ship'. Consul Tritonj interpreted the money provided by Arif for the transit tax as evidence that the Ottoman state owned the ship. The Ottoman government had indeed paid 18,000 Turkish lira to Mihran, although this was not unequivocally a purchase transaction. When the Italian warship *Piemonte* made use of the canal's neutrality to surveil the *Kayseri*, the latter hesitated to depart, reopening the canal's neutrality dispute. Tritonj reasserted that the 'hospital ship' was a farce and that the Ottomans used the *Kayseri* to smuggle soldiers and war material into Egypt. The Italian Navy eventually seized the vessel in December 1911, an action considered legal according to international law only if a belligerent state owned the ship or if it carried war contraband.[40] A note issued by the Ottomans claimed that the Italians had acknowledged the transformation of the *Kayseri* into a hospital ship through the mediation of the German Embassy in Rome. The *Kayseri* was formally not owned by the state and therefore not exposed to a seizure according to the Hague convention.[41] Without further hesitations, the Italian Maritime Seizure Commission (*Commissione delle Prede*) confirmed the legitimate capture of the *Kayseri* and notified Mihran as 'affected party' in the lawsuit. Although Ottoman officials tried to delay the sanction, the *Kayseri* remained in the hands of the Italian Navy after the end of the conflict, and was turned with a royal decree into an Italian ship called *Eritrea*. [42] Within a year and a half, this ship passed from private to state ownership and from serving one state to serving its enemy. Ironically, it also changed its name from its owner Mihran's birthplace to an Italian colony which, in order to be reached from Italy, would have to pass through the same Suez Canal that saw its seizure as a smuggling ship.

[39] Huber (2013, 86–94).

[40] Tritonj to Sangiuliano, June 8, 1912, USMAR, 219.

[41] Copy of a note by the Ottoman Ministry of Foreign Affairs, December 28, 1911, USMAR, 246 3 20.

[42] *Gazzetta Ufficiale del Regno d'Italia*, n.119, May 20, 1912; The Sublime Porte to the Ministry of Foreign Affairs, May 16, 1912, BOA BEO, 4045 303343; *Gazzetta Ufficiale del Regno d'Italia*, n.300, December 21, 1912.

Conclusion

This chapter revolved less about territorial achievements and failures of Italian and Ottoman imperialism than about their capacity to play with extraterritoriality. Decentering the war from the battlefield allows for a search for a personal, infrastructural, and economic geography of Mediterranean contraband. In the end, we can re-center the Italo-Ottoman War through its short-term explosion in the production of reports related to smuggling. The narrative barely set foot in Libya and mostly moved in a space ranging from Suez to Sollum. This borderland was permeable to Mediterranean and Red Sea circuits which intensified smuggling during the conflict of 1911–12. Especially the Suez Canal experienced a densification of Mediterranean connections in this period. Maritime movement from Lebanon, Crete, and the Aegean reflected a web of economic and political relationships at the threshold of illegality in Egypt's main cities. Alexandria and Port Said were neuralgic points where smuggling was both organized and countered. Although the Sultan could still claim suzerainty over Egypt, neutrality and extraterritoriality were intensely discussed during the war by Ottoman and Italian state agents. They negotiated with Egyptian and British representatives as well as traders, brokers, and laborers involved in smuggling. What Simon Kitson has called 'alternative hierarchies' between the state and the 'underworld' equally holds true for such transnational framework.[43] Egypt and the Suez Canal's multilayered institutions left interstices open for extraterritorial actors moving between legality and illegality.

This story cannot be quantified or told as a linear chronological sequence. We will never know exactly how many rifles and ammunition arrived in Libya or were intercepted, when this trafficking precisely began or ended. Aware of these limits, studying smuggling is an exploration of a fragmented history, in which documents reveal rumors, suspicions, incertitude, improvisation and risk, as well as lack of information. The examples of ships changing owners, name, routes, and nationality show the importance of (mis)representations and detailed communication. We only know what the institutions discovered. They admitted that their findings were only the tip of the iceberg and were often wrong or inaccurate. What we do know is that negotiations around smuggling are a symptom of an Italian advance and an Ottoman retreat from the Southern Mediterranean

[43] Kitson (2014, 50).

in a crucial decade of imperial warfare. The Italo-Ottoman war is a multi-faceted snapshot inviting us to investigate earlier and later ebb and flows of smuggling economies linking Egypt and the Suez Canal with the rest of the Mediterranean.

References

Abbreviations

ASDMAE, Archivio Storico Diplomatico del Ministero degli Affari Esteri, Rome
    AAPP, Affari Politici
BOA, Başbakanlık Osmanlı Arşivi, Istanbul
BEO, Bab-ı Ali Evrak Odası
    DH, Dahiliye
        H, Hukuk
        ŞFR, Şifre Kalemi
    HR, Hariciye Nezareti
        İD, İdare
        SFR 3, Londra Sefareti
SYS, Siyasi Kısım
    Y PRK ASK, Yıldız Askeri Maruzat
USMAR, Ufficio Storico della Marina (Militare)

Bibliography

Blumi, Isa. 2003. Thwarting the Ottoman Empire: Smuggling Through the Empire's New Frontiers in Yemen and Albania, 1878–1910. *International Journal of Turkish Studies* 9 (1–2): 251–270.

Carminati, Lucia. 2017. Alexandria, 1898: Nodes, Networks, and Scales in Nineteenth-Century Egypt and the Mediterranean. *Comparative Studies in Society and History* 59 (1): 127–153.

Childs, Timothy W. 1990. *Italo-Turkish Diplomacy and the War over Libya: 1911–1912*. Leiden: Brill.

Ellis, Matthew H. 2018. *Desert Borderland: The Making of Modern Egypt and Libya*. Stanford: Stanford University Press.

Fortna, Benjamin C. 2016. *The Circassian: A Life of Eşref Bey, Late Ottoman Insurgent and Special Agent*. Oxford: Oxford University Press.

Huber, Valeska. 2013. *Channelling Mobilities: Migration and Globalisation in the Suez Canal Region and beyond, 1869–1914*. Cambridge: Cambridge University Press.

Kitson, Simon. 2014. *Police and Politics in Marseille, 1936–1945*. Leiden: Brill.
Kozma, Liat. 2011. Cannabis Prohibition in Egypt, 1880–1939: From Local Ban to League of Nations Diplomacy. *Middle Eastern Studies* 47 (3): 443–460.
Labanca, Nicola. 2012. *La guerra italiana per la Libia: 1911–1931*. Bologna: Il Mulino.
McCollum, Jonathan. 2015. Reimagining Mediterranean Spaces: Libya and the Italo-Turkish War, 1911–1912. *Diacronie* 23: 3.
Öztan, Ramazan Hakkı. 2017. Tools of Revolution: Global Military Surplus, Arms Dealers and Smugglers in the Late Ottoman Balkans, 1878–1908. *Past & Present* 237 (1): 167–195.
Schayegh, Cyrus. 2011. The many worlds of ʿAbud Yasin; or, what narcotics trafficking in the interwar Middle East can tell us about territorialization. *The American Historical Review* 116 (2): 273–306.
Simon, Rachel. 1987. *Libya between Ottomanism and Nationalism: The Ottoman Involvement in Libya during the War with Italy (1911–1919)*. Berlin: K. Schwarz.
Tagliacozzo, Eric. 2005. *Secret Trades, Porous Borders: Smuggling and States along a Southeast Asian Frontier, 1865–1915*. New Haven: Yale University Press.
Vandervort, Bruce. 2018. A Military History of the Turco-Italian War (1911–1912) for Libya and its Impact on Italy's Entry into the First World War. In *Italy in the Era of the Great War*, ed. Vanda Wilcox. Leiden: Brill.
Wright, John. 1989. *Libya, Chad and the Central Sahara*. London: Hurst.

PART IV

Imperial Strategies in the Mediterranean, from the Mid-1860s to World War II

CHAPTER 14

Italian Sea Power and Suez—From Opportunity to Obsession (1861–1943)

Fabio De Ninno

Introduction

The opening of the Suez Canal (1869) changed global maritime communications forever.[1] The new mobility provided for by the canal also marked a 'strategic revolution' that changed the geopolitical value of the enclosed sea for the major European naval powers. For Britain, until the Great War, controlling the Mediterranean (and the canal) was pivotal, being part of a 'superhighway to the world wide web' to its Empire.[2] During the interwar years, Britain continued to dedicate significant resources to defend the Mediterranean and the strategic choke point of Suez, during the rise of Fascist Italy as a Mediterranean antagonist.[3] The second Mediterranean naval power was France, which maintained important trade connections

[1] Abulafia (2013, 4).
[2] Simpson (1998, 51).
[3] Morewood (2004).

F. De Ninno (✉)
Università di Siena, Siena, Italy
e-mail: fabio.deninno@unisi.it

B. Curli (ed.), *Italy and the Suez Canal, from the Mid-nineteenth Century to the Cold War*,
https://doi.org/10.1007/978-3-030-88255-6_14

with the Indian Ocean and the Far East through Suez. Initially, for the French *Marine Nationale*, British control over the waterway was a major concern, but tensions faded after the Fashoda Crisis (1898) and with the formation of the Anglo-French Entente (1904) and the Anglo-French Naval Agreement of 1912. Instead, the rising Italian Navy, due to the advantageous geographical position of the Italian peninsula, was a potential threat for French maritime communications.[4]

Italian diplomatic and naval relations with Britain and France determined the importance of Suez for the Italian sea power. For Italy, Suez and Gibraltar were the only possible access points to global maritime communications. Moreover, the two choke points were interdependent: Italian access to the Atlantic via Gibraltar made access to Suez pivotal in case of interdiction of the Italian communications in the Western Mediterranean or the Atlantic.

The chapter discusses this evolution in relation to the development of the Italian Navy, its strategy and plans, and how they were influenced by the more general political and economic aspects of the Italian relation with Suez. The chapter also divides the relationship between Italian sea power and Suez according to three periods: accessing the canal (1869–1918); securing the canal (1919–1935); blocking or seizing the canal? (1935–1943).

Accessing the Canal (1869–1918)

At the inauguration ceremony of the Suez Canal, the Italian Navy (Regia Marina) was going to be well represented. A squadron of six warships, including the powerful battleships Castelfidardo and Roma, under the command of vice-admiral Amedeo of Savoia, reached Alexandria. However, the squadron returned to Italy before the ceremony, being their commanding officer requested at home by a sudden illness of King Vittorio Emanuele II. Six Italian merchant ships participated in the inauguration.[5] Italian naval presence reflected the aspiration of the new political élites to take advantage of the return of the Mediterranean at the centre of global

[4] Ropp (1978, 90, 240, 317, 320), Roksund (2007, 102).

[5] Appunto dell'ufficio storico della Marina, *Navi della Marina militare italiana che furono dislocate ad Alessandria per tenersi pronte a partecipare all'inaugurazione del canale di Suez*, AUSMM, Rdb b. 106.

communications.[6] Indeed, during the late 1860s and early 1870s, sharing this illusion, the *Rivista Marittima*, the journal of the Italian Navy, published various articles claiming the necessity to connect the peninsula with the Indian and Pacific oceans via Suez, making Genoa the trade link between Germany and the Far East by replacing in this role the Austro-Hungarian port of Trieste.[7]

The opening of Suez, instead of starting a new golden age, marked the beginning of a decline for the Italian maritime economy. British, French, and German shipping companies rushed into the Mediterranean with their more advanced technology and corporate capacities, leading to a reduction of the Italian merchant fleet.[8] Only at the beginning of the twentieth century things started to change. With the industrialisation of north-western Italy, trade expanded: between 1900 and 1915 imports grew by 214% and exports by 81%: around 90% travelled by sea. The merchant fleet also expanded, reaching 1,28 million tons in 1914, but the tonnage was insufficient for the commercial needs of the Italian economy and two thirds of Italian trade travelled on foreign shipping. However, Suez remained marginal in this growth. The largest part of the Italian maritime trade was with Europe and North America via Gibraltar and between 1908 and 1913 trade with India and the Far East, via Suez, accounted only for 9,1% of imports and 4,2% of exports.[9]

The limited importance of Far Eastern trade affected the relation of the Regia Marina with the canal until the First World War. In the Mediterranean, Italy was involved in a fierce naval competition with France and Austria-Hungary and its primary task was the defence of Italian coastlines, projecting Italian naval power mainly in the Adriatic and Tyrrhenian seas. Nevertheless, the second most relevant mission of the Regia Marina was to support the Italian oceanic projection and colonial ambitions in East Africa.[10] Naval support to colonial enterprises in the Red Sea, via Suez, started already in the first decade after the Unification. The first Italian colony, the settlement in the Assab Bay (1869), was a coaling station established by the Rubattino company on its route to India. The Navy

[6] Frascani (2012, 34-35).

[7] Cronaca, *Rivista marittima*, ottobre 1874, Il Lloyd Italiano, p. 157; Storia del bacino di Sanpierdarena, *Rivista marittima*, n. 1, 1878, p. 118.

[8] De Courten (1989).

[9] Annuario Statistico italiano. 1914. Serie II, Vol. IV. Istat: Roma, p. 224.

[10] Gabriele (1973).

selected the location, subsidised the construction and offered logistical support.[11]

However, Italian naval presence was initially unwelcomed: in 1879, when Italy attempted the acquisition of Socotra, an island dominating the southern access to the Red Sea, Britain opposed the project.[12] In 1879 and 1880, the sloop *Rapido* was sent to the Red Sea to provide military assistance for the expansion of the colony. Cesare De Amezaga, commander of the warship, went as far as to envisage that the establishment of an Italian colony in Eritrea could end British predominance in the Red Sea and that Suez could partially fall under Italian influence.[13] Ambitions and realities could not be more different. Italy also entered the Triple Alliance (1882) to reinforce its power projection capacity in the Mediterranean. The alliance with Germany and Austria-Hungary allowed to free Italian naval forces from the Adriatic, but within the Alliance Italian stance remained strongly pro-British and anti-French, refusing the idea of confrontation with the British Empire.[14] Indeed, Italy needed British support against France in the Mediterranean, and London's dominant position over the canal allowed to control the Italian communications with the Red Sea. Domenico Bonamico, Italy's leading sea power theorist of the time, argued that an alliance with Britain was necessary to secure Italian communications in the Mediterranean and the Red Sea, but Italy lacked the naval power necessary to be a credible partner for the British.[15]

Naval assistance to colonial expansion continued. The Eritrean colony was established in 1882 and in 1885 the Regia Marina deployed a division in the Red Sea made up by one battleship, one cruiser, one frigate and two corvettes. At a strategic level, its mission was to secure Italian communications with the motherland via Suez. The division had also to provide logistical support and mobility for the troops deployed in Eastern Africa and tactical assistance for landing operations along the Eritrean coast.[16] That year marked the peak of Italian naval presence in the canal until the Great War. However, the growth trend of the Italian warships passages through Suez continued until 1890 (Table 14.1).

[11] Labanca (2002, 49).
[12] Cadorna to Visconti Venosta, 3 maggio 1872, DDI, II, Vol. III.
[13] De Amezaga to Cairoli, Aden, 3 January 1880. DDI, II, Vol. XII, n. 524.
[14] Sondahaus (2001, 113).
[15] Bonamico (1899, 73-4).
[16] Gabriele (1982, 53-70).

Table 14.1 Italian warships stop at Port Said, calculation based on AUSMM, Rdb, b. 119, Navi della Marina Militare che dal Novembre 1869 a tutto l'anno 1920 fecero sosta a Port Said

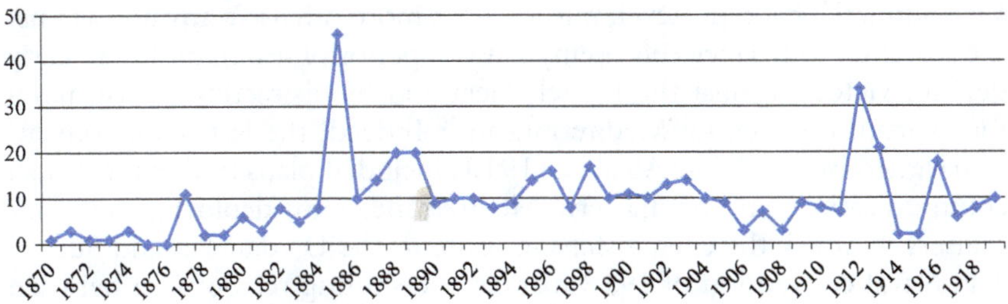

Colonial expansion slowed down in the 1890s, but naval presence grew again during the years of the first Italo-Ethiopian war (1895–1896). After the defeat at Adwa (1896) and during the first decade of the twentieth century it decreased again. A second peak occurred in 1911–1912, when the Italo-Turkish war for Libya extended to the Red Sea and the Regia Marina destroyed the limited Turkish naval presence in the area.[17] This conflict marked a decisive strategic shift for the Italian position in relation to Suez. The conquest of Libyan ports and the Italian occupation of the Dodecanese Islands gave Italy the control of both shores of the Central Mediterranean and a naval base in the Aegean Sea. The British Admiralty regarded this as a threat:

> [T]he possession by Italy of the Naval bases in the Aegean Sea would imperil our position in Egypt, would cause us to lose our control over the Black Sea and the Levant trade sources, and would in a war expose our route to the East via the Suez canal to operations of Italy and her allies.[18]

British worries were premature because the Italian naval position was precarious. At that time a global naval arms race was also investing the Mediterranean, with France, Italy and Austria-Hungary rapidly expanding their fleets. The forming of an Anglo-French alliance partially shifted the Royal Navy from the Mediterranean to the North Sea, making a naval war between the Triple Alliance and Britain a realistic possibility. The Regia

[17] Gabriele (1998, 111-30).

[18] Memorandum by the Admiralty, 20 June 1912, TNA, CAB 37/111/77.

Marina, despite being unwilling to fight a war with Britain, urged its German and Austrian allies to strengthen naval cooperation, and a naval convention was signed in 1900.[19] After 1912, the Italian naval staff, led by Vice-admiral Thaon di Revel, pushed for a more offensive attitude against the Entente, to balance the enemy naval superiority searching for an early decisive victory against the French Fleet and by distracting British naval forces from the Central Mediterranean.[20] Indeed, the last naval convention signed by the Triple Alliance (1913) included plans to disrupt British communications in the Eastern Mediterranean by deploying auxiliary cruisers to attack the enemy ships in route to the Dardanelles and Suez.[21] The Naval convention of 1913 marked the shifting relationship between the Italian Navy and Suez: from a waterway to access colonial territories to a commercial and strategic highway capable of affecting the Mediterranean balance of power.

Changes became even more evident during the First World War, a global economic conflict centred around blockade and counter-blockade strategies implemented by both the Entente and the Central Powers. The Entente's control of strategic choke points, like Suez and Gibraltar, was decisive for the outcome of the war, and significantly affected Italian decision-making before and during the conflict. In August 1914, the Italian choice for neutrality was also due to the Navy insistence that in a war with Britain Italy risked the loss of its maritime communications.[22]

The war at sea reshaped also the Italian maritime economy relationship with the canal. In 1913, 93% of Italian coal, 47.5% of iron and 28.16% of wheat were imported, almost entirely by sea. While Britain was Italy's main supplier of Italian coal, Germany's isolation and the closure of Dardanelles, caused by the Ottoman Empire entering into war, deprived the peninsula of its main markets for iron (Germany) and wheat (Russia and Romania), forcing the research for new suppliers.[23] Outside Europe,

[19] Gabriele (1969, 352-95).

[20] Memoria n. 1 a S.E. il Capo di stato maggiore della Marina per la cooperazione con la flotta A.U., marzo 1913, AUSMM, Rdb., b. 295, f. 1, sf. CN. 3.

[21] Convenzione Navale addizionale per il Mediterraneo, 23 giugno 1913, in USMM, 1935, 170.

[22] Promemoria del capo di stato maggiore della marina in occasione della proclamazione della neutralità italiana, 1 August 1914. ACS, Carte Salandra, b. 2, f. 16.

[23] ASI, Serie II, Vol. IV, 1914, Tipografia nazionale: Rome: 157, 227-233: in 1913, Britain accounted for 81% of Italian supplies of coal; Germany accounted for 26% of iron; Russia for 48.6% of wheat and Romania for 17.5%.

the United States, Argentina, Australia, and India were the largest producers: this increased the importance of Italian communications with the Indian Ocean via the canal. Already in February 1915, protests erupted in Parliament because the import of wheat supplies met with increasing shipping difficulties on both the Gibraltar and Suez routes.[24]

Italy's entry into the war (24 May 1915) against Austria-Hungary further enhanced the importance of access to oceans via the Mediterranean choke points. During the initial phase of the war, Italian imports came mostly from the Atlantic through Gibraltar, passing the western Mediterranean and reaching the Italian ports of the Tyrrhenian Sea: in 1916, passages through Suez accounted only for 7% of the traffic directed to the Italian peninsula.[25] During 1917 the situation changed dramatically. Coal still arrived from the Atlantic, but during the peak of the U-boat offensive (spring 1917), lack of shipping in the Atlantic, high losses in the Western Mediterranean and reorganisation of the Entente shipping system shifted Italian imports of food from the Americas to India and Australia.[26] In 1917, the two countries accounted for 40,7% of all Italian food imports[27] and were Italy's main source of wheat (Table 14.2).

Table 14.2 Italian wheat imports July 1916–December 1918 (in tons), by origin

Period	*North America*	*South America*	*India*	*Australia*
July–December 1916	914,274	313,303	4,156	16,750
January–June 1917	350,605	265,699	306,013	381,451
July–December 1917	261,640	63,867	639,550	
January–June 1918	840,303	369,666	420,046	31,953
July–December 1918	539,968	635,392	413,840	10,482

Sources: AUSMM, Rdb, b. 499, Ministero per i trasporti marittimi e ferroviari, Dati e notizie relativi alla situazione e all'impiego del tonnellaggio nazionale ed estero sotto il controllo italiano, Situazione di tutte le merci importate via mare durante gli anni 1915, 1916, 1917, 1918, 1919, ripartite per qualità e provenienza.

24 Discussioni, 22 February 1915, APC XXIV Legislatura, p. 6256.

25 AUSMM, Rdb, b. 610, f. 6, Condizioni del problema antisommergibile, Studio del capitano di fregata Alfredo Baistrocchi, 28 novembre 1916: 4-5.

26 Salter (1921, 96).

27 Bachi (1925, 132).

As a result, imports via Suez were extremely important to relief Italian food scarcity.[28] However, access to this vital route was hindered by German submarine warfare. During 1916, in order to avoid to provoke the United States in the Atlantic, German submarine attacks in the Western Mediterranean intensified, and continued during 1917–1918. Meanwhile, in the Mediterranean, the Entente antisubmarine effectiveness remained poor until late 1917. Causes were the Italo-French rivalry over the control of the Entente naval forces, limited effectiveness of antisubmarine technology, and delay caused by the British in the introduction of convoys. Finally, the Regia Marina opposed to shift resources from the Adriatic to support the antisubmarine war effort in the Mediterranean.[29]

In summer 1917, losses of merchant ships had just passed their peak (April 1917) and remained high, but Revel continued to refuse to send Italian warships to patrol sea lanes passing south of Crete and directed to Italy from Suez.[30] More important, during 1917, Germans acknowledged Italian supplying difficulties, and U-boats intensified the attacks in the Eastern Mediterranean, precisely when traffic from the India had become preponderant in feeding Italy.[31] Even in 1918, with the decreasing effectiveness of German attacks, a constant number of U-boats was destined to the Aegean Sea and Egyptian waters, attacking ships exiting from Port Said and along the sea lane between Egypt and Italy.[32]

During 1917–1918, for the first time, accessing Suez became vital for the survival of the Italian economy and this 'trauma' significantly affected Italian naval strategic thinking after the war, reshaping the relationship between the canal and the Fascist vision of Italian sea power.

[28] Dentoni (1995, 133-147).

[29] Halpern (2012, I, 429-88; II, 21-107, 167-81, 418-31); too much optimism about the italian results in Marcuzzi (2018).

[30] Il capo di stato maggiore della Marina al Comandante in capo dell'Armata, 26 agosto 1917. AUSMM, Rdb, b. 743.

[31] Relazione sull'attività e funzionamento dell'ispettorato per la difesa del traffico marittimo nazionale dal 31 dicembre 1917 al 31 dicembre 1918, p. 154, AUSMM, Rdb, b. 922.

[32] Azioni dei sommergibili nemici nel Mediterraneo nel 1918, p. 141. AUSMM, Rdb, b. 922.

Securing the Canal (1919–1934)

The rise of Fascism (1922) marked a radical change in Italian foreign policy. Mussolini pursued the project to make the Mediterranean an Italian vital space, putting the country on a collision course with France and Britain. Mussolini geopolitical project was partly inspired by the geopolitical and strategic conceptions of the Italian nationalist movement and of the Italian Navy. According to their geopolitical thinking, Italy's position in an enclosed sea required to secure the Adriatic, gain access to the oceans (via Gibraltar and Suez) and gain a predominant position in the Red Sea, while building a merchant fleet large enough to lessen Italian dependence on British shipping.[33] As one article of the *Rivista Marittima* argued in 1922:

> Italy [...] is a slave, because its central position in the Mediterranean means that its maritime communications, in addition to the two choke points of Gibraltar and Suez, are dominated by major naval bases of other nations. The blockade of our coasts is the most effective weapon in case of war to force our surrender, if we consider that 9/10 of our supplies arrive from sea.[34]

Again, during the interwar years, the symbiotic relationship between Gibraltar and Suez was self-evident. With American markets reachable again, commerce with the Indian Ocean and Asia loss importance and Suez became less pivotal for the Italian maritime trade (Table 14.3).

Table 14.3 Italian commerce with Asia and Oceania in percentage (1925–1928)

Year	*Imports*	*Exports*
1925	15.20%	6.50%
1926	12.64%	9.20%
1927	13.20%	9.30%
1928	11.50%	9.90%

Sources: ASI, Serie III, Vol. IV, 1930, Commercio speciale distinto per paesi di provenienza e destinazione, p. 279.

[33] Knox (1992, 17).
[34] Accinni (1922, 811).

Table 14.4 Merchant traffic passing Suez in 1934

Nationality	*No. of Ships*	*Tonnage*
British	3,071	17,238,128
German	509	2,976,451
Netherlandese	378	2,559,182
Italian	435	2,089,003
French	305	1,976,285

Source: Sandiford 1935, 150

Indeed, until the second Italo-Ethiopian War, the tonnage of Italian merchant ships passages through the canal continued to be relatively inferior compared to other European powers, even to Germany, that had no significant colonial interest East of Suez (Table 14.4).

However, it was essential to secure access to Suez in case of war. Italy's naval rivals were France and its allies: Yugoslavia and Greece. The Marine Nationale's predominance over the Western Mediterranean isolated Italy from Gibraltar and the Atlantic. In 1921, the Italian Chief of Naval Staff acknowledged this situation, stating that 'Italy is locked in a maritime prison whose gates are in the hands of others'. In case of war with France, shipping in the Western Mediterranean had to be limited or suspended and 'defence of communications in the Eastern Mediterranean is vital for Italy'. Therefore, the Regia Marina should be able to exercise its control over the sea lanes directed to the Dardanelles and Suez, protecting Italian trade from Suez and the Red Sea.[35]

Safeguarding the access to Suez contributed to the shift in war planning of the Regia Marina from an Adriatic-Tyhrrenian perspective to a Mediterranean dimension. According to the 1925 naval plans against France, Yugoslavia and Greece, in the Eastern Mediterranean the sea lane Port Said-Tobruk had to be maintained open, although it could be threatened by enemy submarines and surface forces based in Beirut.[36] While the largest part of the fleet had to control the Central Mediterranean, two naval divisions, composed of light units and based at Tobruk in Cyrenaica and Leros in the Dodecanese, had to protect Italian traffic from Suez. Finally, forces in the Red Sea had to defend the Italian traffic in the Strait

[35] Promemoria sull'assetto marittimo dell'Italia, June 1921. AUSMM, Rdb, b. 1574.

[36] Documento L.G. 1, 1925, Italia contro Francia, Jugoslavia, Grecia, Mediterraneo orientale, p. 10. AUSMM, DG, DG 1 B,C,D, pacco 12, busta 2, D4.

of Perim and the Gulf of Aden.[37] Nevertheless, the Regia Marina was still recovering from the restraints of the postwar period, and available warships were insufficient to meet such plans: only two prewar armoured cruisers were available to be deployed in Tobruk.[38]

During the next decade, defence of communications in the Eastern Mediterranean remained one of the cornerstones of Italian naval policy, and access to Suez and the Dardanelles to guarantee Italy's supplies was put forward as a justification for naval expansion and request for naval parity with France.[39] Indeed, the possibility to secure access to the Dardanelles and Suez depended on a fleet numerically strong enough to compete with the Marine Nationale. The necessity to build such fleet influenced the development of the force-structure of the Italian Navy. Luckily for the Regia Marina, the Washington naval conference (1922) ratified the parity with the French Navy in capital ships. By 1934, naval armaments limitations and massive naval spending allowed the Regia Marina to have a surface fleet greater in number of non-capital ships than the Marine Nationale and slightly inferior in terms of total tonnage.[40]

Increased naval power would improve the defence of Italian access to Suez. In 1929, revised naval plans centred around a conflict with France and Yugoslavia, relying on Greek neutrality. Access to the Aegean Sea guaranteed imports from the Black Sea, and access to both choke points of Suez and the Dardanelles was considered 'preeminent and vital in time of war'.[41] In 1932, it was planned that even shipping from South America had to be redirected towards Suez, via the Cape of Good Hope, merging into the Red Sea with the traffic coming from the Indian Ocean, before entering the Mediterranean via the canal. In case of war Italy would have to assure the passage through the canal of at least 12 steamships per week, out of a total of 30 necessary to resupply the country.[42] The main threat to

[37] Documento L.G. 1, 1925, Italia contro Francia, Jugoslavia, Grecia, Allegato N. 7, 8, 10 al documento n. 7. AUSMM, DG, DG 1 B, C, D, pacco 12, busta 2, D4.

[38] Documento L.G. 1, 1925, Italia contro Francia, Jugoslavia, Grecia, Allegato 1, Allegato al documento 4, Composizione e dislocazione iniziale delle forze navali. AUSMM, DG, DG 1 B,C,D, pacco 12, busta 2, D4.

[39] Riassunto della politica navale delle potenze firmatarie dei trattati di Washington e di Londra 1922-1930-1934, p. 54. AUSMM, Rdb, b. 2322.

[40] Sullivan (1998, 238-39).

[41] Promemoria sulle isole italiane dell'Egeo, 23 ottobre 1931, p. 4, AUSMM, DG, DG-9, pacco 27, f. A-1.

[42] Verbale della Riunione tenuta presso il capo reparto operazioni il 14 gennaio 1932, Ipotesi est-ovest, Apprezzamento della situazione, p .2 AUSMM, DG, DG 1 B, C, D, pacco 12, busta 2, D-2.

this traffic were French submarine and surface forces based at Beirut[43] and Cattaro in Yugoslavia.[44] Convoys had to be formed in the Eastern Mediterranean, involving large scale Italian naval activity between Suez and the Italian ports. Indeed, in 1933 an important training exercise was held in the waters south of Crete, to test the Navy capacity to destroy French forces attacking Italian convoys moving from Egypt to the Ionian Sea. Involving the entire 2nd Naval Squadron, the exercise showed that a large portion of the Italian fleet could operate in those waters, securing access to Suez.[45]

Suez was not just a getaway into the Mediterranean for the Italian traffic, but also an access point to the Red Sea and the Indian Ocean. Since the early 1920s Italian foreign policy aimed to undermine British control over the eastern coast of the Red Sea and the Persian Gulf, with the direct involvement of the Italian Navy.[46] The 1929 naval plans called for offensive action in the Red Sea, and envisaged operations against Djibouti[47] and the deployment of auxiliary cruisers to attack the French traffic outside Diego Suarez and Saigon.[48] Plans for the Red Sea were further developed in the following years, on the assumption of possible attacks of French naval forces against Italian traffic in the area. According to Italian plans, the Regia Marina could deploy in the Red Sea, via the canal, a naval force numerically superior to the French, to destroy the local enemy naval forces. The naval plans also called for the use of auxiliary cruisers, sailing from Italian East African colonies, to attack enemy shipping in the Indian Ocean.[49]

Later plans shifted towards the deployment of submarines, tested in late 1933 and early 1934 when the boats *Sciesa* and *Toti* circumnavigated the

[43] Arielli (2010, 21).

[44] Direttive per le operazioni in Mediterraneo orientale, Operazioni contro l'Egitto e in Cirenaica, Basi e punti di appoggio in Egitto. AUSMM, DG, DG 9, D-2; Promemoria per S.E. il Capo di stato maggiore, Difesa del traffico, Ottobre 1933, p. 3. AUSMM, DG, DG-5, A,B,C,D, pacco 19.

[45] Esercitazione E.N. 2 della 2ª squadra, 18 December 1932. AUSMM, Rdb., b. 1719.

[46] Fiore (2012, 122-32, 166-67).

[47] Gooch (2007, 154-155).

[48] Apprezzamento della situazione iniziale in un'ipotesi di conflitto Est-Ovest, p. 31. AUSMM, DG 0-1, B,C,D, Pacco 12, busta 2, D-3, Libro di guerra, Documento n. 2. 1929.

[49] Promemoria n. 21, Reparto O, Ufficio I, attacco al traffico francese, 21 dicembre 1931, AUSMM, DG-10, F. 5, SF. A3.

African continent from east to west, also passing through the canal.[50] By 1934, the Naval staff planned to deploy a naval division in the Red Sea to protect Italian shipping, and cooperate with the Army to the conquest of Djibouti and attack French traffic in the Indian Ocean with submarines.[51] In the framework of Fascist expansionism, access to Suez turned out to be the key for the oceanic projection of Italian sea power and to secure communications in case of war with France. However, this will not last for long because of the emerging confrontation with Britain.

Blocking or Seizing the Canal? (1936–1940)

In the early 1930s, Japanese expansionism in China and the rise of Nazi Germany favoured a more active phase of Fascist foreign policy, starting with the conquest of Ethiopia. During the spring of 1935, when it became clear that Britain would oppose the invasion, a diplomatic and naval crisis broke out in the Mediterranean. The Italian Navy faced the possibility that the canal could be closed to ships transporting Mussolini's Army in East Africa.[52] The Italian Naval Staff, supervised by the newly appointed Undersecretary of the Navy, Admiral Domenico Cavagnari, analysed the question from a legal perspective. The access to Suez was regulated by the 1888 Constantinople Convention, which allowed all powers to use the canal even in case of war, but during the First World War Britain had closed the passage to enemy shipping. Moreover, while Egypt had become formally independent in 1922, Britain maintained relevant military forces there and the right to secure its control over Suez in case of necessities connected to imperial defence. Cavagnari informed Mussolini that Britain could legally close the canal to Italian shipping if the League of Nations backed the action.[53]

The possible closure of Suez made necessary to study a new set of diplomatic options and naval plans for a confrontation with Britain. By May 1935, plans for a naval war against Britain were ready, but the Naval Staff

[50] Relazione finale sulla crociera intorno al continente africano dei Smg. A.Sciesa ed E. Toti. AUSMM, Rdb, b. 2568, f. 1.

[51] Ufficio del capo di stato maggiore, direttive di massima per l'ipotesi di guerra su due fronti valevoli per l'anno 1934, p. 35-36. AUSMM, DG, DG 1-B,C,D, Pacco 12, busta 2, D-3.

[52] Mallet (2015, 101, 165, 173).

[53] Sul transito del canale di Suez, 18 maggio 1935, seen by Mussolini, p. 5. AUSMM, Rdb., b. 2330.

excluded the possibility of direct attacks against Alexandria and Suez: they were too far from the operational range of the Italian Navy and Air Force.[54] The new plans also acknowledged that without access to Suez and Gibraltar, Italian traffic had to come entirely from the Dardanelles and from ports in Southern France and Spain, reducing total possible imports to just 3,83 million tons per year against the necessary 10 million tons. The closure of Suez counted alone for 1,9 million tons of lost imports.[55]

The crisis also induced Britain to strengthen its military links with Egypt. The Mediterranean fleet was now based in Alexandria and in August 1936 the Anglo-Egyptian treaty was signed. During 1937–1939, the British Admiralty started to fear that the Italians could block the canal in case of war, in a moment when the British Empire was threatened at the same time by Germany, Italy and Japan and needed fast strategic redeployment of its naval forces.[56] Indeed, a plan for a possible blockage had been studied by the Italian Navy at least since May 1936, but conclusions were that a permanent obstruction 'was difficult to obtain' because the canal Company had the necessary means to provide a swift reopening of the waterway. The employment of aircrafts for minelaying was also excluded, due to the distance from Italian airbases.[57]

The plans of 1935–1936 were a temporary solution, because they assumed that Italy had to face only the British Empire. Instead, after the end of the Ethiopian campaign (May 1936), the Italian armed forces policy and strategy were directed against both Britain and France as primary enemies, and assumed Germany as an ally, also because of the Rome-Berlin Axis (October 1936) and common involvement in the Spanish Civil War.[58] Now Italian Admirals had to plan for a war in which both Suez and Gibraltar were closed to Italian traffic. The possible options produced a split in the Italian Admiralty. Many senior leaders asked for a more ambitious naval policy, supporting the construction of an offensive escape fleet, with aircraft carriers, capable of an undisputed control of the Mediterranean and to re-open access to the Mediterranean choke points. Cavagnari was of a different opinion, thinking that the crisis of 1935–1936 showed that

[54] Mallett (1998, 25-6).

[55] Studio sul rifornimento della nazione in guerra, aggiornato al giugno 1935, XIII, Importazione del traffico italiano: 1-4. AUSMM, DG, DG-5, A, B, C, D.

[56] Morewood (2004, 86-96, 191-193).

[57] Notizie riguardanti il canale di Suez ed opere portuali in Egitto, 14 maggio 1936. AUSMM, DG, DG 9, pacco 27, D-1.

[58] Gooch (2007, 318-30).

Italy already had the possibility of closing Suez, with a distant blockade. Indeed, the Italian Navy could close the Sicilian Narrows to enemy shipping, blocking British traffic to India via the Mediterranean. The naval bases of Leros and Tobruk could also attack the enemy traffic in the Eastern Mediterranean and the Italian bases in Eritrea and Somalia could do the same in the Red Sea and the Indian ocean.[59] This opinion was shared also by Mussolini and other high-ranking officers of the armed forces. The Regia Marina would have to focus on the control of the Central Mediterranean and on the defence of Italy from possible Anglo-French joint operations, while closing the Mediterranean to enemy traffic and controlling the Red Sea, where Italian domination over Ethiopia guaranteed a 'capital' position.[60]

Italo-British relations over Suez remained tense in the years leading up to the Second World War. The Royal Navy rapid rearmament was perceived as a threat to the Italian access to oceans. The Gentlemen's agreement of 2 January 1937 (confirmed by the Easter accords of April 1938) apparently settled the matter, stating the necessity of free access to Gibraltar and Suez for both powers. However, Italian foreign policy was aligned with Germany and Japan, as its adhesion to the Anti-Comintern Pact (6 November 1937) showed. Despite British appeasement, British control over Gibraltar and Suez and British alliance with Egypt remained the key problems of Italo-British relations. Thanks to its predominant strategic position, Britain could impose a blockade of Italian sea lanes and dominate the Mediterranean, even though the Italian control over the Central Mediterranean could deny the free use of the Gibraltar-Suez route to British shipping.[61]

Therefore, between 1936 and 1940, the Regia Marina continued to develop plans for a distant blockade of Suez, by controlling the Red Sea, eventually seizing Djibouti and the Strait of Bab-el-Mandeb and attacking the traffic ashore of Aden.[62] In April 1937, plans called for the construction of a fleet of 3 oceanic cruisers, 4 flotilla leaders, 4 destroyers, 8 torpedo boats and 12 submarines to operate against the British communications in

[59] Il Trattato Anglo-Egiziano, 10 settembre 1936, p. 5. AUSMM, Rdb., b. 2346.

[60] Dall'Atlantico all'Oceano Indiano attraverso il Mar Rosso, Ufficio Trattati, Articolo per l'ISPI dell'Ammiraglio Rainieri Brescia, Roma, 23 gennaio 1938. AUSMM, Rdb, 2330.

[61] Considerazione del nostro problema strategico in base alla situazione politica del momento presente, 26 ottobre 1938. AUSMM, DG, DG Studi politici, busta 2, P2, A8.

[62] Mallett (1998, 107).

the Indian Ocean and block the Red Sea.[63] In November 1938, Cavagnari stated that were Britain involved in a war with Japan, Italy could attack Suez with Japanese support.[64] All these ambitious plans were undermined by Italian limited economic resources available for naval expansion and the necessity to concentrate such resources in the Central Mediterranean. Nevertheless, still after the outbreak of the Second World War in Europe and during Italian non-belligerence (1 September 1939–10 June 1940), Cavagnari ordered to plan attacks against enemy ports in the Red Sea up to Port Said. The local commander, Rear Admiral Carlo Balsamo, prepared for a more cautious approach, envisaging the impossibility to replace its losses, due to the isolation of Italian East Africa from the Mediterranean.[65]

Conclusion: The Unreachable Obsession

Suez was a vital objective in the war against Britain and for the Italian claims of hegemony over the *mare nostrum*, but it was an unreachable goal for the Regia Marina. The Naval policy of the second half of 1930s renounced the design to project Italian naval power in the Eastern Mediterranean, leaving to the Army in North Africa the task to seize the canal.[66] Nevertheless, the importance of Suez continued to stand at the centre of Mussolini's ambitions, becoming a dominant *leitmotiv* of Fascist foreign policy and preparations for war.[67] A 'window on the oceans', either Gibraltar or Suez, was the main objective of Italy's war declared on 10 June 1940.[68]

Italian control over the Sicilian Narrows and the southern entrance of the Red Sea would close the Mediterranean route to British shipping, with important effects on Britain's global logistical system: in early 1943, the longer route via Cape costed 365,000 tons per month of extra tonnage.[69] Nevertheless, the Italian distant blockade of the canal could not exercise decisive strategic results in the Italo-British confrontation during the Second World War. Indeed, British control over Suez allowed British pre-

[63] Documento di guerra L.G. N.2, Dislocazione dei mezzi, Ufficio Piani operazioni, 20 aprile 1937: 17-18. AUSMM, DG, DG-8, A,B,C, pacco 14, busta 1, f. C-1.

[64] AUSMM, Rdb., b. 2688, f. 5, Cavagnari to Badoglio, 18 November 1937.

[65] Mallett (1998, 168-69).

[66] Minniti (2000, 145).

[67] Gooch (2020, 108).

[68] De Felice (1981, 774).

[69] Leighton and Coakley (1995, 671).

dominance in the Eastern Mediterranean, securing the Middle East and its oil fields from the Axis powers.[70] On the contrary, for Italy, the closure of Gibraltar and Suez meant the impossibility to access markets outside the Mediterranean and the Black Sea. Moreover, the first day of war (10 June 1940), British control of the two exits of the Mediterranean left 1.26 million tons of Italian merchant ships trapped outside the Middle Sea, out of 3.3 million tons available, decisively affecting later scarcity of Italian merchantmen to resupply the Axis forces in North Africa.[71]

Suez was the only Mediterranean choke point within the range of the Italian armed forces and became the main target of Italy's 'parallel war', once the fall of France allowed the Italian Navy to control the Central Mediterranean and supply the Italian forces in Libya. In September 1940, Mussolini ordered an offensive against Egypt, aiming at Suez. In December, a British counteroffensive led to a massive Italian defeat, while the Italian troops were repulsed in Greece and three Italian battleships damaged by the British carrier-borne strike at Taranto (11–12 November 1940). These defeats marked the end of Italy's parallel war and its possibility to seize Suez without German assistance. By March 1941, also the Italian Naval presence in the Red Sea was annihilated, securing British access in the area. In early 1941, German air, naval and armoured units started arriving in the Mediterranean mitigating the Italian military crisis, allowing systematic minelaying operations against the canal by air, but not being able to interrupt the enemy traffic.[72] During the spring of 1942, after Japan's entrance into the war, the Italians hoped that the Imperial Japanese Navy was going to push into the Western Indian Ocean and the Red Sea, joining the Axis forces in North Africa and aiming at Egypt and the canal.[73] However, the Japanese were stopped by the British reinforcements in the Indian Ocean, and the Axis fortunes in North Africa fluctuated, with Hitler rejecting the Kriegsmarine plans to seize Suez.[74] After Allied victory at El-Alamein (October–November 1942) and the final Axis defeat in North Africa (May 1943), the Mediterranean was reopened to Allied shipping, finally allowing faster reinforcement of the British forces in the Far East.[75]

[70] Smith (1996, 30), Morewood (1991, 91-2).
[71] Ferrante (1993, 134-38), Hammond (2020, 27).
[72] Massignani and Greene (1998, 140).
[73] Supermarina, Promemoria n. 22, March 1942: 3-5. AUSMM, Patto Tripartito, 1942.
[74] Wegner (2001, 122-27).
[75] Boyd (2017, 347).

The end of the Fascist war in Africa marked the sunset of Italian ambitions of domination over Suez. Since unification, Italian sea power relation with the canal shifted from utility to necessity. Naval presence was important for supporting Italian colonial expansion, but the access to Suez was not vital for the Italian economy or naval policy during this period. The First World War showed that access to oceanic sea lanes, provided by Suez, was vital for the Italian economy and could be put at risk if a hostile power controlled the Western Mediterranean. During Fascism, Italy set on a collision course with France and Britain and coveted direct control over the canal to ensure its maritime and naval security and oceanic projection. However, the Italian Navy did not develop the necessary means to achieve such ambitious goals, making Suez an unreachable obsession and marking the end of a historical cycle in Italy's relationship with the canal.

References

Archival Sources and Abbreviations

ACS, Archivio centrale dello Stato, Rome
ASI, Annuario Statistico Italiano
APC, Atti parlamentari, Camera dei deputati, Rome
AUSMM, Archivio dell'Ufficio storico della Marina Militare, Rome
DG, Direttive Generali
Rdb, Raccolta di base
b., busta
DDI, Documenti diplomatici italiani
TNA, The National Archives, London
CAB, Cabinet papers
USMM, Ufficio storico della Marina Militare, Rome

Bibliography

Abulafia, David. 2013. *Il Grande Mare, Storia del Mediterraneo*. Milan: Mondadori.

Accinni, Francesco. 1922. 'La Conferenza di Washington e la Nostra Marina da guerra'. *Rivista marittima*, June, 807-834.

Arielli, Nir. 2010. *Fascist Italy and the Middle East, 1933-1940*. London: Palgrave Macmillan.

Bachi, Riccardo. 1925. *L'alimentazione e la politica annonaria in Italia*. Bari: Laterza.

Bonamico, Domenico. 1899. *Il problema marittimo dell'Italia*. La Spezia: Tip. Lega Navale.

Boyd, Andrew. 2017. *The Royal Navy in Eastern Waters, Lynchpin of Victory, 1935-1942*. London: Seaforth.

De Courten, Ludovica. 1989. *La marina mercantile italiana nella politica di espansione (1860-1914), Industria, finanza e trasporti marittimi*. Rome: Bulzoni.

De Felice, Renzo. 1981. *Mussolini e il fascismo, Lo stato totalitario, Vol. 2, 1936-1940*. Turin: Einaudi.

Dentoni, Maria Concetta. 1995. *Annona e consenso in Italia 1914-1919*. Milan: Franco Angeli.

Ferrante, Ezio. 1993. La marina mercantile italiana e la lotta per le comunicazioni marittime nel secondo conflitto mondiale. In *L'Italia in guerra, il 1° anno 1940*, ed. Antonello Biagini and Raimond H. Rainero, 133–138. Rome: CISM.

Fiore, Massimiliano. 2012. *Anglo-Italian relations in the Middle East, 1922-1940*. Farnham: Ashgate.

Frascani, Paolo. 2012. *Il Mare*. Bologna: il Mulino.

Gabriele, Mariano. 1969. *Le convenzioni navali della Triplice*. Rome: Usmm.

———. 1973. *La flotta come strumento di politica nei primi decenni dello stato unitario italiano*. Rome: Usmm.

———. 1982. *La politica navale italiana dal 1885 al 1915*. Rome: Usmm.

———. 1998. *La Marina nella Guerra-italo turca, Il potere marittimo strumento militare e politico (1911-1912)*. Rome: Usmm.

Gooch, John. 2007. *Mussolini and his Generals, The Armed forces and Fascist foreign policy 1922-1940*. Cambridge: Cambridge University Press.

———. 2020. *Mussolini's War, Fascist Italy from triumph to collapse, 1935-1943*. London: Penguin.

Halpern, Paul. 2012. *La grande guerra nel Mediterraneo*. Gorizia: Leg.

Hammond, Richard. 2020. *Strangling the Axis. The Fight for Control of the Mediterranean during the Second World War*. Cambridge: Cambridge University Press.

Knox, MacGregor. 1992. L'ultima guerra dell'Italia fascista. In *Annali della Fondazione Luigi Micheletti, 5, L'Italia in guerra 1940-1943*, ed. Bruna Micheletti and Paolo Poggio, 17–32. Fondazione Micheletti: Brescia.

Labanca, Nicola. 2002. *Oltremare, Storia dell'espansione coloniale italiana*. Bologna: il Mulino.

Leighton, Richard M., and Robert W. Coakley. 1995. *Global Logistics and Strategy 1940-1943*. Washington: Center for Military history.

Mallett, Robert. 1998. *The Italian Navy and Fascist expansionism, 1935-1940*. London: Frank Cass.

———. 2015. *Mussolini in Ethiopia, The Origins of Fascist Italy's African War*. Cambridge: Cambridge University Press.

Marcuzzi, Stefano. 2018. 'From the Adriatic to the Mediterranean: Italy in the Allied Naval Strategy (1915–1918)'. *War in History.* December. https://doi.org/10.1177/0968344518778307

Massignani, Alessandro, and Jack Greene. 1998. *The Naval War in the Mediterranean, 1940-1943.* Annapolis: Naval institute press.

Minniti, Fortunato. 2000. *Fino alla guerra, Strategie e conflitto nella politica di potenza di Musssolini, 1923-1940.* Naples: ESI.

Morewood, Steven. 1991. Protecting the jugular vein of empire: The Suez canal in British defence strategy, 1919-1941. *War & Society* 10 (May): 81–107.

———. 2004. *The British defence of Egypt, Conflict and crisis in the Eastern Mediterranean, 1935-1940.* London: Routledge.

Roksund, Ale. 2007. *The Jeune Ecole, The Strategy of Weak.* Leiden-London: Brill.

Ropp, Theodor. 1978. *The Development of a Modern Navy, French Naval Policy 1871-1904.* Annapolis: Naval institute press.

Salter, J.A. 1921. *Allied Shipping control, An experiment in International Administration.* Oxford: Oxford University Press.

Sandiford, Roberto. 1935. 'Il canale di Suez e il suo regime internazionale', *Rivista marittima*, November, 133.

Simpson, Michael. 1998. Superhighway to the World Wide Web: The Mediterranean in British Imperial Strategy 1900-45. In *Naval Policy and Strategy in the Mediterranean*, ed. John B. Hattendorf, 51–77. London: Frank Cass.

Sondahaus, Lawrence. 2001. *Naval Warfare, 1815-1914.* London: Routledge.

Smith, Kevin. 1996. *Conflict over convoys Anglo-American logistics diplomacy in the Second World War.* Cambridge: Cambridge University Press.

Sullivan, Brian. 1998. Italian Naval power and the Washington disarmament conference of 1921-1922. In *The Washington Conference, 1921-22*, ed. Erik Goldstein and John Maurer, 87–101. London: Frank Cass.

Wegner, Bernd. 2001. Hitler's Grand Strategy between Pearl Harbor and Stalingrad. In *Germany and the Second World War, Vol. VI, The Global War*, ed. Horst Boog, Werner Rahn, Reinhard Stumpf, and Bend Wegner, 112–144. Oxford: Oxford University Press.

CHAPTER 15

Connecting the Two Seas: Negotiating an International Modus Vivendi—Italian and Ottoman Diplomacies in the Suez-Red Sea Area

Giorgio Ennas

Diplomatic Neutralisation of the Suez Canal Between the 1860s and the 1870s

This chapter analyses the search for a new *modus vivendi* between conservative and interventionist powers and the negotiations behind the transformation of the political status of the Eastern Mediterranean, Suez Canal and Red Sea areas between the second half of the nineteenth and the beginning of the twentieth century. It shows, in particular, how the opening of the Suez Canal produced an effective expansion of the Eastern Mediterranean geopolitical area, which gradually included the Red Sea.

G. Ennas (✉)
Franklin University Switzerland, Lugano, Switzerland
e-mail: Giorgio.Ennas@eui.eu

B. Curli (ed.), *Italy and the Suez Canal, from the Mid-nineteenth Century to the Cold War*,
https://doi.org/10.1007/978-3-030-88255-6_15

After centuries of essentially utopian projects, by the end of 1840s the Suez Canal project was finally ready for completion thanks to the new cultural positivist atmosphere and new material technologies.[1] Between 1854 and 1856, de Lesseps obtained two firmans (orders) from the Egyptian *vâlî* (governor) for the Suez Canal Company to construct the canal, for which the explicit authorisation of the Sublime Porte was to be eventually obtained.[2] His initial attempts to proceed with only the authorisation of the Egyptian governor had faced strong opposition from the Porte and the British government. Despite the fact that the new Ottoman cabinet, led by Mehmed Emin Aali Pasha and Keçecizade Mehmed Fuad Pasha, was apparently willing to give its official support to the Suez Canal project, the Ottoman *chargé d'affaires* in London, Musurus Bey, requested the opinion of the British cabinet several times on their behalf before the officialisation of the construction of the canal.[3]

On 4 January 1860, the Ottoman foreign minister, Fuad Pasha, wrote that the Porte officially blocked the work that had begun without Ottoman sanction and the administrative procedure for the concession of the imperial firman. Moreover, he listed three fundamental conditions that were to be met in order to obtain permission from the Porte to complete any construction, guarantees concerning the conservation of the Ottoman suzerainty over the Egyptian province, and guarantees by France and Britain regarding the neutrality and security of the canal.[4]

The Porte informed the Egyptian *vâlî*, Ismail Pasha, that the previous agreement had two unacceptable propositions for all the powers effectively interested 'in the progress of civilisation in the Orient', and reserved the right to block the company's enterprise.[5]

In April 1863, Mehmed Emin Aali Pasha addressed two separate dispatches to the Ottoman representatives in Europe regarding the Suez Canal question. In the first, he officially invited the French and the British cabinets to collaborate with the Porte towards the development of a sustainable *modus vivendi* that respected both Ottoman neutrality and its territorial integrity. In the second, he gave indications to the *chargés*

[1] Bilici (2019, 45).

[2] Bilici (2019, 80–1).

[3] De Lesseps to Lange, 31 March 1858; Foreign Office to Musurus Bey, 11 March 1858, London, BOA, HR, SYS, D. 00970.

[4] Fuad Pasha to the Ottoman representants in London and Paris, 4 January 1860, Constantinople, BOA, HR, SYS, D. 00970.

[5] Aali Pasha to Ismail Pasha, 3 April 1863, Constantinople, BOA, HR, SYS, D. 00970.

d'affaires in the other European capitals to obtain support for the Anglo-French-Ottoman agreement, including the preservation of Ottoman control over the Egyptian *eyâlet* (Ottoman province) and the neutrality of the Suez Canal.[6]

In the meantime, the Ottoman *chargé* in Turin, Rüstem Bey, reported his conversation with the Italian foreign minister, Emilio Visconti Venosta, regarding the question of the neutralisation of the canal.[7] Apparently, Visconti Venosta affirmed that the interest of the Italian government in the Suez Canal question was focused only on commercial aspects and he reported his positive attitude concerning the recognition of the canal's neutrality. Finally, after a decade of diplomatic negotiations, the British-Ottoman faction obtained the official neutralisation of the Suez Canal. The sultan ratified the previous firman on 22 February 1866 and sanctioned the project with the promulgation of an official firman on 19 March.[8] Specific institutions aimed at the management of infrastructure were then established, such as an Administrative Council of the Company, which was created to negotiate the questions related to navigation through the canal.[9] Finally, on 17 August 1869, a provisional regulation of the canal's navigation was printed. It imposed the taxation standard for cargo passing through the Suez Canal for three years following its inauguration on 17 November 1869.

In the 1870s, de Lesseps' activism questioned the Ottoman suzerainty and the neutrality of the canal. Without the Porte's approval, the Canal Company increased the taxation on vessels passing through the canal, completely ignoring the claims of powers such as the Kingdom of Italy.[10] On 15 November 1872, the new Ottoman foreign minister, Halil Şerif Pasha, underscored that the Porte considered the tonnage question as an issue that regarded the respect of 'the laws and the traditions of the Empire'.[11]

Starting its work on 6 October 1873, the International Tonnage Commission established a specific system of calculating tonnage. Despite

[6] Aali Pasha to the Ottoman representants in Europe, 6 and 9 April 1863, Constantinople, BOA, HR, SYS, D. 00970.

[7] Rüstem Bey to Aali Pasha, 14 May 1863, Turin, BOA, HR, SYS, D. 00970.

[8] Bilici (2019, 277).

[9] Report of the Commission, BOA, HR, SYS, D. 00980.

[10] Italian Legation in Constantinople to Rashid Pasha, 20 June 1872, Constantinople, BOA, HR, SYS, D. 00979.

[11] Halil Pasha to de Lesseps, 15 November 1872, BOA, Constantinople, HR, SYS, D. 00980.

the success of the diplomatic negotiations, on 14 April 1874 the foreign minister, Mehmed Rashid Pasha, reported to the Ottoman representatives in Europe the Suez Canal Company's refusal to acquiesce to the international plan for taxation.[12] The Ottoman government refused the possibility of taking direct control of the canal, in order not to risk a possible global escalation.[13]

The Italian government officially confirmed its support to the Porte as executor of the decision previously taken by the International Commission. Nevertheless, in 1875 the Ottoman ambassador to Rome, Alexander Caratheodory Efendi, reported Italian interest in negotiating an agreement between the Company and the Commission.[14]

Moreover, between the 1870s and the 1880s the Egyptian crisis accelerated the rivalry between Italy and the Ottoman empire in the Eastern Mediterranean. In 1875 the economic-political crisis of the Khedivate of Egypt and its consequent inability to ensure security and geopolitical stability on the western side of the Red Sea extended the question of suzerainty from the Suez Canal to the Red Sea.[15] The crisis of the *status quo* gradually persuaded the British government of the need to abandon the policy of preservation and of neutralisation of the Eastern Mediterranean, favouring the constitution of a new system or *modus vivendi* in the Suez-Red Sea area. At the same time, after the parliamentarian triumph of the Historical Left in 1876, the Italian government was finally ready to seize the opportunity offered by the opening of this new area of expansion.

The Italian-Ottoman 'Silent' Conflict in the 1880s and 1890s

As highlighted by the Italian and Spanish claims for the development of commercial stations on the west coast of the Red Sea, the inauguration of the canal opened the Red Sea space to powers interested in enhancing

[12] Report of the Commission, BOA, HR, SYS, D. 00981; Rashid Pasha to the Ottoman representants in Europe, 14 April 1874, Constantinople, BOA, HR, SYS, D. 00981.

[13] Grand Vizier to the Khedive, 25 April 1874, Constantinople, BOA, HR, SYS, D. 00982.

[14] Artom Efendi to the Ottoman Foreign Minister, 16 April 1874, Rome, BOA, HR, SYS, D. 00981; Caratheodory Efendi to Safvet Pasha, 20 August 1875, Rome, BOA, HR, SYS, D. 00982.

[15] Gelvin (2009, 101–2); Assim Pasha to Musurus Pasha, 13 January 1882, Constantinople, BOA, HR, SYS, D. 00987.

their international position.[16] As a matter of fact, the 'illegal' occupation of the west coast created a new boundary between expansionist powers and the Ottoman state. This new geographical proximity created the need to negotiate a new equilibrium. At the same time, the clash between the interests of the Ottomans and of these expansionist powers exacerbated the polarisation of international relations in the Red Sea.

Between 1879 and 1882, the direct transfer of the ownership of the Bay of Assab to the Italian government, as well as the intensification of the Italian military activities in the region, strongly irritated the Ottoman *Hariciye Nezâreti* (foreign ministry). Italy's unauthorised occupation of the *kaymakamlıks* (governorships) deeply concerned imperial diplomacy, which rapidly focused its efforts on containing the Italian threat: firstly, through diplomatic pressure on the other European Powers; secondly, by forcing the Italian government to recognise Ottoman suzerainty over these territories; thirdly, by reporting the cargo of all those vessels sent to territories where the Italian expansionist policy was directed, for example, Beilul and Massawa.[17]

Moreover, Italian expansionism in the Red Sea area received the endorsement of the British government for the occupation of the geopolitical *vacuum* created by the Egyptian crisis. However, even though the Italian-British agreement of 23–28 February 1882 entailed recognition of Italian ownership of the Bay of Assab, Ottoman suzerainty was confirmed over the north side of the Red Sea coast to Assab. The informality of this first agreement aimed to lessen the risk of an open conflict between the two powers in the Red Sea. At the same time, in the second part of the 1880s it created a period of 'silent' conflict between the Ottoman and the Italian diplomacies.[18]

The Italian occupation of Massawa in February 1885 created a potentially explosive situation. In spite of this act of hostility, on 9 February 1885, the Italian foreign minister, Pasquale Stanislao Mancini, highlighted Italy's desire to recognise Ottoman suzerainty over the territories around Massawa. Notwithstanding the Italian guarantees, the Ottoman foreign minister, Assim Mehmed Pasha, wrote to the Ottoman ambassadors to the other European powers protesting against aggression towards 'a

[16] Ottoman dispatches from Madrid, BOA, HR, SYS, D. 00987.

[17] Mihran Efendi to Assim Pasha, 16 December 1884, Rome, BOA, HR, SYS, D. 00990. Arzemy Efendi to Said Pasha, 27 March 1886, Naples, BOA, HR, SYS, D. 00989.

[18] Lenci (1990, 33, 37).

civilisation' that 'based its international relations on the respect of the treaties' and indicating how the Italian case represented a dangerous precedent for international diplomacy. To avoid the possibility of international complications, Mancini unequivocally confirmed the recognition of Ottoman sovereignty on several occasions and underlined the temporal boundaries of the Italian military occupation due to the inability of the Egyptian authorities to preserve local order.[19]

In August 1885, the Ottoman ambassador to Rome, Mihran Efendi, wrote to his *Hariciye Nâzırı* (foreign minister) describing the eventuality of a British-Italian plan to 'pacify' Sudan. Moreover, he reported that both the Italian and the British governments were interested in the development of a *modus vivendi* in the Red Sea region, and that St. James had proposed an agreement between the Italian and Ottoman authorities.[20] British diplomacy pushed for a compromise based on mutual recognition: for the Italians, this entailed accepting the principle of Ottoman suzerainty, whereas for the Ottomans it required the acknowledgement of the Italian *de facto* military occupation of their territories.

In February 1888, Rüstem Pasha, now ambassador to London, complained in a conversation with the prime minister, Lord Salisbury, about British support to the Italian occupation of the territories under Ottoman-Egyptian administration. The British prime minister responded by complaining about the unresolved question of the recognition of Ottoman suzerainty. At the same time, he blamed the apparent lack of interest of the Ottoman government to cover the *vacuum* created by the crisis of the Egyptian khedivate, militarily occupying the west coast of the Red Sea.[21]

Between the 1880s and the 1890s the Italians adopted an imperialist attitude 'humouring' and 'ignoring' the Ottoman authorities with the long-term objective of justifying Italian *de facto* occupation of territories that were under international protection;[22] on the other hand, the Ottomans 'obstinately' chose to protest against the occupation of their

[19] Farah (2002, 80–1); Assim Pasha to the Ottoman ambassadors in Europe, 10 February 1885, Constantinople, BOA, HR, SYS, D. 00989.

[20] Mihran Efendi to Assim Pasha, 18 August 1885, Rome, BOA, HR, SYS, D. 00984.

[21] Rüstem Pasha to Said Pasha, 29 February 1888, London, BOA, HR, SYS, D. 00988.

[22] Lenci (1990, 50–1). Since the Congress of Paris of 1856, the integrity of the Ottoman territories was put under international protection.

territories 'without their knowledge' and 'insisting' on delegitimising the Italian occupation.[23]

The international situation in the Red Sea was made more complex by the necessity for both Italian and Ottoman authorities to impose customs administrations against illegal activities such as piracy, slave trading and contraband between the two sides of the sea. The 'silent' friction regarding the Ottoman-Italian borders on the Red Sea continued through the official diplomatic channels. An illustrative incident is the arrest of the Eritrean-Italian dhows named '*Dukhan*' and '*Feth-ül Djévad*' under the auspices of the Yemenite-Ottoman authorities in 1892.

In a telegram from Constantinople to the Ottoman ambassador in Rome, Mahmoud Nedim Bey, the foreign minister, Said Pasha, described another 'incident' that had occurred with a dhow from Massawa which was blocked by the imperial navy. Although the foreign minister remarked upon the inevitability of the incident due to the peculiarity of Massawa's situation, the Italian government maintained its intention to send a military ship to the city of Hodeida to put pressure on the Ottoman governorship. In a second dispatch, Said Pasha mentioned the need 'to solve the question not engaging the matter of principle' (the question of the Ottoman suzerainty over Massawa), and ordered Mahmoud Nedim Bey to negotiate with the Italian foreign ministry to obtain the 'immediate' suspension of the expedition of the military ship to Hodeida and its return to Massawa.[24] Despite Mahmoud Nedim Bey's efforts, the Italian government aimed to push the Porte to allow a pecuniary compensation as soon as possible, in order to assert Italy's superior status as a Great Power.

In August, Mahmoud Nedim Bey wrote that the negative attitude of the Eritrean-Italian authorities was due to the refusal of the *vâlî* of Hodeida to negotiate the requests of the Italian consul.[25] The refusal of the Ottoman-Yemenite governor to negotiate with the Italian-Eritrean authorities risked compromising the Great Power status of the Italian state. For this reason, the imposition of a compensation through an act of 'gunboat diplomacy' became the only valid option to protect Italy's international status.

[23] Musurus Pasha to Assim Pasha, London, BOA, HR, SYS, D. 00987; Assim Pasha to the Ottoman representants in Paris, Berlin and Vienna, 2 February 1885, Constantinople, BOA, HR, SYS, D. 00989.

[24] Said Pasha to Mahmoud Nedim Bey, dispatches of 15, 16, and 18 February 1892, Constantinople, BOA, HR, SYS, D. 00098.

[25] Mahmoud Nedim Bey to Said Pasha, 19 August 1892, Rome, BOA, HR, SYS, D. 00098.

On 22 August, Mahmoud Nedim Bey confirmed that the Italian naval expedition was definitively cancelled. Although the Ottoman diplomatic agent in Rome obtained guarantees concerning the cancellation of the expedition, at the end of August the lack of communication between the Ottoman and Italian governors risked the issue being reopened.[26] In fact, modifying the course of the '*Feth-ül Djévad*' dhow from Massawa for contraband activity, the Ottoman authorities of Mocha risked creating a new diplomatic incident with the Italian-Eritrean governorship.[27] Apparently, the Eritrean dhow was stopped by a ship with an Ottoman flag and was forced to change its course for Mocha, where it was quarantined and sequestered without any evidence of contraband. The Italian consular authorities forcibly requested compensation for the economic losses due to the irregular management by the Ottoman authorities. The Italian general secretary of the foreign ministry, Isacco Malvano, and the Italian government did not intend to create a question against the Ottoman authorities, but the Ottoman governor of Hodeida had to offer reparation for the losses.[28]

In September, the foreign minister, Said Pasha, confirmed the desire of the Porte to open negotiations with the Italian consul to achieve a compromise regarding the reparation. Due to the hesitation of the Ottoman-Yemenite authorities for economic compensation, on 3 October the Italian foreign minister, Benedetto Brin, informed Mahmoud Nedim Bey of a military ship expedition to Hodeida supporting the request of the Italian consular authorities. Mahmoud Nedim Bey obtained a new postponement of the expedition when he informed Brin that the delay of the Yemenite authorities was due to the reformulation of the Italian claims regarding the reparation.[29]

Despite Said Pasha's justifications of the legitimate activity of the imperial authorities and his intention to continue with the investigations,

[26] Mahmoud Nedim Bey to Said Pasha, dispatches of 22 and 31 August 1892, Rome, BOA, HR, SYS, D. 00098.

[27] Ottoman report of the incident (BOA, HR_SYS__00098_00002_042_001-002).

[28] Mahmoud Nedim Bey to Said Pasha, 22 August 1892, Rome, BOA, HR, SYS, D. 00098; Mahmoud Nedim Bey to Said Pasha, 31 August 1892, Rome, BOA, HR, SYS, D. 00098.

[29] Said Pasha to Mahmoud Nedim Bey, 10 September 1892, Constantinople; Brin to Mahmoud Nedim Bey, 3 October 1892, Rome; Mahmoud Nedim Bey to Said Pasha, 4 October 1892, Rome, BOA, HR, SYS, D. 00098.

especially regarding the recognition of the crew's nationality,[30] Mahmoud Nedim Bey reported the opinion of the Italian foreign minister, who was not effectively interested in further prolonging the question of the national identity of the crew, insisting on immediate compensation as the only way to successfully conclude the negotiations without any military intervention.[31]

Undoubtedly, the Ottoman's final concession of reparation to the cargo owner can be considered a success of the so-called gunboat diplomacy. However, what clearly emerges from this lengthy negotiation is the Ottoman desire not to jeopardise diplomatic relations with its neighbour and to thus preserve stability on the Red Sea.[32]

Ottoman-Italian Relations: Between 'Turkish Piracy' and the Fight Against Smuggling (1899–1902)

Between 1898 and 1899, increased Ottoman control of Eritrean-Italian vessels produced a 'global' confrontation between the Ottoman and Italian governments. The decision made by the Italian authorities to interpret these sequestrations as acts of 'Turkish piracy' against 'Italian subjects' caused an immediate deterioration of relations between the two states. Moreover, it reawakened the question of legitimacy-sovereignty for the Ottoman-Italian 'unofficial condominium'.[33]

In March 1900, the Ottoman authorities responded to the Italian attitude by intensifying their controls over Eritrean vessels with Italian flag that were probably involved in contraband activities on the Yemenite coast. In this case the Italian authorities preferred to avoid controversy and decided to accept Ottoman intervention. Taking advantage of the situation, the Ottoman authorities declared that, considering the Ottoman identity of the Eritrean populations, the owners of the ships should avoid

[30] Said Pasha to Mahmoud Nedim Bey, 12 November 1892, Constantinople, BOA, HR, SYS, D. 00098.

[31] Mahmoud Nedim Bey to Said Pasha, 26 November 1892, Rome, BOA, HR, SYS, D. 00098.

[32] Mahmoud Nedim Bey to Said Pasha, 25 December 1892, Rome, BOA, HR, SYS, D. 00098; Said Pasha to Mahmoud Nedim Bey, 24 December 1892, Constantinople, BOA, HR, SYS, D. 00098.

[33] Lenci (1990, 46); By reference to the official British-Ottoman condominium between 1899 and 1955.

every consular interference and ask the Ottoman administration directly for the restitution of their vessels and goods.[34]

In so doing, the Ottoman authorities reopened the question of the legitimacy of Italian occupation and highlighted the ambiguity of the informal compromise. This time the challenge was brought to the diplomatic domain, where the Ottomans had the chance to openly engage the Italians concerning the question of Ottoman suzerainty over Massawa.

Italian hesitation to recognise the Ottoman citizenship of Eritreans caused imperial central and provincial authorities to react. Their refusal to recognise the 'Italianness' of the Eritrean side brought out the question of the lack of recognition of the Italian administration of Massawa.

Italian documents testify that Italy's concern stemmed from the possible consequences of an official Ottoman policy to force those vessels with an Italian flag to hoist the Ottoman one and impose the use of Ottoman documents. Indirectly threatened by the Ottomans with the possibility of a direct clash, in November 1901 the Italian government requested a diplomatic roundtable to achieve a peaceful solution to the controversy.[35] Finally, Ottoman diplomacy had the possibility to officially discuss the question oflegitimate 'sovereignty' over Eritrea with its Italian counterpart.

The Diplomatic Development of an International *Modus Vivendi* (1902)

In the negotiations of January 1902, Ottoman diplomacy immediately focused its efforts on two main objectives: the acceptance of consular protection for vessels flying an Italian flag and arriving from Massawa and all other Italian Eritrean territories; and the international official recognition of Ottoman requests concerning the legitimacy of Italian occupation of the imperial territories.[36]

Refusing the Italian request for a simple '*entente de fait*', which for the Ottoman government could imply the recognition of Italian sovereignty over its *kaymakamlıks*, the Sublime Porte offered an official agreement on most of the problems which had emerged in almost 20 years of forced cohabitation in the Red Sea: on the basis of international law, vessels officially registered with an Italian flag would be treated as official Italian ships

[34] Lenci (1990, 54–5).
[35] Lenci (1990, 60–1).
[36] Lenci (1990, 72).

by the imperial Yemenite authorities; Ottoman law would be applied to those vessels with regular documentation; Ottoman ships would receive the same treatment as official Ottoman vessels. The 'ordinary application of European international law' excluded the application of the system of Capitulations, which made this proposal 'absolutely unacceptable' to Italian diplomats. In an effort to create a *modus vivendi* acceptable for both parties, the Ottomans suggested an official system that could put the Ottoman and the Italian governments in a position of equality, based on reciprocal flexibility. Therefore, the Italians unilaterally interrupted the diplomatic negotiations with the Porte, avoiding the Ottoman request.

Due to the unilateral interruption of diplomatic negotiations, Italian diplomacy forced the Ottoman-Yemenite authorities to apply the sanctions planned in the imperial *irâde* decree. In the summer of 1902, the Ottoman navy blocked several Eritrean vessels, providing the ships' officials and the crews with Ottoman passports as well as navigation documents. The perfectly legal Ottoman initiative obliged the Italian government to 'impose' the re-opening of negotiations in order to reach an agreement, as the Ottoman government desired to end this matter as quickly as possible. Finally, on 26 October, the Ottoman and the Italian governments found an agreement. The Porte proposed once more to consider the Eritrean vessels under the Italian 'protection', specifying that their crews would be treated in accordance with the international law and not with the system of Capitulations. The Italian government agreed.[37] But, at the end of October, the situation was further complicated on the Ottoman-Italian border by the expedition of an Italian military ship to Hodeida.[38]

The Ottoman foreign minister, Tevfik Pasha, wrote to the ambassador in Rome, Reshid Bey, describing the situation. Moreover, he wrote to the Italian government to remind them of the efforts of the imperial government against piracy, to emphasise the traditional conciliatory attitude of the Porte regarding acts of piracy by its formal subjects and to admonish the Italian authorities for police acts on the territories of a 'friend state',

[37] Tevfik Pasha to Reshid Bey, 26 October 1902, Rome, BOA, HR, SYS, 00098.

[38] Lenci (1990, 72–5); Tevfik Pasha to Reshid Bey, 15 October 1902, Constantinople, BOA, HR, SYS, D. 00098.

which were formally against international law. The good faith of the Porte was the main shield against Italian gunboat diplomacy.[39]

During the negotiations, Tevfik Pasha confirmed the Ottoman acceptance of Italian requests for the extradiction of the pirates. In exchange, the Ottoman foreign minister requested the immediate retreat of the *Piemonte* and the nullification of the Italian ultimatum, which should be considered 'null and void'.[40]

The increasing tension and the ambiguity produced by the prolongation of the negotiations caused the bombardment of Midi on 31 October 1902, which sought to impose the unilateral acceptance of Italy's requests to the Ottoman-Yemenite authorities.[41] The Ottoman government immediately considered this decision a crime that went against international law and was a direct act of hostility that could not be justified by the legalist attitude adopted by the imperial government. The British cabinet supported the Italian position, although it was evidently an open violation of international law, and the bombardment of Midi was officially labelled as an incident of piracy.[42]

In 1902, the result of the Italian-Ottoman negotiations was ratified by an unofficial exchange, to 'not create a precedent of agreements with the Sublime Porte not based on the Capitulations', and because of the need for the Italian government not to re-open the question of sovereignty with the Ottoman government.[43] Moreover, by being more interested in the configuration of an official *modus vivendi* on the Red Sea than in initiating military escalation with the Italian state, the Ottoman government agreed to define its previous claims around the question of imperial suzerainty as simple 'enunciations'.[44]

By acting with a mixture of diplomatic-administrative aggressiveness and simultaneously using international pressure as well as the principle of

[39] Tevfik Pasha to Le Marquis de Malaspina, 17 October 1902, Constantinople, BOA, HR, SYS, D. 00098.

[40] Tevfik Pasha to Reshid Bey, dispatches of 21 and 31 October 1902, Constantinople, BOA, HR, SYS, 00098.

[41] Tevfik Pasha to Reshid Bey, 1 Nov. 1902, Constantinople, BOA, HR, SYS, D. 00098.

[42] Governor of Yemen to the Commissar of Hodeida, 2 August 1902; Tevfik Pasha to Reshid Bey, 17 Nov. 1902, Constantinople, BOA, HR, SYS, D. 00099.

[43] Lenci (1990, 75). The informality of the exchange guaranteed the preservation of the system of Capitulations, not creating an official precedent. DDI, Prinetti to Giacchetti, 21 October 1902, Rome.

[44] Lenci (1990, 75).

'good-faith', Ottoman diplomacy achieved the objective of temporarily stabilising the new international boundary between its Yemenite territories and the rising Italian colonial empire. The mutual need of Italian-Eritrean and Ottoman-Yemenite authorities to combat Arabic piracy and smuggling on the Red Sea played a pivotal role for both sides also in light of the stabilisation of the region.[45]

Conclusion

From the 1850s to the 1860s, the opening of the Suez Canal produced an expansion of the Eastern Mediterranean area towards the Red Sea and of the policies of the European powers interested in preserving or modifying the international equilibrium sanctioned in Paris in 1856. Throughout the second half of the nineteenth century, confronting each other in the diplomatic arena, the European powers created new equilibrium and *modus vivendi* on the Red Sea. In particular, this demonstrates that, in the newly opened Red Sea theatre, the Kingdom of Italy and the Ottoman sultanate progressively developed a new *modus vivendi* by pursuing different foreign policies.

At the same time, the question of sovereignty in the Red Sea deeply affected their diplomatic relations. Their gradual deterioration was in part due to the Italian project of 'taking over' the position of the Ottoman dynastic state in Africa and to the Ottoman government's failure to impose respect for its suzerainty over parts of its territories in the area.[46]

References

Archival Sources and Abbreviations

BOA, Başbakanlık Osmanlı Arşivi (The Ottoman Archives of the Prime Minister), İstanbul
HR, Hariciye Nezareti (Foreign Ministry)
SYS, Siyasi (Political Correspondence)
DDO, Documents Diplomatiques Ottomans
DDI, Documenti Diplomatici Italiani
D., dossier
d. dispatch

[45] DDI, Prinetti, to Malaspina, 7 Nov. 1902, Rome/Hamid Bey to Tevfik Pasha, 11 Nov. 1902, London, BOA, HR, SYS, 00991.

[46] DDO, d. 15 355/223, Munir Pasha to Tevfik Pasha, 26 June 1901, Paris.

Bibliography

Bilici, Faruk. 2019. *Le canal de Suez et l'Empire ottoman*. Paris: CNRS. Editions.

Documenti Diplomatici Italiani. 2010. *III serie*. Vol. VII. Roma: Istituto Poligrafico Zecca dello Stato.

Farah, Cesar E. 2002. *The Sultan's Yemen. Nineteenth-Century Challenges to Ottoman Rule*. London: I.B.N. Tauris & Co. Ltd.

Gelvin, James L. 2009. *Storia del Medio Oriente moderno*. Torino: Einaudi.

Kuneralp, Sinan. 2010. *ODD Sur l'Empire Ottoman et l'Europe III, Documents diplomatiques ottomans sur la politique des Puissances au Maroc (1859–1912)*. Istanbul: Isis Press.

Lenci, Marco. 1990. *Eritrea e Yemen. Tensioni italo-turche nel mar Rosso 1885–1911*. Milano: Ed. Franco Angeli Libri S.R.L.

CHAPTER 16

When Two Worlds Collide: Britain, Italy and Suez in the Fascist Era

Steven Morewood

For Britain, the Suez Canal created a short cut to India, jewel of empire, and the lucrative Far East, becoming a vital highway for military and commercial traffic. British ships topped canal user figures year-after-year. Anthony Eden called Suez 'the swing-door of the British Empire, which has got to keep continually revolving if our communications are to be what they should'.[1] For Italy, the Suez Canal took on a greater significance once Benito Mussolini assumed power in October 1922. After the Great Depression, Italy, alongside Norway and Japan, became one of only three states to enhance their canal tonnage. Indeed, by the start of the war with Abyssinia, Italy stood second only to Britain on the list of canal users.

[1] *Hansard*, Commons, Vol.233, 23 December 1929, c. 2047.

S. Morewood (✉)
University of Birmingham, Birmingham, UK
e-mail: S.MOREWOOD@bham.ac.uk

B. Curli (ed.), *Italy and the Suez Canal, from the Mid-nineteenth Century to the Cold War*,
https://doi.org/10.1007/978-3-030-88255-6_16

Suez and Abyssinia

British decision-makers tended to dismiss *Il Duce's* speeches referencing a desire to reconstitute the Roman Empire in the Eastern Mediterranean—only achievable at British and French expense—as theatrical posturing designed for internal consumption. Aside from the Corfu incident, throughout the 1920s fascist Italy conformed with the international system, signing up to peace treaties and negotiating friendship pacts with regional powers. On the surface, it appeared as though the traditional friendship between Britain and Italy was enduring. Austen Chamberlain, Britain's foreign secretary for much of this period, established a warm personal rapport with Mussolini, who kept his photograph on his desk.

The international scene darkened at the start of the new decade, clouded by the Great Depression and the emergence of Adolf Hitler to lead Germany with another fascist regime, but British governments hoped to remain on good terms with Rome. In November 1933 Italy, alongside France and the United States, was designated as a power against whom no defensive military preparations were required.

On 28 September 1932, King Victor Emmanuel III travelled through the Suez Canal bound for Massawa, returning on 12 October, escorted by an Egyptian cruiser. There was, however, a more sinister aspect to the trip. Accompanying the monarch was the minister of colonies, Emilio de Bono, charged by Mussolini with investigating the viability of invading Abyssinia from Eritrea. He reported that a surprise attack was out of the question because of the necessity to send substantial reinforcements through Suez.[2] Subsequently, Mussolini would protest that he was acting defensively in the wake of the Wal Wal incident, but the sheer scale of the military build-up traversing the canal exposed the absurdity of this claim.

It was in 1932 that Mussolini gave serious consideration to absorbing Abyssinia into his burgeoning empire. Clearly the merchant shipping network needed to be transformed in advance so it could handle the logistical underpinnings of the military expedition. That year four large combines were created: Lloyd Triestino based on Trieste, the Italia based on Genoa, the Adriatica based on Venice and the Tirrenia based on Naples. From 1933 onwards the Italia began to figure prominently in canal shipping. The Tirrenia took care of much of the colonial trade and mail, managing to make the journey from Naples to Massawa in nine days, with the Lloyd

[2] De Bono (1937, 10).

Triestino handling long-distance services to Africa and Asia. Italian Somaliland exported a growing volume of bananas. It too was visited by the King, again accompanied by de Bono, on 25 October 1934, on a trip that lasted until 26 November.[3]

On 7 January 1935 Mussolini met Pierre Laval, French Foreign Minister, in Rome, seeking a free hand in Abyssinia, which he thought he got. The same day, de Bono left unostentatiously for Massawa.[4] During January the Italian air forces in East Africa began to be strengthened through bomber and reconnaissance aircraft. Some were sent via Suez while others were flown from Tripoli and across Egypt and the Sudan.[5] From 23 February onwards a stream of reinforcements started pouring into Eritrea and Italian Somaliland, proceeding there through the canal.[6]

The sheer amount of troops, weapons and supplies transiting the canal could not escape the notice of the British. The senior naval liaison officer at Port Said began sending the Admiralty details of Italian vessels making passage and their passengers and cargoes. For example, in February he reported *Caffaro* and *Gange* had made the journey from Naples (the usual embarkation port) to Massawa in Eritrea, passing transiting Suez on the 18th and 21st respectively. The first transport carried 283 fascists, who were not in uniform, plus military vehicles and war materials. The second had 2030 troops aboard and an indeterminate cargo.[7]

'Up to the present', Mussolini claimed in a speech on 14 May 1935, 'the number of workmen who have left is perhaps greater than the number of soldiers'. The British military attaché in Rome was not sure whether to believe *Il Duce*, since Italian press reports barely mentioned this aspect.[8] However, the returns of the British naval liaison officer at Port Said do bear this out. From Rome, British ambassador Sir Eric Drummond sent confirmatory reports to the Foreign Office. On 20 February 1935 he drew attention to 4000 skilled workmen leaving Naples for Eritrea that day.[9] Moreover, the British embassy had gleaned insights through

[3] Ibid., 139–40.

[4] Ibid., 56.

[5] British Air Attaché Rome, 'Aircraft reinforcements for Italian East African colonies', 16 February 1935, TNA CO 535/107/4.

[6] Farnie (1969, 595).

[7] Naval liaison officer Port Said to director of naval intelligence, 22 February 1935, TNA CO 535/107/4.

[8] Military attaché Rome, memorandum, 15 May 1935, TNA CO 535/107/4.

[9] Drummond to Simon, 20 February 1935, TNA CO 535/107/4.

W.E. Steele, the representative in Italy of the British firm Huston-Horsby Limited, who revealed that the firm Società Anonima Italiana Lavori Pubblici, which undertook public works in the colonies, was in charge of developments. In February they sent out to Eritrea a large number of labourers, surveyors and engineers, together with rails, dump-wagons, picks, shovels and air compressors. These were conveyed in steamships and transports via Suez in order to build military roads. The contract was reportedly worth 180 million lire.[10]

The line spun by the fascist regime was that troops were being sent to East Africa in a purely defensive role lest Emperor Haile Selassie used his superior numbers to attack along Eritrea's exposed 400-mile-long border with Abyssinia. This rationale became increasingly dubious over coming months as reinforcements kept pouring in. Early in February, the 19th (Gavinana) Division at Firenze and the 29th (Peloritana) Division at Messina were mobilised, becoming the first reinforcements to pass through Suez. The liner *Conte Biancamano* left Naples on the evening of 26 February carrying 73 officers and 2600 troops, commencing the deployment.[11]

Mussolini became increasingly nervous at the prospect of League of Nations opposition. Accordingly, he planned to win the upcoming war as soon as possible and looked to deploy a massive force to ensure this eventuality. On 8 March 1935, Mussolini wrote to de Bono, who was to command operations: 'You have asked me for three divisions by the end of October; I intend to send you *10*, ten I'm telling you.'[12] De Bono had transformed this order into a deployment plan by 7 May 1935. The potential threat of League sanctions and the need to be ready to invade Abyssinia by the autumn gave the matter some urgency, which was reflected in the plan. De Bono planned to transport the reinforcements to Massawa in two main components. Prior to the rainy season, commencing in June, two regular divisions would arrive at the Eritrean highlands minus their pack horses. With the rainy season drawing to a close by early September, the second phase would be activated, involving two more divisions, the *Gran Sasso* and the *Sila* plus five *MVSN* divisions. At the same time, the equipment, supplies and pack horses for all divisions would be shipped, the total

[10] Drummond to Simon, 11 February 1935, TNA CO 535/107/4.

[11] Military attaché Rome, 'Italo-Abyssinian Relations', 26 February 1935, TNA CO 535/107/4.

[12] Mussolini to de Bono, 8 March 1935, *DDI*, 7, XVI, number 707.

deployment to be complete by the end of the year.[13] An added complication was mounting opposition in the British Parliament to Italian designs on Abyssinia, leading to de Bono receiving orders, on 18 May 1935, to begin stockpiling three years' worth of munitions and victuals against the worst case scenario: the closure of Suez.[14]

Marshal Badoglio, Italian Army Chief of Staff, doubted whether this was a realistic timetable. The deployment would involve shipping seven regular divisions, *MSVN* units, 8000 pack horses, ammunition, supplies and combat aircraft over 2000 miles from Naples to Massawa via Suez. It was fortunate that some two hundred bombers and over a thousand air crew and technicians had already been sent, given the anticipated role of air power in subduing the Abyssinian tribesmen warriors. Even at this stage, Air Chief of Staff Giuseppe Valle was considering employing outlawed chemical weapons such as mustard gas and phosgene to ensure success.[15]

The volume of Italian traffic passing through Suez increased exponentially between February and September 1935. To 1933 Italian canal usage only amounted to 5–6% of total net tonnage, ranking Italy fifth among users, reflecting Italy's demand for oil from the Persian Gulf and the salt trade from the salines of Assab in Eritrea.[16] During 1934, when Italian empire-building took on renewed vigour, the share almost reached 7%, pushing Italy up to fourth place. Not surprisingly, in 1935, Italy jumped into second place behind Britain.

The Suez Canal Brigade commander 'Tim' Pile reckoned that 45,000 Italian troops had traversed Suez since October 1934 to 1 April 1935. Pile's impression was that Italy would start war with Abyssinia in September with three divisions.[17] He was not far wrong. On 11 June 1935 an Italian auxiliary, *Città di Siracusa*, arrived at Port Said. As was usual practice, a British constable requested the captain to fill out entry forms. But he became enraged, tore up the forms and threw them overboard 'as there is a war on'. The Italian consul needed to intervene and eventually the forms were completed.[18] The febrile atmosphere was assisted by members of the

[13] Mallett (2015, 61).
[14] De Bono, 161.
[15] Ibid., 162.
[16] Farnie (1969, 95).
[17] Pile to Basil Liddell Hart, 13 May 1935, LHCMA, LH 1/575. The *Capona* basis reference probably refers to gangster-like cheating after Al Capone.
[18] RAF intelligence, 1935, TNA AIR 8/138.

Italian colony at Port Said, whose younger generation wore black shirts, perched on the breakwater, singing the fascist national anthem to troopships, whose passengers responded with the salute, '*Duce! Duce!*'[19]

From August 1935 onwards, following the failure of the Tripartite Talks in Paris and the end of the East Africa rain season, Italian military traffic escalated markedly as the final countdown to war began. The invasion commenced on 3 October. Between 1 February and 2 October 1935 Italian vessels had carried 193,200 passengers south through Suez to East Africa, paying around £750,000 in canal dues.[20]

The Failure to Close Suez

If anything, once the conflict commenced, the importance to Italy of Suez deepened and became integral to the success of the war effort against Abyssinia. In fact, during the opening month of the campaign the volume of Italian traffic was such that the record tonnage of January 1929 was surpassed.[21]

Rome was nervous lest the British closed the canal to their transports, a move being called for by some British parliamentarians, sections of the press and supporters of Abyssinia such as the League of Nations Union. To counter this prospect, articles in the Italian press began appearing from June 1935 which emphasised British obligations under the 1888 convention. The Italian Consul at Port Said Mario Conti was also charged with making strong representations to the British and Egyptian authorities on the spot. Indeed, British intelligence intercepted a communication from the Italian foreign ministry to the Alexandria legation, ordering the establishment of an intelligence service 'to enable us to be quickly and accurately informed of whatever may happen or be prepared on the banks of the canal'.[22]

From Rome, Drummond cautioned that 'however foolish it may sound … Italy might in such an eventuality [closure] consider seriously the possibility of war with Great Britain.'[23] Italian concern at this prospect was again indicated when the Italian minister in Cairo Emilio Pagliano saw

[19] Farnie (1969, 600).
[20] Ibid., 599.
[21] Ibid., 600.
[22] Italian foreign ministry to legation, Alexandria, 20 August 1935, TNA HW 12/194.
[23] Drummond to Hoare, 18 June 1935, TNA FO 371/19112.

Egyptian prime minister Nessim Pasha and pointedly enquired: 'Does Egypt intend to shut the Suez canal?' Pagliano's sudden recall to Rome in July was regarded by British high commissioner Sir Miles Lampson as reflecting his lack of success in this endeavour. This then produced Italian scare tactics, with the Italian government issuing a chilling warning through the Egyptian minister in Rome that 'it would be inadmissible that closing would be decided without Egypt being heard. Such closing would have as an immediate effect, a declaration of war against Egypt.' Mussolini's reinforcement of Libya in September 1935 gave substance to this threat. It certainly succeeded in frightening the Egyptians.[24]

In fact, it was in British hands as to whether or not Suez remained accessible to Italian vessels feeding the vicious onslaught against spear-wielding Abyssinian warriors. The National government of Stanley Baldwin never seriously contemplated closure. In June 1935 head of the foreign office Sir Robert Vansittart directed its legal adviser to study the legal aspects. He concluded the canal could only be denied to Italian ships as part of general League action, whose efficiency would be impaired should Italy send supplies in neutral vessels.[25] Anthony Eden, despite being a League stalwart and minister for League of Nations affairs, agreed. 'This does not arise at present anyway', he wrote. 'I trust we may never have to consider it.'[26]

In the autumn general election the Opposition mounted a close Suez campaign, declaring it the ultimate sanction. Foreign Secretary Sir Samuel Hoare put out an official statement accusing Clement Attlee (Labour leader) and Lloyd George of being 'warmongers': following their policy would destroy the League's basis—collective security. Hoare repeated the contrarian reasoning of his September address to the League's general assembly: 'We will take our full share boldly and loyally in collective action, but we refuse to take isolated action.'[27] Soon afterwards Chamberlain weighed in again, suggesting that imposing military sanctions 'would be to reduce the League to a farce'.[28]

Both Admiralty and Foreign Office were sensitive to the possibility of war flaring up accidentally as a result of Italy misconstruing defence

[24] Kelly to Hoare, 1 September 1935, TNA FO 371/19074.
[25] Beckett, 'Legalities of closing Suez', 11 June 1935, TNA FO 371/19117.
[26] Minute, 17 June 1935, ibid.
[27] Ibid., 11 November 1935.
[28] Ibid., 14 November 1935.

measures at Suez. The Suez Canal Defence Plan was conceived in 1934 to deal with a Far Eastern war. In the event of Japan becoming hostile, a fleet would be sent to Singapore via Suez, the fastest route, and the canal taken over for two weeks to ensure its safe passage. With the outbreak of what is tellingly referred to in British records as 'the Italo-Abyssinian Emergency', the plan needed to be adapted to meet the situation. There was an acute consciousness that while defence preparations elsewhere in the Mediterranean could be presented as purely precautionary, any at Suez could be misread as preparing to deny the canal to Italy. Thus the Admiralty repressed the War Office's proposal to create a new post of 'fortress commander, canal zone'[29] and Lampson was debarred from discussing the defence of Suez with the leak-prone Egyptian government.[30] Moreover, only moderate elements of the defence plan were instigated, more drastic steps held back lest Mussolini turned 'mad dog'. Thus garrisons for terminal ports and armed guards for merchant ships remained in Cairo; guns and port examination batteries stored at Alexandria where two destroyers were on standby to proceed to Suez (one at either end); four boom ships to guard the entrances requisitioned but not deployed; and two RAF squadrons stood by to prevent air attacks to sink block ships.[31]

Such was Admiralty nervousness at the prospect of war after Canada tabled an oil sanction at Geneva, that it overturned the previous assumption that Italy would not attempt to block Suez in response to its imposition. Since an oil sanction would probably thwart Italian plans to overcome Abyssinian resistance before the next rain season, Naval plans division warned that

> the blocking of the canal by Italy as part and possibly the first act of a sudden violent offensive policy against Great Britain, cannot be disregarded. If Italy has in fact contemplated … blocking … it must be assumed that the operation has been carefully thought out and will be resolutely prosecuted. A ship or ships can be … prepared in an Italian port and sailing at a judiciously chosen time, not necessarily by direct route, can enter the canal with little chance of detection … The initiative therefore lies with the enemy.[32]

[29] Naval plans division, 'Defence of the Suez canal', 11 December 1935, TNA ADM 116/3833.

[30] Hoare to Lampson, 14 October 1935, TNA FO 407/213.

[31] Admiralty, 'Suez canal area', July 1936, TNA ADM 116/3476.

[32] Naval plans division, 'Defence of Suez', 11 December 1935, TNA ADM 116/3833.

Taking no chances, Britain aligned itself with France in kicking an oil sanction into the long grass, embracing craven appeasement through the infamous Hoare-Laval Pact. One assumption underlying British calculations was that Italy would become bogged down in Abyssinia, thereby becoming increasingly amenable to reason. Until 11 February 1936, when Italy began a new offensive, this assumption had some merit. However, on 22 January British intelligence reported that an unknown quantity of mustard gas had passed through Suez and gas cylinders distributed for bomber aircraft.[33] The spraying of gas from the air onto the hapless Abyssinian warriors proved a turning point. Moreover, the assumption that the rain season would halt the Italian advance in April was belatedly realised to be false—heavy impeding rains did not fall until June.[34]

On 6 April Eden, Hoare's successor, conceded in Cabinet: 'The only effective action at the present stage would be the closing of the Suez canal.'[35] Public opinion certainly favoured such a course. Winston Churchill advocated delivering Italy an ultimatum: Suez would be denied unless it made terms with the League. A body within the foreign office propounded closure, or at least formally tabling a motion at Geneva so that, after the anticipated French rejection, blame for League defeat could rest with France.[36] Profoundly worried by these developments, Admiral Chatfield, chief of naval staff and chairman of the chiefs of staff, wrote a note to Sir Maurice Hankey, secretary to the cabinet and committee of imperial defence reminding him: that it was a tenet of imperial strategy that 'the canal should always remain open and free', concerned that if Suez were closed over Abyssinia this would set a dangerous precedent and 'weaken our international position' should the waterway have to be seized to facilitate the passage of a fleet to Singapore.[37] Chatfield wanted a rejoinder in Parliament and got it the same day from Eden: 'If honourable Gentlemen wish to take military action I must warn them that you cannot close the canal with paper boats.'[38] Such an insult to the capabilities of the Royal Navy would have been unthinkable in earlier times.

As it happened, Italian troops marched into Addis Ababa on 5 May 1936, leading Mussolini to proclaim victory. Germany's remilitarisation of

[33] Intelligence, 22 January 1936, TNA WO 106/283.

[34] Ellington to Brooke-Popham, 20 March 1936, LHCMA, BP II/5/32.

[35] Cabinet meeting, 6 April 1936, TNA CAB 23/83.

[36] See TNA FO 371/20472.

[37] Chatfield to Hankey, 6 May 1936, TNA FO 954/6A/140.

[38] *Hansard*, 5th Series, *CCCXI*, 6 May 1936, c.1736.

the Rhineland two months prior had concentrated British and French minds even harder on the rising Nazi menace. There was no inclination to continue sanctions, let along close Suez. They were lifted in July 1936, drawing the Italo-Abyssinian crisis to a close. In all, from February 1935 through July 1936 Italy spent 950 million lira on ships and canal dues, ferrying 595,204 men, 634,900 tons in supplies, 10,084 vehicles and 40,859 animals via Suez to achieve tainted victory.[39]

Suez in the Lead-up to the Mediterranean War

In Milan, on 1 November 1936, Mussolini made an overture to Britain: 'If for others the Mediterranean is a route, for us Italians it is life. We have said a thousand times ... that we do not intend to threaten this route.'[40] Eden's favourable response led Mussolini to propose a 'gentleman's agreement'. Against his better judgement, Eden was pushed by the chiefs of staff to follow up. The resulting Anglo-Italian declaration of 2 January 1937 recognised that British and Italian Mediterranean interests were compatible and agreed to respect the status quo.[41]

In May 1937 Chamberlain became Prime Minister. Describing himself as a 'go-getter' for peace, he was soon writing to Mussolini seeking to restore the traditional friendship. In these efforts, the status of Suez came up for discussion. An internal government debate saw Eden lead doubters while the chiefs of staff, their ministers and supporters believed that, despite the Rome-Berlin axis of November 1936, Italy could still be brought back on side. Hankey weighed in by emphasising that the great Italian fear was 'that we shall interrupt their communications with Abyssinia at the Suez canal'. Given this threat, he believed that 'it is still possible to come to terms with them'.[42]

Ultimately, the debate turned in favour of the military, who Chamberlain backed, bringing Eden's resignation. This paved the way for talks in Rome. The Suez issue was introduced by the Italian foreign minister, Count Ciano, causing consternation in Whitehall which feared the Italians might seek a director on the Suez Canal Company Board. In 1937, Italy retained second place among canal users, with 13.6% of traffic, behind Britain

[39] Gooch (2020, 17).
[40] *The Times*, 2 November 1936.
[41] Italy No. 1 (1937), CMD 5348.
[42] Hankey memorandum, 19 July 1937, TNA PREM 1/276.

(48%) but ahead of Germany (9.5%), Holland (8%), France (5%) and Japan (2.7%). Yet only Britain, France and Holland held directorships. An Italian director was opposed on a number of grounds: the admiralty worried they could become privy to defence plans; the treasury, given Britain's 51% shareholding, believed he might seek reduced dues; the board of trade contended canal usage did not confer automatic representation; the foreign office saw such an appointment as a step towards internationalising the waterway.[43] It came as a relief, therefore, when the Italian proposal was received, which merely reconfirmed the 1888 Constantinople Convention guaranteeing a neutral canal.

With this hurdle cleared, the Anglo-Italian agreement was signed on 16 April 1938. It did not, however, guarantee Italian neutrality. In defining Italy's probable wartime status Lord Halifax, Eden's successor, warned that 'Italian policy is likely to remain opportunist and that in any ... war she might be expected to throw in her lot with what she considers likely to be the winning side'.[44]

The Sudetenland crisis threatened to unleash another world war. On 26 September 1938 the admiralty authorised the initiation of 'preparation for canal control' phase of the Suez Canal Defence Plan.[45] Three ships were earmarked to act as boom ships and the canal company began distributing dredgers and salvage craft to clear any potential blockage. Naval control service staffs were also activated.[46] Although the defence plan considered Italian landings improbable, the man on the spot feared Italy might attempt to control both ends of the canal with a small military force ostensibly bound for Eritrea. This was a valid point. Ciano had suggested landings at Port Said and Suez and chief of the general staff, general Alberto Pariani, promised to undertake a technical investigation.[47] Mussolini, however, knew Italy was unready for war and played a key role in facilitating the Munich conference, leading to the suspension of defence preparations on 30 September and their rescinding three days later.[48]

[43] Interdepartmental meeting, 19 March 1938, TNA FO 371/21980.

[44] Halifax to Chamberlain, 2 June 1938, TNA FO 800/318.

[45] Admiralty to naval liaison officer, Port Said, 26 September 1938, TNA ADM 116/3835.

[46] Senior British naval officer, 'Report on the Suez canal defence plan', 1 November 1938, TNA ADM 116/3835.

[47] Ciano diary, 14 February 1938.

[48] Ismay note, 30 September, India Office to Government of India, 3 October 1938, TNA CAB 21/519.

On 26 October 1938 the Cabinet decided to bring into force the Anglo-Italian Agreement.[49] Chamberlain was pleased by Mussolini's recent intervention and the Foreign Office, seeking Mediterranean stability, was willing to accept Italy's withdrawal of 10,000 'volunteers' from Spain, met the precondition for implementation.[50] Britain and Egypt followed France in recognising Italian sovereignty over Abyssinia. In a message to Perth, Halifax spoke of 'a new era in the relationship between our two countries'.[51]

At Munich Mussolini invited Chamberlain for talks in Rome, which were to take place in January 1939. Complications arose after Italy made claims against France, including a canal director and reduction in dues. The same objections remained valid on the British side, added to which was a concern that to concede an Italian directorship would pave the way for German and Japanese claims for similar status.[52] The French shared this concern when Chamberlain and Halifax visited Paris, making it known that their directors would refuse to back an Italian nominee. Nor did they want to risk Italian representation on an important highway to French colonies.[53] Against this background, the Rome summit (11–14 January 1939) achieved nothing beyond pomp and ceremony. Neither side had a programme. Armed with umbrellas, Chamberlain and Halifax failed to impress Mussolini and Ciano, convincing them the German horse was the one to back. 'The British', Ciano recorded in his diary, 'do not want to fight.'[54]

The international situation deteriorated after Germany invaded Czechoslovakia in March 1939. That month, the Foreign Office supported the admiralty and war office in wanting batteries mounted at port Suez because of the exposed position of oil refineries there, which could be targeted by a submarine before defence plans became operational. This was despite the fact that the convention debarred permanent fortifications on either canal bank. To ameliorate the Italians, for now Egyptian gunners manned the batteries. Should Italy complain, the response would be that the guns were not mounted on either bank and not intended to interfere

[49] Cabinet, 26 October 1938, TNA CAB 23/96.
[50] Foreign office memorandum, 21 October 1938, TNA FO 434/5.
[51] Halifax to Perth, 26 October 1938, TNA FO 434/5.
[52] Foreign office memorandum, 18 November 1938, TNA FO 371/21944.
[53] Anglo-French conversations, November 24 1938, 6 December 1938, TNA CAB 24/280.
[54] Ciano diary, 12 January 1939.

with free navigation. Halifax admitted these were not convincing answers, but with appeasement floundering received Cabinet approval.[55]

Three weeks later, Italy invaded Albania and quickly formed a military alliance, the Pact of Steel, with Germany. Thereafter it became uncertain whether or not Italy would join Germany if it came to war over Poland, which Britain and France had guaranteed against external aggression. At dawn on 1 September 1939 German forces invaded Poland. At 2.30 p.m. the Suez Canal 'control telegram' was cabled to Cairo. Simultaneously, the British ambassador in Rome, Sir Percy Loraine, was instructed to inform Ciano that precautionary measures to defend Suez would be taken but were not directed against Italy.[56] Upon receiving the cable, the commander-in-chief Mediterranean fleet, admiral Andrew Cunningham, instructed canal authorities to avoid giving any impression that Suez was about to be closed to Italian vessels.[57] At 5.30 p.m. came the news that Italy had declared her neutrality, or 'non-belligerence' as an embarrassed *Duce* deigned to call it.

Until summer 1940, Mussolini sat on the fence, biding his time to see which way the war went. Despite the allied farce in Norway, he still refused to name a date of entry to Hitler. His hesitancy evaporated once Germany unleashed its devastating blitzkrieg campaign in the west. British intelligence now picked up increasing indications that Mussolini was on the verge of taking the plunge. This scenario revived fears that Italy might attempt to block Suez. The cabinet learned that Cunningham intended to stop Italian traffic through the canal once the situation became critical. In the meantime, there was the matter of the Italian aircraft carrier *Maraglia*. This 5000-ton warship had left Tobruk and was due at Suez on 25 May 1940, ostensibly heading for Massawa from there.

It emerged that *Maraglia* was following a programme drawn up months before, making it unlikely it was intent on a blocking mission. Loraine reported it was Italy's solitary aircraft carrier, with thirty planes aboard, hardly a willing sacrifice. Halifax came up with the solution adopted: that if it arrived at Port Said laden with aircraft the *Maraglia* be allowed through; if they were offloaded at Benghazi then it should be stopped.[58]

[55] Cabinet, 29 March 1939, TNA CAB 23/98.
[56] Admiralty to Loraine, 1 September 1939, TNA ADM 1/9878.
[57] Cunningham to canal authorities, 1 September 1939, TNA ADM 1/9878.
[58] War cabinet, 23 May 1940, TNA CAB 65/7/30.

Cunningham was given the option of blocking passage, but in the event decided to take no action.[59]

At 12.30 p.m. on 10 June 1940, the war cabinet convened not realising that in the evening Mussolini would declare war against Britain and France. But if the timing was obscure, there was enough evidence to warrant extreme precautions. The day before an Italian ship, *Verbania*, arrived at Port Said carrying munitions and cement, ideal for rapid sinking. Cunningham refused to allow it through the canal unless if first offloaded its cargo. The ship was being held pending a decision from London. Lampson strongly supported Cunningham, and the War Cabinet agreed to detain *Verbania* on a pretext concerning contraband control.[60] By the following day, all subterfuge surrounding the detention of Italian ships[61] became unnecessary as the European war engulfed the Mediterranean. For the duration of its belligerency as part of the axis, Italy would see Suez denied to its ships. As Italian east Africa withered on the vine, British, Commonwealth and imperial armed forces launched an offensive in January 1941, which led to the fall of Addis Ababa within three months and the restoration of Emperor Haile Selassie to his throne.[62]

Overview

Britain's failure to deny Suez to Italian warships, transports and supply ships during the Italo-Abyssinian crisis ranks as one of the 'what if' moments. 'Had we stooped the passage of Italian transports through the Suez canal, and the import of oil fuel into Italy', wrote Cunningham in his memoirs, 'the whole subsequent history of the world might have been altered.'[63] Unquestionably the measure could have been effected had the Baldwin government held its nerve. As Neville Chamberlain recognised in private: 'By putting his great army on the other side of the Suez canal Mussolini has tied a noose round his own neck and left the end dangling for anyone with a noose to pull.'[64] Britain was in the best position to play hangman. Expecting the role, the commander-in-chief Mediterranean

[59] War cabinet, 25 May 1940, TNA CAB 65/7/33.

[60] War cabinet, 10 June 1940, TNA CAB 65/7/55.

[61] Others were being held at Gibraltar and Port Sudan under the guise of contraband control.

[62] See Stewart (2016).

[63] Cunningham (1951, 177).

[64] Neville to Ida Chamberlain, 8 December 1935, CRL, NC 18/1/941.

fleet, admiral Sir William Fisher, who was supremely confident of defeating the Italians, wanted to act aggressively but his hands were tied by the Admiralty and his government.

Briefly, early in September 1935, Fisher's flagship, the aptly titled *Resolution*, with eight 15-inch guns, was moored at Port Said alongside the *Despatch*, a light cruiser, having come from Malta, which the Admiralty deemed too close to Italian bombers based on Sicily. Italian troopships came close by as they entered the canal. Cunningham recalled:

> The troops generally cheered and shouted 'Duce! Duce!' on passing, and for our benefit. One day during the dinner-hour a large transport crammed with two or three thousand soldiers arrived, all cheering defiantly at our two ships. As they passed they broke into the Fascist anthem 'Giovinezza'; but were considerably put out by the loud shouts of 'Encore!' from the hundreds of sailors on the forecastles of the *Despatch* and *Resolution*. It is impossible to describe the withering contempt the British bluejacket can put into his applause if he dislikes the entertainment or entertainer, and on this occasion their sarcastic shouts penetrated even the thick skins of the Italians.[65]

The British warships only remained at Port Said from 2 to 15 September, before moving to their new base at Alexandria. But the demeanor of their crews was symptomatic of a gung-ho attitude, which does not appear misplaced in view of Italy's subsequent military performance in the Second World War.

References

Sources

CRL, Cadbury Research Library, University of Birmingham
LHCMA, Liddell Hart Centre for Military Archives, Kings College, London
TNA, The National Archives, Kew, London

Bibliography

Cunningham, Viscount. 1951. *A Sailor's Odyssey*. London: Hutchinson.
De Bono, Emilio. 1937. *Anno XIIII. The Conquest of an Empire*. London: Cresset.

[65] Cunningham (1951, 174).

Farnie, Douglas. 1969. *East & West of Suez. The Suez canal in History 1854–1956.* Oxford: Clarendon Press.

Gooch, John. 2020. *Mussolini's War. Fascist Italy from Triumph to Collapse 1935–1943.* London: Allen Lane.

Mallett, Robert. 2015. *Mussolini in Ethiopia 1919–1935.* New York: Cambridge University Press.

Morewood, Steven. 2014. *The British Defence of Egypt 1935–1940: Crisis and Conflict in the Eastern Mediterranean.* London: Routledge.

Muggeridge, Malcolm, ed. 1980. *Ciano's Diary 1937–1943.* London: Phoenix.

Pratt, Lawrence. 1976. *East of Malta, West of Suez. Britain's Mediterranean Crisis, 1936–1939.* Cambridge: Cambridge University Press.

Stewart, Andrew. 2016. *The First Victory.* New Haven: Yale University Press.

PART V

Work and Migration

CHAPTER 17

The Suez Canal in Conversations Through Time: Evangelical and Waldensian Transnational Mobility in the Mediterranean

Paola Schellenbaum

Introduction

Egypte: Le vice roi a entrepris un voyage en Europe. On l'attend à Venise, puis à Florence et à Marseille, d'où il se rendra à Paris, à Berlin et à Vienne afin d'inviter les souverains de l'Occident à l'ouverture solennelle de l'isthme de Suez.

L'Echo des Vallées, 20 May 1869[1]

In the *Chronique politique*, brief political world news in *L'Echo des Vallées*, the Waldensian newspaper written in French and printed in

[1] [Egypt: The viceroy has made a trip to Europe. He is awaited in Venice, then in Florence and Marseille, from where he will go to Paris, Berlin and Vienna in order to invite the kings of the West to the grand opening of the isthmus of Suez.]

P. Schellenbaum (✉)
Società di Studi Valdesi, Torre Pellice (TO), Italy

B. Curli (ed.), *Italy and the Suez Canal, from the Mid-nineteenth Century to the Cold War*,
https://doi.org/10.1007/978-3-030-88255-6_17

Piedmont since 1848, the Suez Canal is mentioned several times. On 20 August 1869, it is reported that the waters of the Mediterranean joined the Red Sea while its grand opening resonated in the Waldensian valleys on 26 November 1869. The following month, on 5 December 1869, the newspaper writes about the initial construction of a Protestant church in Cairo, in the presence of the Prince of Prussia, with criticisms by the Catholic press that feared competitors in missionary work.[2]

L'Echo des Vallées and other evangelical newspapers, such as *La Buona Novella* (1851–1862 in Italian, printed in Turin) and *Le Témoin* (1875–1896), were part of Protestant transnational networks. Bibles and foreign printed material had started to circulate in limited numbers after the royal edict of 1730, obtained through the mediation of the English ambassador in Turin. In 1848 the freedom of press was introduced in Piedmont with civil and political rights for religious minorities, before the Albertine Statute inserted Italy into the current of European liberalism.[3] Waldensians, a Calvinist denomination, were mainly poor peasants but some families started to live at the outskirts of Pinerolo (*Pignerol*), on the French border and at the frontier between the valleys and Turin, for the sake of commerce or textile manufacturing.[4] After emancipation, they published newspapers to enlarge evangelization efforts and gradual diffusion of Protestantism in 'Italy and the Levant', as Edward Clarke (1820–1912), an English Baptist pastor based in La Spezia, defined his mission in 1866. British missionary societies reinforced a transnational perspective, opening new migration routes.[5] According to Census data of 1861 Protestants in Italy were 33,000, mainly in Piedmont, but they progressively expanded geographically in the peninsula and elsewhere, as a consequence of the Awakening.[6]

These newspapers are now preserved in the Waldensian Library in Torre Pellice (Turin).[7] The historical archives document the life and work of the Waldensians, after the Glorious Return (1690), as well as of other Protestant denominations in Italy.[8] Family papers are preserved in the Waldensian Studies Society's collections, founded in 1881 to study the

[2] *L'Echo des Vallées*, 20 May 1869; 20 August 1869; 26 November1869; 10 December 1869.

[3] Isabella (2012).

[4] Romagnani (1999), Barbero (2008).

[5] Raponi (2014).

[6] Ward (1992), Vogel (2011).

[7] Papini and Tourn (2005). See www.studivaldesi.org and www.fondazionevaldese.org

[8] See http://patrimonioculturalevaldese.org

Reformation and other religious movements, including Waldensian history and circular migrations, within a frame of a global history and religious pluralism. Nowadays, the Waldensian Church (Union of Methodist and Waldensian Churches) is a member of the World Communion of Reformed Churches and is active in Italy, Rio de la Plata and Switzerland. The American Waldensian Society and the English Committee in Aid of the Waldensian Church Mission are a consequence of a 'glocal' heritage.[9]

Crossing the established boundaries of regional or national studies, the aim of this chapter is to enlarge the thinking of the Mediterranean with its complex forms of hybridity, exploring mobility circuits through social networks, trade, migration and missionary encounters in a global history of the Suez Canal, and combining history and anthropology in 'archival itineraries'.[10]

Multi-sited Ethnography

The Suez Canal provides us with a privileged viewpoint in order to overcome essentialist and homogenizing representations, both in space and in time. Its relevance was echoed while interviewing Egyptian migrants in Port Said, Suez and Ismailia during a multi-sited ethnography on transnational families in the 1990s.[11] They wanted to share thoughts on the relevance of the canal in global and trade history, crossed by racial inequality and stratification. They showed interest in the involvement of Italians, thus exploring the new cultural forms to which changes in colonial subaltern situations had given rise, telling complementary stories.[12] While walking with them in the historic centre of Ismailia, founded and named after Pasha Ismail Khedive of Egypt in the 1860s, questions arose on ethnic, gender and race relations in different multicultural contexts, as well as on human mobility and long-distance travels. Before the canal, few reached the Far East and had to adapt to local culture in hybrid ways, while long-distance travels' extended boundaries meant to go beyond biblical geography, as expressed in Ferdinand de Lesseps's motto 'Diviser deux continents pour réunir deux mondes'.[13] In the Suez Canal Company, the division of

[9] Gosso (2015).
[10] Stoler (2010), Evans (2016).
[11] Schellenbaum (1999).
[12] Marcus (1995).
[13] Bonnet (1950), Karabell (2003).

labour along national lines—*européen* versus *indigènes*—according to a colonial segregated model, added gender as later organizing principle, with priority to the wives, sisters and daughters of Europeans, in 'multi-ethnic Egypt' until 1950s.[14]

Subsequently, these field-notes were developed in a historical ethnographic research on Waldensians and Evangelicals in the Mediterranean, with migrant biographies of the past. How was North Africa transformed into a migratory frontier throughout the nineteenth century? First, the development of trade, migration and tourism between Europe and the Levant, after the opening of the Suez Canal in 1869, was of strategic importance and brought to Egypt a flourishing expatriate community. Second, the wide missionary activity helped at first Europeans with their religious duties, education and medical care, and then also served the local population, both Muslim and other, through the translation and distribution of literature.

Instead of keeping these flows separate, seasonal and circular displacements—with floating taxonomies of race/ethnicity, religion and nationality—represent a rather complex phenomenon.[15] Yet, current attention on borders highlights their role as symptoms of growing ambivalence, since contact and conflict coexist.[16] Showing historically distinct, although overlapping, patterns of trans-Mediterranean movements, such larger currents would include exiles, missionaries and migrants with unexpected consequences on both sending and receiving countries. Current debate on transnationalism and cosmopolitanism in anthropology is inspiring also for framing biographical itineraries of the past.[17] How did missionary activity—both Catholic and Protestant—influence migratory flows and, on a broader level, how did local education in missionary endeavours prove to be decisive moments in the history of North Africa, 'not as yet another episode in European empire'?[18]

A global history perspective brings to the fore relationships through time and empires appear as 'dynamic sites articulated through intricate webs of connection, across and between binaries of power'.[19] Drawing on recent research on imperial encounters, ethnocentric vocabulary used by

[14] Curli (2014, 554).

[15] Cohen and Sheringham (2016), Bauman (2016).

[16] Graziano (2018).

[17] Appiah (1997), Driessen (2005), Amelina et al. (2013), Glick Schiller and Irving (2014).

[18] Clancy-Smith (2011, 249).

[19] Haggis et al. (2017, 107).

missionaries such as 'pagan' and 'barbarian' emphasizes their 'civilizing' effort, while creating more nuanced and shared contact zones of mutuality.[20] Difference was not only a language barrier.[21] It was rather a way to conceive the Mediterranean as imagined and culturally constructed as 'Europe was made by its imperial projects, as much as colonial encounters were shaped by conflicts within Europe itself'.[22]

Mobility Circuits and Religious Transnational Networks

Descending from a medieval itinerant religious movement, although at times segregated within an Alpine ghetto, Waldensians migrated to different destinations (i.e. France, South America, Mediterranean shores, United States), as reported in Synod debates over the years.[23] In the wake of the resumption of persecutions of the Jews and of the Waldensians just before 1848, Protestants would also go on to play a role in the liberal *Risorgimento* out of all proportions to their small numbers, as international factors became central to Italy's unification.[24] *Casa Monnet*, the first evangelization station in Pinerolo, near Turin, illuminates the frontier between the Waldensian valleys, Italy and the Levant. Built in 1704, and adjacent to a Catholic cloistered monastery, the villa was bought in 1850 by David Monnet (1808–1902), an exiled medical doctor, who returned after graduating from the University of Montpellier and upon recognition of his degree by the University of Turin.[25] His mother was French and married to a pastor involved in the Waldensian Bible Society (1816–1829), aimed at bibles and books' circulation.[26] Casa Monnet was a 'crossroad of biographies'.[27] The *Société évangélique de Genève*, where Waldensian pastors received their theological education, would initially fund the station in Piedmont. Missionaries sent in the 1840s by the American Bible Society,

[20] *La Buona Novella*, 16 January 1857; 31 January 1857; 14 February 1857.
[21] Mitchell (1988), Sharkey (2008).
[22] Cooper and Stoler (1997, 1).
[23] Pons (1948).
[24] Spini (1998 / 1956), Romagnani (1999).
[25] Register, 1853, AS Pinerolo, n. 1479. Letters Patent July 21, 1848, ASU, University of Turin, Faculty of Medicine, 1848 IX.B1, n. 5702 (Correspondence). See www.studivaldesi.org/dizionario/xixsecolo.php
[26] Maselli (2006).
[27] Schellenbaum (2014).

such as Robert Baird (1798–1863), had previously toured this remote borderland as well as other English travellers since 1810s.[28] On January 1853 Lord Shaftesbury (1801–1885) came as president of the English committee in Aid of the Waldensian Church Mission, operating since 1825. Also Methodists joined Protestants missions for political pressure in support of religious freedom and William Arthur (1819–1901), a Wesleyan minister advocate of modernization in Italy and India, in 1860 wrote *Italy in Transition* on his return from the peninsula, exciting 'intense interest'.[29]

Soon after becoming a British colony, Malta had emerged as a centre for evangelization in the Mediterranean, closely tied to abolition movements. Since 1815 the Church Missionary Society had established on the island, and missionaries arrived in Egypt in 1819.[30] Founded in London in 1804, the British and Foreign Bible Society was also based in Malta and operated with moral and spiritual aids at the port of Alexandria, among sea travellers and sailors. After the opening of the Suez Canal, it opened a permanent depot in Alexandria in 1871 and started to distribute printed material in the Nile Delta and in Port Said, among seamen and travellers sailing to the Holy Land or to the Far East. These societies' Christian mission, built on the 'transformative capacity' of Bible reading, was also an economic enterprise as their agents sold, albeit cheaply, printed and translated material. They never sponsored a church and their audience was in great flux.[31]

A minister of the newly formed Free Church of Scotland, Robert Stewart (1812–1887), visited the Waldensians in 1844 to support their educational institutions and the following year established the first missionary chaplain of the Scottish Church in Livorno, a Tuscan flourishing free port and business community with exchanges across the Mediterranean.[32] Migratory movements from Pinerolo went to Switzerland, France and to the Mediterranean basin, as Islam was not regarded as a threat to Europe.[33] Stewart made a four months' journey to Egypt through Malta in 1853, writing a travelogue *The Tent and the Kahn*, while Waldensian pastor Giovanni Turino (1824–1909) had been sent to Constantinople to spiritually lead the French-speaking community, assist-

[28] Villani (2014).
[29] *The Waldensian Review*, No. 120, Summer 2012, 6.
[30] Clancy-Smith (2011).
[31] Howsam (2002), Sharkey (2011, 441-43).
[32] Brown (1890).
[33] Evans (2016).

ing war injured people from the carnage of the Crimean war. Mobility networks also emerge from other sources: in letters addressed to his eldest son Louis, living in Alexandria since 1840s, Jacques Vinçon (1794–1869) sent pastoral news from *Casa Monnet* writing about a new place for Sunday worship, schooling and a charity for poor people that also hosted a group of young women—both Catholics and Protestant—with two foreign teachers who had previously been in Saint Petersburg where they had met Jacques's daughter, Joséphine Renée, on her way back to Pinerolo.[34] In 1857, David Monnet left Pinerolo, after his marriage to Helen Walker (1813–1879), an English woman who was living in Turin.[35] Correspondence with Quintino Sella (1827–1884) shows their social networks in the city.[36]

The Suez Canal in Marsh's Diary

Social networks included Caroline Crane (1816–1901) and her husband George P. Marsh (1801–1882), the first US Minister to the newly formed Italian Kingdom, who arrived in Turin when Cavour had just died.[37] Helen Walker worked for the French Embassy, where Monnet also served as a medical doctor. In 1859, she was engaged in an international Women's Committee for the help of injured people at war with Anna Koppmann (1819–1885) and Felicita Bevilacqua (1822–1899).[38] Subsequently, the *Associazione Nazionale Filantropica delle Donne Italiane* (National Philanthropic Association of Italian Women) opened under the auspices of Giuseppe Garibaldi and lasted from 1861 to 1864.[39] Chair of the Committee was Anna Pallavicino Trivulzio, cited in Marsh's diary[40] for her engagement in the modern medical assistance to war injured people. In a brief note to Felicita Bevilacqua, Helen Monnet shared with her friend Marsh's benevolence towards the Association.[41] The Suez Canal was under

[34] Vinçon to Louis, 21 January 1851 and 6 October 1851, ASSV, Vinçon-Turin Family papers.

[35] Marriage Act, 28 July 1857, ASTV, Turin Consistoire.

[36] Walker to Sella, 10 August 1861, Sella Foundation (Correspondence).

[37] *La Buona Novella*, 15 June 1861, no. 11.

[38] Pallavicino Trivulzio to Caroline Marsh, 2 April 1864, UVSC-GPM Collection (Correspondence).

[39] Folders 1067, 1082, 1091, 1737, 1741, BCVr Bevilacqua Papers (Correspondence).

[40] Diary, 29 January 1862, UVSC-GPM Collection. The seventeen notebooks are being digitalized.

[41] Folder 1091, BCVr Bevilacqua Papers (Correspondence).

construction in those years and these women's activity was looking south. Through transnational solidarity networks it was intended to support schools for the poor in Southern Italy where migration, ethnic and religious segregation was common. It shows how 'culture, private life and political action were intertwined'.[42] In Marsh's words, it made life abroad 'so full and satisfactory'.[43]

Marsh's diary is preserved at the University of Vermont, because her husband was a Vermont-born lawyer and legislator, long-serving diplomat, geographer, philologist and historian.[44] She was 'sixteen years her husband's junior but neither the age gap nor Marsh's awesome erudition made her passive or subordinate [...] she kept up her own literary pursuits and contacts'.[45] They were both raised in Christian orthodoxy but her faith was evangelical.

Being a traveller, Caroline Marsh had started to write a travelogue in Constantinople in 1850, sharing thoughts on both public and private life, when the American couple visited the Gulf of Suez during their forty days' trip to the Holy Land. Years later, the Suez Canal is mentioned in Marsh's diary.[46] In 1861 and in 1863 the couple met Ferdinand de Lesseps. Their curiosity and interest about the construction of the canal led them to different comments: English opposition meant for the French to be in charge of the entire financing, becoming a national interest instead of a transnational one and slowing down the terms of the construction.[47] Two years later, the canal is again commented upon: the British 'are hated by the entire world' for their opposition to the digging works.[48]

When not travelling, the American couple lived in Turin and in Piobesi, near Pinerolo. From this small village, they visited the Waldensian valleys, extending the mountain trip to the Monviso ('the most magnificent sight') and proceeding to France following the Durance, a river draining towards the Mediterranean, a popular migration route.[49] Some years before, well-known European novelist Fredrika Bremer (1801–1865), a feminist from

[42] Sodini (2006, XV).

[43] Daniels in Ducci (2012).

[44] Lowenthal and Quartermaine (2004).

[45] Lowenthal (2000, 45-46).

[46] Diary, 23 October 1861, UVSC-GPM Collection.

[47] Diary 3 November 1861, UVSC-GPM Collection.

[48] Diary 17 November 1863, UVSC-GPM Collection.

[49] Diary 21-24 July 1863, UVSC-GPM Collection. These pages have not been included in the Italian translation and I was able to consult them in Vermont in 2017.

Scandinavia, who had been in Palestine and in the United States, had also come to this remote border region, with in-depth spiritual encounters described in *Two years in Switzerland and Italy*, translated in 1861. As mentioned in Evans, she travelled in Europe, arriving in Piedmont before going south.[50]

In Piedmont George P. Marsh wrote pioneering insights on environmentalism and anthropogenic change and his book *Man and Nature*, published in 1864, would describe social interdependence, a unique concept of human geography at the times.[51] In a later edition, he wrote about the benefits of the Suez Canal against adverse ecological effects, with both admiration and concern about practical consequences of free trade and 'unwanted interchanges of marine life'.[52] Partly, his critical view was due to Italy's structural weaknesses: the living conditions among the rural population, where poverty was endemic, would worsen when crop prices plummeted because of cheap rice and silk arriving from the East.[53] Caroline Marsh would instead envision growth favoured by the opening of the canal and increased circulation of liberal principles, supported by Italian democrats led by Cavour. In brief, Marsh's question was 'whether man is of nature or above her', and it resonates even today.[54]

Migrations in the Mediterranean

As Curli notes, 'the Suez Canal has recently returned to the attention of historical research, either in the framework of the new global history of migrations and imperial connections or from the perspective of economic and business history'.[55] While smaller in numbers and little studied, as compared to larger trans-oceanic migration waves, migratory circuits in the Mediterranean in the nineteenth century—the so-called 'silent revolution'[56]—highlight the extent to which the legal and social privilege of capitulations attracted Italians and Greeks. New migration routes and biographical itineraries created a sort of cultural hybridity with multi-layered

[50] Evans (2016, 444).

[51] Marsh (1864), Lowenthal (2016). The Waldensian library preserves a signed copy of the book, donated in 1864.

[52] Lowenthal (2000, 390).

[53] Lowenthal (2000, 337).

[54] Lowenthal (2000, 289).

[55] Curli (2014, 553).

[56] Barone (2018, 401).

identities, with migrants rooted both in Europe and in North Africa or Ottoman Empire. Older generations were different from newly arrived people, thus diasporas are conceived in their complexity as contact zones and mobile space.[57] Chain migrations offer new insights on migrant agency, including working-class women, at the intersection of micro and macro-historical changes and dynamics,[58] including globalisation issues of forced migrations, especially for women travelling alone.[59]

Waldensians came mainly from the Alpine border zone and the first reported migrants arrived in Egypt in 1879. Since then, almost every year people from the valleys continued to depart until 1929.[60] According to Cairo Census data, 37 percent of the total population was immigrant in 1848 and all neighbourhoods were mixed.[61] Italians in Egypt were around 2000. By 1907, they had grown to 34,926, both as permanent residents in large cities and smaller sub-urban and countryside areas and as temporary workers on the seashore. Between 1902 and 1907 emigration reached its peak with 12,000 persons departing per year from Southern Italy to Northern Africa: islanders preferred circular migrations, as often observed on frontiers, to work in sea-related jobs.[62] During the war the British tried to further discipline working-class migration through their consulate in Naples but by 1917 Italians had grown to 40,198 persons in Egypt, becoming one of the many diasporas.[63]

The permeability of domains observed by anthropologists in situated or diaspora cosmopolitanism actively place migrants in processes of cross-border participation and practices of translation, putting into question a nation-state-centred approach. Being a dynamic historical construct, the ethnic/religious lens further complicates issues of cultural identity and social dynamics. After 1882, the European migrant population in Egypt exploded not simply as 'displaced communities of European origin', but as 'unique cultural configurations', thus offering 'new constructions of European-ness': experiential categories of cosmopolitanism and

[57] Isabella and Zanou (2016, 4-7).
[58] Biancani (2019).
[59] Clancy-Smith (2011).
[60] Bodrero (1995).
[61] Alleaume and Fargues (1998).
[62] Fauri (2015, 161-167).
[63] Gabaccia (2000).

nationalism were not mutually exclusive, but were instead co-dependent processes.[64]

Christian missionaries represented a 'polyglot assortment of churches, orders, and societies; they came from many countries; they spread out in cities, towns, and villages'.[65] Moreover, members of French or Swiss Reformed churches went mainly to Tunisia, while British residents in Egypt would attend the Anglican Church in Alexandria since 1839 and in Cairo since 1876 where there were also other denominations. The first American Presbyterian missionaries came to Egypt in 1854 and embraced a strict anti-slavery line.[66] For the sake of Bible reading, they opened schools, medical facilities and public libraries carrying out literacy campaigns among peasantry in Luxor and in Assiut. With their motto 'Each One Teach One', they differentiated themselves from Catholics while reading the Bible in Arabic and giving lay members leading roles in the church. From 1863 the Evangelical Theological Seminary on the houseboat *Ibis* served as a floating learning centre until 1926. In 1871, Anna Young Thompson (1851–1932) arrived to engage in women's rights and literacy, as reported in her diaries and correspondence.[67] In the same year, the British and Foreign Bible Society had opened a permanent depot in Alexandria with colporteurs[68] travelling to the Delta and to Upper Egypt. Protestants exerted an important social impact towards Coptic Orthodox and Muslims, as many people profited from educational opportunities for women and gender roles reconfiguration, without becoming converts.[69]

[64] Clancy-Smith (2011, 48-49).

[65] Sharkey (2017, 200).

[66] The Evangelical Presbyterian Church of Egypt became autonomous in 1926: https://www.oikoumene.org/en/member-churches/evangelical-presbyterian-church-of-egypt-synod-of-the-nile

[67] Anna Young Thompson Papers (1868–1931): https://www.history.pcusa.org/collections

[68] They were holy books' vendors.

[69] Sharkey (2008, 25-30).

Waldensians and Evangelicals in the Suez Canal Zone

In the Waldensian historical archives, records from Egypt are preserved.[70] Two colporteurs had been trying to create an Italian-speaking evangelical congregation since 1858, when digging works of the Suez Canal were to start, but only in 1873 an international Committee of different Christian denominations mobilized for the creation of *Chiesa evangelica italiana* in Alexandria.[71] No Waldensian pastor was available and the congregation started to flourish under the leadership of Giovanni Cerioni, a convert who had arrived from Malta, with weekly meetings, public conferences and books' sales, sent by Claudiana, the Italian Protestant publishing company.[72] In previous years, the nearly 250 Protestants of Alexandria would worship in the Prussian Embassy. In 1864 a Lutheran congregation was formed in Cairo and in 1873 the Calvinist *Communauté Evangélique de langues française et allemande* was also established. In 1878 Khedive Tawfiq officially recognized the Protestant community of Egypt as Christian and established regulations for marriages and other personal status affairs of the Egyptian Evangelical Church (former American Presbyterian Mission). Since 1895 the Swiss Ernest Cramer (1837–1923) started giving funds to the French-speaking Protestant congregation of Cairo, supported by the *Société évangelique de Genève*. Keen on Egyptology, he was connected to Edouard Naville (1844–1926), who supported the French-speaking Protestants while in the field from 1883 to 1912. In those years, Ernesto Schiaparelli (1856–1928) was involved with Catholic missionaries during the Italian Archaeological Mission (1902–1920) that was to enlarge the Egyptian Museum's collections in Turin. They had both studied in Paris with Gaston Maspero. After 1909, Cramer also supported the Waldensian Church (1907–1920) that had moved for worship from the American Presbyterian Mission, where it had been hosted, to Sh. Kantaret el Dekka, with a school and book depository.

[70] *Carte d'Egitto* is a conference held in Turin on 27 November 2019; see https://archiviodistatotorino.beniculturali.it/giornata-di-studi-carte-egitto-2019/

[71] Gamble to Prochet, 21 March 1870. Reinhardt to Prochet, 2 April 1872, on international Committee. Genta to Prochet, 29 March1872; 22, 29 April 1872; 7 May 1872; 7 July 1872, ASTV, Egypt (Correspondence); see: http://patrimonioculturalevaldese.org/en/scheda?id=f5751246-5df0-3c3f-bdbd-65c80dafc9b0

[72] Prochet to Cerioni, 25 January 1875, ASTV, Egypt (Correspondence).

In a letter to Matteo Prochet (1836–1907), president of the evangelization committee, the director of a refuge for the poor in Cairo and Alexandria had again asked in 1904 to send a Waldensian pastor for Italian migrants[73] while Cerioni was preaching at the Scottish missionary school. In those years, the Italian consulate had also opened a school, in addition to religious Catholic orders whose schools for *élites* were in French and Italian, but Protestant schools taught Arabic and Italian in order to reach the larger needy indigenous population, especially girls.[74] Waldensians were a diversified population. In the register, one can trace members' names from Alpine valleys and Southern Italy, but there were also some new converts.[75] Their social integration was good, even during the cholera crisis in 1902, when they would act in inter-faith and social solidarity.

On 3 January 1906, Thomas Hall, a Presbyterian of New York and member of the American Waldensian Aid Society, wrote a letter from Naples to Matteo Prochet, just before embarking for Alexandria, assuring 'loyal affection'.[76] When the young pastor Rinaldo Malan arrived in Egypt in 1907, with a document for full recognition of the Waldensian Church, he was welcomed.[77] Some members came from Southern Italy and their strong bond became evident in the aftermath of the Messina earthquake (1908). Fund-raising in Cairo and Alexandria among Waldensians allowed them to express a sort of 'diffused patriotism' compatible with a profound social disunity that migrations would worsen. Pastor Chauvie and his wife Marcella Eynard, descendent of doctor David Monnet, had died in that earthquake and in Egypt 727.41 francs were collected in their memory,[78] building on solidarity networks around tragedy. Dickie argues that the patriotic narratives in that period constructed imagined communities, but he does not mention transnational faith-based solidarity.[79] Gratitude to Arturo Muston, the new head of the evangelization committee, was expressed by Mr. Colucci, who was the treasurer of the congregation, for

[73] Petition to Prochet, 16 February 1904. Rudolph to Prochet, 19 February 1904, 23 March 1904, ASTV, Egypt (Correspondence).

[74] Russell (2004, 110).

[75] Membership register, ASTV, Egypt (Correspondence).

[76] Naples Waldensian Church, ASTV (Correspondence). Relationships with the American Protestant Churches had started in 1853 when Waldensian Moderator Jean Pierre Revel (1810–1871) had made his first trip to New York for networking.

[77] Annual Report 1908–1909, Document, ASTV, Egypt (Correspondence).

[78] 20 February 1909, ASTV, Egypt (Correspondence).

[79] Anderson (1983), Dickie (2006).

supporting the missionary activity in Egypt,[80] and Ernest Cramer would personally welcome him in 1911, after sending the train timetable from Alexandria to Cairo.[81]

Multiethnic and polyglot population of the Suez Canal zone was growing diverse and since 1912 the British and Foreign Bible Society had opened in Port Said an administrative headquarter for Egypt, building a Bible House named 'Shaftesbury Building', with local, regional, and global distribution, while the canal would also provide easy access to travellers to the Holy Land and other destinations. In 1914 colporteurs sold Bibles in 57 different languages.[82] The Waldensian annual reports offer details about pastoral visits to Waldensian families not only in larger cities but also in Ismailia, Port Said, Suez, Mansura, Zeitun. Conferences in Sh. Kantaret el Dekka in Cairo were given by lay people to promote dialogue with Catholics, Jews, Orthodox and Muslims.[83] Clerical reactions were soon to appear with other rivalries due to war times but more people attended and support increased. The Waldensian Church in Cairo was dismissed in 1920 as the inter-war period required to divert evangelization efforts to Southern Italy.

Conclusion

Current debate on borderlands, transnational encounters and sociability, both as public narrative and as social practice, challenges the dominant sedentary paradigm. The creation of difference—even when racialized—was not natural but it was a fact, sometimes an opportunity, considering European consular agents or missionaries' complex transnational backgrounds that arrived to Egypt as a consequence of the opening of the Suez Canal. Such multiethnic diversity significantly problematizes spatial theory based on national or religious affiliations, that is, Europeans versus Muslims as homogeneous categories, as they were not the sole source of belonging.[84] Mobility flows were not separate phenomena and race/ethnicity, religion and nationality would intertwine in complex ways with

[80] Colucci to Muston, 24 October 1908, ASTV, Egypt (Correspondence).
[81] Cramer to Muston. 19 January 1911, ASTV, Egypt (Correspondence).
[82] Sharkey (2011, 446).
[83] Annual Report 1914–1915, ASTV, Egypt (Correspondence).
[84] Burbank and Cooper (2011, 12).

larger currents including exiles, missionaries and migrants, and their relations with the local population, both Muslim and other.

For the international missionary movement, Egypt was strategic.[85] In this period of greatest expansion of the British occupation, work intensified as travelling missionary women grew in numbers and became outspoken critic. They were given freedom of movement with a career outside the home and this would liberate them from social restrictions as well as, reciprocally, the women they encountered in the field: 'over the years, Egyptian Copts and Muslims alike found many of the missionaries' ideas and practices compelling, and that is perhaps why missionaries seemed so subversive and threatening to Coptic Orthodox patriarchs and to men like Hasan al-Banna, the founder of the Muslim Brotherhood'.[86] The Missionary Conference in Cairo (1906) was about education and shed light on the premises of the World Missionary Conference in Edinburgh (1910) that was to overcome denominational barriers. Yet, stereotypes against 'backward and primitive south-easterners Europeans' remained.[87] People of faith would also grasp alternatives of amity and coexistence, beyond binary visions of contrasting civilizations and 'within but resistant to imperial contact zones'.[88] Their general attitude was Christian universality, religious liberty and belief as a matter of individual choice.

While representing an important but ignored chapter of global migration history, migrants to Egypt imagined a culturally constructed space through notions of borderland, with social exchanges and missionary encounters. Waldensians and Evangelicals were an active part in this process. Since 2015, this perspective has inspired an ecumenical project named 'humanitarian corridors' that links the two shores of the Euro-Mediterranean region, as a message of peaceful cohabitation. After the so-called refugee crisis, safe passages were offered to more than 3000 vulnerable people.[89] One of the premises was the adoption of a human-rights framework, in the difficult effort of transforming Europe into a genuinely 'diverse' society.

[85] Sharkey (2008).

[86] Sharkey (2008, 219).

[87] Borutta and Gekas (2012).

[88] Haggis et al. (2017, 2).

[89] Zanfrini (2019). Mediterranean Hope is a project funded by the Waldensian Board and run by the Federation of Protestant Churches in Italy and the Community of S. Egidio (www.mediterraneanhope.com). Similar programmes are also sponsored by the Catholic Church.

References

Abbreviations

AS, Archivio storico Pinerolo
ASTV, Archivio storico Tavola Valdese, Torre Pellice
ASSV, Archivio Società di Studi Valdesi, Torre Pellice
BV, Biblioteca Valdese Torre Pellice
BCVr, Biblioteca Comunale Verona
Sella Foundation, Biella
PHS, Presbyterian Historical Society, The National Archives of the PC (USA)
ASU, Archivio storico, Università di Torino
UVSC-GPM Collection, University of Vermont

Bibliography

Appiah, Kwame A. 1997. Cosmopolitan Patriots. *Critical Inquiry* 23: 617–639.

Alleaume, Ghislaine, and Philippe Fargues. 1998. *'Voisinage et frontière: résider au Caire en 1846' in AA.VV. Urbanité arabe*, 77–112. Paris: Sindbad-Actes Sud.

Amelina, Anna, Thomas Faist, and Devrimsel Nergiz. 2013. *Methodologies on the Move. The Transnational Turn in Empirical Migration Research.* London: Routledge.

Anderson, Benedict. 1983. *Imagined Communities. Reflections on the Origin and Spread of Nationalism.* London: Verso.

Barbero, Alessandro. 2008. *Storia del Piemonte.* Turin: Einaudi.

Barone, Giuseppe. 2018. *Storia mondiale della Sicilia.* Bari-Rome: Laterza.

Bauman, Zygmunt. 2016. *Strangers at Our Door.* Cambridge: Polity Press.

Biancani, Francesca. 2019. *Sex Work in Colonial Egypt. Women, Modernity and the Global Economy.* London: I. B. Tauris.

Bodrero, Giuseppe. 1995. 'The Emigration Phenomenon in the Waldensian Valleys between 19th and 20thC', B.A. Thesis, University of Turin.

Bonnet, E. George. 1950. Ferdinand de Lesseps. *Revue des deux mondes* 6: 1–29.

Borutta, Manuel, and Sakis Gekas. 2012. A Colonial Sea: the Mediterranean, 1798–1956. *European Review of History* 19: 1–13.

Brown, Wood J. 1890. *An Italian Campaign, or the Evangelical Movement in Italy, 1845-1887.* London: Hoddwer and Stoughton.

Burbank, Jane, and Fredrik Cooper. 2011. *Empires in World History. Power and the Politics of Difference.* Princeton: Princeton University Press.

Clancy-Smith, Julia A. 2011. *Mediterraneans.* Berkeley: University of California Press.

Cohen, Robin, and Olivia Sheringham. 2016. *Encountering Difference. Diasporic Traces, Creolizing Spaces.* Cambridge: Polity Press.

Cooper, Fredrik, and Anna L. Stoler. 1997. *Tensions of Empire. Colonial Cultures in a Bourgeois World*. Berkeley: University of California Press.
Curli, Barbara. 2014. Dames Employées at the Suez Canal Company: the "Egyptianization" of Female Office Workers, 1941-56. *International Journal of Middle East Studies* 46: 553–576.
Dickie, John. 2006. Timing, Memory and Disaster. *Modern Italy* XI: 147–166.
Driessen, Henk. 2005. Mediterranean port cities: Cosmopolitanism reconsidered. *History and Anthropology* 16: 129–141.
Ducci, Lucia. 2012. *George P. Marsh Correspondence: Images of Italy, 1861-1881*. Lanham: The Rowman & Littlefield Publishing Group.
Evans, Richard J. 2016. *The Pursuit of Power: Europe, 1815-1914*. London: Allen Lane Penguin.
Fauri, Francesca. 2015. *Storia economica delle migrazioni italiane*. Bologna: Il Mulino.
Gabaccia, Donna. 2000. *Italy's Many Diasporas*. London: UCL Press.
Glick Schiller, Nina, and Irving Andrew, eds. 2014. *Whose Cosmopolitanism? Critical Perspectives, Relationalities and Discontents*. London: Berghahn Press.
Gosso, Elisa. 2015. 'Glocal Heritages. Memories, Places and Waldensian Transnational Routes', Ph.D. dissertation, University of Turin.
Graziano, Manlio. 2018. *What is a Border?* Stanford: Stanford University Press.
Haggis, Jane, et al. 2017. *Cosmopolitan Lives on the Cusp of Empire. Interfaith, Cross-Cultural and Transnational Networks, 1860-1950*. London: Palgrave Macmillan.
Howsam, Leslie. 2002. *Cheap Bibles. Nineteenth-century publishing and the British and Foreign Bible Society*. Cambridge: Cambridge University Press.
Karabell, Zachary. 2003. *Parting the Desert. The Creation of the Suez Canal*. London: John Murray.
Isabella, Maurizio. 2012. Freedom of the Press, Public Opinion and Liberalism in the Risorgimento. *Journal of Modern Italian Studies* 17 (5): 551–567.
Isabella, Maurizio, and Zanou Constantina. 2016. *Mediterranean Diasporas. Politics and Ideas in the Long 19th Century*. London and New York: Bloomsbury.
Lowenthal, David, and Quartermaine Luisa. 2004. *Un'americana alla corte dei Savoia (translation of Caroline Crane Marsh's Diaries)*. Turin: Allemandi.
Lowenthal, David. 2000. *George Perkins Marsh: Prophet of Conservation*. Seattle: University of Washington Press.
———. 2016. Origins of Anthropocene Awareness. *The Anthropocene Review* 3: 52–63.
Marcus, George E. 1995. Ethnography in/of the World System. *Annual Review of Anthropology* 24: 95–117.
Marsh, George (1864) 2003, *Man and Nature: Or, Physical Geography As Modified by Human Action*, ed. David Lowenthal, Seattle: University of Washington Press.

Maselli, Domenico. 2006. Storia della Società biblica valdese (1816-1829). *Bollettino Società Studi Valdesi* 198: 191–211.

Mitchell, Timothy. 1988. *Colonising Egypt*. Cambridge: Cambridge University Press.

Papini, Carlo, and Giorgio Tourn. 2005. *Claudiana (1855-2005). 150 anni di presenza evangelica nella cultura italiana*. Turin: Claudiana.

Pons, Theophile J. 1948. Actes des Synodes des Eglises Vaudoises (1692-1854). *Bollettino Società Studi Valdesi*. 88.

Raponi, Danilo. 2014. *Religion and Politics in the Risorgimento: Britain and the New Italy, 1861-1875*, 73–111. London: Palgrave Macmillan.

Romagnani, Gianpaolo. 1999. Italian Protestants. In *The emancipation of Catholics, Jews and Protestants. Minorities and the nation state in nineteenth-century Europe*, ed. Liedtke Rainer and Wendehorst Stephan. Manchester: Manchester University Press.

Russell, Mona L. 2004. *Creating the New Egyptian Woman*. London: Palgrave Macmillan.

Schellenbaum, Paola. 1999. 'Bab El-Had (The door's limits): multiple identities, social boundaries and communication networks in Egyptian migrant families between local and global', Ph.D. dissertation, University of Turin.

———. 2014. Casa Monnet a Pinerolo: un crocevia di biografie. *La Beidana-Cultura e Storia nelle Valli valdesi* 80: 4–27.

Sharkey, Heather J. 2008. *American Evangelicals in Egypt. Missionary Encounters in an Age of Empire*. Princeton: Princeton University Press.

———. 2011. The British and Foreign Bible Society in Port Said and the Suez Canal. *The Journal of Imperial and Commonwealth History* 39 (3): 439–456.

———. 2017. *A History of Muslims, Christians, and Jews in the Middle East*. Cambridge: Cambridge University Press.

Sodini, Elena. 2006. *Il buon nome della famiglia e l'amor per la patria*. In *Famiglia e nazione nel lungo Ottocento italiano*, ed. Ilaria Porciani, 107–129. Rome: Viella.

Stoler, Anna L. 2010. *Along the Archival Grain. Epistemic Anxieties and Colonial Common Sense*. Princeton: Princeton University Press.

Spini, Giorgio. 1998. *Risorgimento e protestanti*, Milan: Il Saggiatore (I ed. 1956).

Villani, Stefano. 2014. Dal Galles alle Valli: Thomas Sims (1785-1864) e la riscoperta britannica dei valdesi. *Bollettino della Società di Studi Valdesi* 215: 103–171.

Vogel, Lothar. 2011. 'Comunità e pastori del protestantesimo italiano', in *Cristiani d'Italia*, Treccani.it https://www.treccani.it/enciclopedia/comunita-e-pastori-del-protestantesimo-italiano_%28Cristiani-d%27Italia%29/

Ward, Reginald W. 1992. *The Protestant Evangelical Awakening*. Cambridge: Cambridge University Press.

Zanfrini, Laura. 2019. *The Challenge of Migration in a Janus-Faced Europe*. London: Palgrave Macmillan.

CHAPTER 18

'Their Parents Are All Sailors and Blue-Collar Workers': Elementary Education in the Suez Canal Region at the Turn of the Century

Olga Verlato

INTRODUCTION[1]

> Port Said is a new city [and] a big promise for the Mediterranean peoples. [...] It should be possible to bring together in [our] school not only Italian nationals, but also subjects of our descent and foreign subjects, as well as other foreign nationals, easily attracting them under our sphere of influence.[2]
>
> Michelle Guastalla, Industrialist, Port Said, 10 June 1903.

[1] I thank Barbara Curli and all participants and organizers of the conference 'Italy and the Suez Canal' for much helpful discussion, the editors and reviewers at Palgrave Macmillan for their useful feedback, and Federica Onelli and all other archivists and staff in Rome, Cairo, Paris, and Nantes. I also thank my advisor Zachary Lockman and the department faculty at New York University for their continued support, and Joseph Viscomi, Hala Halim, and Will Hanley for their valuable insights and suggestions. Finally, my gratitude to my colleagues and friends Zachary Cuyler, David Kanbergs, Gabriel Young, and Marianna Pecoraro for their comments on various versions of this chapter.

[2] Guastalla, Report on the Schools in Port Said, 10 June 1903, ASDMAE, 72, f. 13.

O. Verlato (✉)
New York University, New York, NY, USA
e-mail: ov308@nyu.edu

B. Curli (ed.), *Italy and the Suez Canal, from the Mid-nineteenth Century to the Cold War*,
https://doi.org/10.1007/978-3-030-88255-6_18

> If it hadn't been for the [Italian] consul of Port Said, who forced me to close this school of mine that guaranteed the daily bread for me and my family, I would have not incurred any damage.[3]
>
> Massoud al-Tawil, School Director, Suez, 28 May 1889.

These quotations offer two very different depictions of what it meant to live and work in the Suez Canal region at the turn of the twentieth century. The European presence in this Mediterranean city assumed diametrically opposite meanings in the eyes of Michele Guastalla and Massoud al-Tawil. Yet the two can be regarded as also having some things in common. Al-Tawil, a teacher from the Levant who directed a small school in Suez which offered instruction in Arabic and French, and Guastalla, an Italian industrialist who in 1899 obtained a concession from the Egyptian government to the saltworks of Port Said and occasionally acted as an advisor for Italian authorities, were not among the important politicians or renowned intellectuals who more often leave traces in the historian's sources.[4] Additionally, they were both immigrants to Egypt, although subject to radically different legal regimes. Al-Tawil, described in the sources as 'a Levantine subject *raʿiyya*', was in fact excluded from the capitulatory privileges granted to Guastalla, an Italian national.[5] Finally, they were both concerned with the same issue: elementary education in the canal region. These shared characteristics encapsulate the three guiding criteria of my analysis: a focus on 'ordinary', virtually unknown historical subjects; a problematization of the history of Mediterranean migration and cosmopolitan life as homogeneous phenomena; and a study of elementary education in the Suez Canal region.

This chapter examines the intricate relations between some of the educational institutions that shaped elementary instruction locally in the Suez Canal in the late nineteenth and early twentieth centuries, focusing primarily on Port Said, but also with references to Ismailia, Suez, and Port Tawfik.[6] I focus on the schools administered and financed (fully or in part)

[3] Ibid., Al-Tawil, Letter to the Italian General Consul in Cairo, 27 May 1889, 32, f. 6.

[4] Ibid., and *Gazzetta Ufficiale del Regno d'Italia*, 1 April 1899, Rome, 2415.

[5] '*Raʿiyya*' was one of the broadest legal definitions used in late nineteenth-century Egypt, referring to 'Ottoman subjects [and] all manner of foreign subjects'. Hanley (2017, 65). In Italian sources, the term usually indicated non-Muslim Ottoman subjects. Margaroli and Battista (1829, 276-7).

[6] I use the term 'elementary education' to refer to general educational provision and distinguish it from primary instruction, which applied to a limited number of schools whose

by the Egyptian, Italian, and French states, and private schools such as al-Tawil's, while also referencing those of religious and charitable organizations. These various schools enrolled students from different communal, linguistic, religious, and socioeconomic backgrounds, and their pedagogical offerings and activities were shaped in large part by the local environment, rather than by official directives alone. By shedding light on the still largely unexplored workings of education at the local level, I argue that the interaction and competition between different local and foreign schools, which in turn relied on various and at times overlapping sources of state, religious, and private support, deeply affected local schooling by creating an uneven educational environment. Examining these schools allows us to reconsider the alleged insularity of different communities, while at the same time exposing the forms of discrimination created and reinforced by the workings of the schools through their pedagogical and administrative choices.

What we may categorize as state, religious, and private schools for the sake of clarity were not, however, isolated typologies. Some of the schools we will examine, such as the Egyptian *amīriyya* (governmental) schools, or the Italian *regie* (royal) schools of Egypt can be generally categorized as state institutions. However, local private schools such as al-Tawil's could also receive the support of foreign nations or local authorities, and therefore cannot be seen solely as private ventures separate from state power. Moreover, since the 1860s, the Egyptian government sought to increasingly bring *kuttāb*s (traditional communal schools) under the control of the Ministry of Education, or *dīwān al-madāris*, while local religious associations routinely administered schools as part of their charitable missions, some of which increasingly embraced political and nationalist agendas.[7] Finally, the schools of various Christian orders received allocations from foreign states, which saw them as a tool to spread their influence through education. Therefore, state, religious, and private schools did not exist as separate from each other, but rather were part of an interconnected, transregional system of funding allocation, pedagogical decisions, and competition for students and prestige.

Additionally, this chapter seeks to offer a reformulation of the notion of cosmopolitanism. 'The cosmopolitics of the eastern Mediterranean port

graduates could access secondary or higher education in the government system, unlike those from elementary schools.

[7] Yousef (2016, 81-82).

cities' are often described as 'involving multilingualism, religious plurality, openness, enterprising ethos, intercultural exchange and at least a weak form of tolerance'.[8] However, this conception risks obscuring the racialized and discriminatory reality and colonial history of many of these locales. Historian Khaled Fahmy, for example, notes that 'the scholarship of cosmopolitan Alexandria [is pervaded by] elitist, exclusionary and static views'.[9] Guastalla's 1903 reference to the 'Mediterranean peoples' of Port Said, itself instrumental to justifying expanding Italian influence in the country through education, constitutes an example of how this rhetoric was already employed at the time. Guastalla's words can in fact be seen as emphasizing 'openness and cosmopolitanism' by, as Fahmy explains, marginalizing 'the Arabic-speaking population' to which someone like al-Tawil belonged, as did his students and the teachers he hired.[10] Why is 'the maid from Upper Egypt who has learned to speak Greek', historian Will Hanley asks, 'never cast as a cosmopolitan, though she has met cosmopolitanism's classic criteria of polyglossia and migration'?[11]

Alternatively, and in line with recent reformulations of this category, I conceive of cosmopolitan relations as always necessarily uneven, that is to say hierarchical, whether in their coercive and discriminatory or adaptive and inclusive materializations depending on both contingent and structural factors shaping a given interaction.[12] In the case of elementary instruction in the Suez Canal, such unevenness manifested on multiple levels, including legal disputes and transactions, funds allocation, pedagogical choices, and educational provision. In approaching cosmopolitanism this way, I seek to allow for al-Tawil and Guastalla's perspectives to be examined together within a single conceptual and empirical framework, along with those of other migrants from the Arab Ottoman provinces, Egyptian shopkeepers, European nationals and, most importantly, the thousands of students who attended elementary schools in the canal region.

The rest of this chapter will take us through a hypothetical turn-of-the-century academic year in elementary schools across the canal. The next section, 'Preparing for the New Year', offers a brief excursus of the school typologies found in the region and examines how schools managed to

[8] Driessen (2005, 138).
[9] Fahmy (2017, 279).
[10] Ibid., 271.
[11] Hanley (2008, 1352).
[12] In addition to Hanley's work, see Fuhrmann (2003) and Halim (2013).

establish themselves through mutual cooperation and competition. Then, 'School Days in the Suez Canal' provides information about the curricula and pedagogical choices of various schools. Finally, 'Taking the Exams' situates the schools within the broader educational system in the country in order to shed light on the uneven professional and educational opportunities available to the students as they left elementary schools.

Preparing for the New Year

For centuries, the *kuttābs* were the almost exclusive providers of elementary instruction across the Egyptian province.[13] Although Mehmet Ali Pasha (r. 1805–1848) sponsored missions to Europe and opened a select number of secondary and specialized schools to train students for his reformed army and bureaucracy, it was not until the 1860s that the Egyptian Ministry of Education, directed by Ali Mubarak Pasha, launched a series of reforms aimed at expanding and centralizing elementary education, including opening a limited number of primary and secondary schools and seeking to standardize *kuttāb* education.[14] With the beginning of the *de facto* British protectorate in 1882, occupation authorities assumed large control of the Ministry of Education (subsequently reduced to a department, *niẓārat al-maʿārif*, in the Ministry of Public Works), cutting spending dramatically.[15]

Additionally, foreign states opened their own schools or financed existing private and missionary ones. For example, starting in the 1860s, and more systematically since the 1890s, Italy worked to create a network of state schools in regions characterized by significant Italian migration, such as Egypt, where Italians constituted one of the largest foreign communities.[16] Italy founded dozens of such schools, classified as *regie* or colonial based on their form of administration, across various Egyptian regions, open to all nationalities and religions.[17] These constituted, at least theoretically, extensions of state schools in Italy in terms of curricula and examinations, although in reality their educational offerings varied greatly from one school to the other. Schools financed by France were categorized as

[13] I am concerned here with 'formal' educational provision. For a history of home schooling and alternative writing and reading practices, see Yousef (2016).

[14] Ibid., 81.

[15] Reid (1977, 355-6).

[16] Viscomi (2019, 355).

[17] Floriani (1974, 9).

either confessional or *laïque* (secular) and offered instruction in line with French curricula. Finally, private individuals like al-Tawil opened their own schools, autonomous, at least in theory, from both state and religious authorities. Since the 1880s, a growing number of private schools funded by single individuals or community groups, and referred to as *ahlī* schools, started to position themselves as serving the Egyptian national cause through education.[18] All these different schools (and additional ones, such as those of the Greek-Orthodox community and American missionary schools) were found in the Suez Canal region, variously catering to the communities that lived there, with student bodies that were oftentimes mixed.

In early twentieth-century Port Said, according to a conservative estimate of the available sources, there were a couple dozen schools dedicated to elementary education. These included an Egyptian primary school;[19] four schools financed by France (one *laïque* and three administered by religious orders);[20] two Italian *regie* primary schools founded and financed by the Italian state (one for boys and one for girls); a private school affiliated with the Italian government; a private Coptic school; a number of private schools not receiving subventions from the Egyptian government (*ahliyya laysa lahā iʿāna*);[21] some *kuttābs*;[22] a Franciscan school; three Greek schools; and two American missionary schools.[23] Such variety and concentration of elementary schools was not necessarily the norm at the time. According to a 1917 report, the overall enrollment rate among children aged 7 through 15 in the *muḥāfaẓa* (governorate) of Suez and the one of the canal was around 36% for boys and 16% for girls, higher than the national average of 29% and 6%, although not as much as in other locales such as Cairo.[24] Moreover, while in Egypt as a whole most students attended *kuttābs* and, increasingly, *ahlī* schools, in the canal region nearly half of the almost 8000 pupils reportedly receiving elementary instruction attended schools described as 'foreign'.[25] Yet despite the higher degree of

[18] Yousef (2016, 82-3).

[19] Sāmī (1917, 64).

[20] Budget Allocations, 1908–1909, MAE, CP, 25, D, c. 223, 1-2.

[21] There were reportedly 11 such schools in the governorates of Suez and the canal at the time. Sāmī (1917, 116).

[22] There were 39 *kuttābs* in the two governorates. Ibid., 115.

[23] Guastalla, Report on the Schools in Port Said, 10 June 1903, ASDMAE, 72, f. 13.

[24] Sāmī (1917, Appendix 2: Table W).

[25] Ibid., Appendix 2: Table D.

educational provision in the canal region compared to other parts of the country, social extraction, legal standing, religion, gender, and geographic location rendered students' access to elementary education highly uneven.

Opening and ensuring the continuous activities of the schools could prove challenging. The first Egyptian primary school in the area opened in 1888 in Suez, followed in 1890 by one in Port Said.[26] Egyptian primary schools had come late to the canal zone when compared to the main centers of Cairo and Alexandria, and other cities in Lower Egypt such as Mansoura, Zagazig, and Damietta, where they had existed since the 1860s. Still, these two schools seemed to have enjoyed good, or at least acceptable, institutional and financial standing, given that they managed to continue their activities for over three decades, unlike other primary schools across Egypt.[27] At the same time, Egyptian state schools suffered from structural underfunding. The British government, which exercised overall control on the Ministry of Education through its appointed advisors, allocated only a fraction of state expenditures to education.[28] As a consequence, and as we will see in more detail shortly, students from certain locales and lower socioeconomic backgrounds could be disproportionally excluded from state primary education.

French schools were another significant source of elementary education provision in the area. In 1910, in addition to the aforementioned four schools of Port Said, there were two French schools in Suez, two in Port Tawfik, and three in Ismailia.[29] These received allocations from the French Ministry of Foreign Affairs, which distributed funding according to the schools' size, local demand, and competition with other educational providers. For example, the French ministry recommended nearly doubling the allocation given to the all-boys school of the Pères de Terre-Sainte in Ismailia for the 1909–1910 academic year due to the fact that 'this school is addressed to a poorer clientele [and] the Italians have founded a school in Ismailia'. The report explained that it was 'necessary to respond to this danger by sustaining the school more efficiently'.[30] This recommendation is one of the numerous indications that one of the key motors behind

[26] Ibid., Chapters 2, 6–7. According to French sources, the school was founded in 1889 rather than in 1890.

[27] Ibid.

[28] Lockman (2002, 146).

[29] Allocations for Egypt, 1910–1911, MAE, CP, 35, D, c. 223, 5.

[30] Ibid., Defrance to Pichon, Letter, 5 March 1911.

establishing and financing schools was competition between different providers.

Italian authorities were extremely concerned with the ability of French schools to divert students away from their own institutions. Guastalla described the state of the *regia* Italian school in Port Said as 'saddening, with a humiliating comparison between [it] and those right in front of it that belong to the French mission, from which everyday ordered lines of students are seen coming out, hundreds of them being Italians, sent there by their parents because there one finds space, air, and light'.[31] Such a dim picture may have been exaggerated in order to attract additional funding from the Italian ministry. Nevertheless, due to repeated cuts to the budget allocated to Italian schools abroad by the Ministry of Foreign Affairs, and to the greater prestige of French education, Italian school directors and administrators systematically struggled to render their schools a prime choice among students in the city. This fact is also confirmed by the lower number of students attending Italian schools in Port Said, around 200 in the 1906–1907 academic year, compared to the more than 600 pupils found in the city's French schools.[32]

Turning to the southern end of the canal, in the city of Suez and the neighboring Port Tawfik (later integrated into the city as its port), the situation appears to have been even worse. In 1899, the small Italian colonial school of Port Tawfik, attended by some 20 students (among them Italian nationals and a few Egyptians, most of whom admitted for free) had been forced to terminate its activities due to insufficient funding, and had been merged with the one in Suez. This choice appears to have placed a heavy burden on the students' families, described in the sources as 'all sailors and blue-collar workers' who claimed they were unable to afford the transportation costs of getting their children to the school in Suez.[33]

The relation between geographical proximity, class, and access to education found in the case of the Italian school of Port Tawfik echoes the concerns of ʿAbd al-Wahhab Suleyman, a member of the Egyptian General Assembly (*al-jamʿiyya al-ʿumūmiyya*) who pushed to open an Egyptian *amīriyya* primary school in Ismailia in 1907, and then again in 1909. The city, a motion submitted to the Council of Ministers complained, 'currently lacks a school that can also serve the various nearby localities. The

[31] Guastalla, Report on the Schools in Port Said, 10 June 1903, ASDMAE, 75, f. 13.

[32] Public and Private Schools in Egypt, 1906–1907, MAE, CP, 102, 7.

[33] Italian School in Suez, 4 December 1889, ASDMAE, 72, f. 4.

richer inhabitants are forced to send their children to the school of Port Said or Zagazig, but the poor children and those of the middle class can do nothing but wait for the solicitude and benevolence of the government.'[34] Despite these attempts, in 1914 Ismailia still lacked a state primary school.[35]

If ensuring acceptable hygienic conditions (or even the very existence) of state schools was at times challenging, when it came to local private schools lacking institutional affiliation, offering elementary education could prove even more difficult, as in the case of al-Tawil's school. In 1889, he wrote a letter to the Italian General Consul complaining that the Italian consul in Port Said had persuaded him to leave the building he was currently renting in Suez, in which his schools was located and where he lived with his family, so that they could open an Italian school in its place. Massoud al-Tawil's school numbered around 50 students, all boys, and hired a number of local teachers. While the school was financed, at least partially, through students' tuition, private schools would often also receive support from local shopkeepers or charitable organizations.[36] At the request of the consul of Port Said, al-Tawil had agreed to cede the space and school material for 640 francs. Having shaken the consul's hand in front of the dragoman (translator), he reportedly fired the teachers working in his school, and compensated the students' parents for having terminated classes earlier. However, a few days later, the consul retracted his offer, claiming that he had been negotiating directly with the owner of the building, one Count Lavison, through his intermediary in Suez, a certain Mr. Costa, and that al-Tawil was not entitled to receive any compensation. The latter complained to the Italian General Consul, hoping he would intercede for him, but his attempts were in vain.[37]

While it could be argued that al-Tawil had been naïve to liquidate his lease to the building without a binding contract, Italian authorities certainly operated from a privileged institutional and legal position. At the same time, al-Tawil proved to be competent in navigating the legal and institutional field of his interlocutors in order to get his demands met. He did this, for example, by rhetorically emphasizing his reasons, the

[34] Motion to open two state primary schools in al-'Arīsh and Ismailia, February 1909, DWQ, Majlis al-Wuzarā', 2-15-2, #0075-016522.

[35] Sāmī (1917, Chapters 2, 6–7).

[36] Yousef (2016, 82).

[37] Al-Tawil, Letter to the Italian General Consul in Cairo, 27 May 1889, ASDMAE, 32, f. 6.

economic damage inflicted upon himself and his family, and his absolute faith in the power of Italian General Consul to intercede with his compatriots on his behalf. Yet despite his moderate social standing as a school director of some means equipped with the knowledge of how to engage with foreign institutions, al-Tawil could not overturn an uneven legal field.

School Days in the Suez Canal

Who were some of the students attending the schools in the canal, and what kind of education did they receive? The statistics produced by the Egyptian Ministry of Education, British officials, and the French and Italian ministries of foreign affairs, which listed students based on religion, sex, and nationality, offer valuable insights into the demographic makeup of different schools. However, they should not be treated as an uncontestable source of objective knowledge. Legal categories at the turn of the century could shift and overlap, and did not necessarily correspond to one's individual or communal form of identification.[38] For example, in some cases we find a distinction between nationality and *protégé* status which is missing in other instances. Moreover, students from Arab Ottoman provinces could be listed as 'Ottoman', or with reference to their specific area of origin, such as 'Syrian'. Therefore, these demographic considerations, while useful to appreciate important differences between the student bodies' makeup across the schools, remain necessarily subject to empirical and theoretical ambivalence.

As mentioned, in the 1914–1915 academic year, nearly half of the elementary-school students in the canal region attended foreign schools, while around 30% of them were enrolled in the *kuttābs*, and 17% in local private schools. Only around 3% of them attended the Egyptian state primary schools of Port Said and Suez, which in 1914 enrolled 123 and 105 students respectively.[39] Due to the high cost of tuition, most of these students likely came from wealthy Egyptian families, although scholarships for poorer pupils were also occasionally granted. The primary cycle lasted for four years, and students were instructed according to a standardized curriculum drafted by the Ministry of Education.[40] While the curriculum underwent some changes during the period of the occupation, it continued

[38] Fahmy (2013), and Hanley (2017).
[39] Sāmī (1917, Appendix 2: Table E).
[40] Belli (2013, 24).

to include among its main subjects the study of the Quran, Arabic, a foreign language (French or English), arithmetic, geometry, geography, and history.[41]

Around 1000 other students in the region attended French elementary schools, while some 300 were enrolled in Italian ones. French and Italian state schools in Egypt operated according to two interconnected principles. On the one hand, they aimed to 'meet the exigencies of the French population' and 'the education and instruction of Italians abroad' respectively, while dedicating only a few hours each week to 'subjects of local interest'.[42] For example, even though local students would at times make up the majority of the student body, history and geography classes followed a Eurocentric curriculum rather than teaching students about Egypt's geography and the history of the country from the pharaonic period to Mehmet Ali's dynasty (as was the case in Egyptian state schools). On the other hand, France and Italy also aimed to attract local students and foreign pupils of other nationalities in an attempt to bring them under their respective spheres of influence and to compete with other educational providers. Italian schools abroad were described as aiming to 'spread the [Italian] national language and culture', and French ones as 'rendering a service to our [French] influence and prestige'.[43] French schools, for example, were attended for the most part by Egyptian, Italian, and Greek students. The fact that a significant turnout from across different segments of the population could in turn be used to justify receiving more funding only made this logic more ingrained.

It is more difficult to gather information about the student body and pedagogical offerings of small private schools. Italian sources only mention that al-Tawil's school offered instruction 'in Arabic and French'.[44] This choice was symptomatic of a context in which monolingual Arabic formation was increasingly deemed as insufficient to secure appropriate schooling and future employment – a trend inextricable from colonialism and Eurocentric cultural production. Crucially, foreign states sometimes sought to finance local private schools as a means of advancing their

[41] Niḍhāra al-Maʿārif al-ʿUmūmiya, *Brūjrām al-Madāris al-Ibtidāʾiyya wa al-Tajhīziyya* (Al-Maṭbaʿa al-Kubrā al-Amīriya bi-Bulāq: 1897), DKM.

[42] Allocations for Egypt, MAE, CP, 35 D. c. 223, 5. And Modification to the *Regie* Schools' Curricula, 20 May 1902, ASDMAE, 75 f. 13.

[43] Floriani (1974, 27). And Geoffray to Pichon, 30 January 1909, MAE, CP, 10, D, c. 210, 1.

[44] Al-Tawil, Letter to the Italian General Consul in Cairo, 27 May 1889, ASDMAE, 32, f. 6.

cultural interests. For example, French authorities sometimes offered to support local private schools teaching French, seeing them as instrumental to counteracting British influence.[45] Therefore, educational choices served as a primary field for inter-imperial competition, and in turn could become a tool for local educational providers to secure additional financial support depending on the kind of subjects they taught.

Taking the Exams

In such an institutionally and pedagogically heterogeneous context of elementary education, final exams and, more generally, the outcome of receiving instruction also varied greatly. In this final section, I examine four of the educational trajectories available to or imposed on the students of the canal, including instances in which students 'switched' between educational paths.[46] Students enrolled in Egyptian primary schools could take the exam to gain the primary certificate, which gave access to state posts and allowed them to continue on to secondary education. The exam, which had a written and oral component, was quite selective. Statistical records indicate that in 1900, out of the 996 candidates from Egyptian state schools across the entire country, only 431 passed the exam. In 1903, the number of pupils succeeding in the exam was even lower, with only 348 out of 1323 candidates from Egyptian state schools receiving the primary school certificate.[47] The fact that candidates needed to pay a fee in order to take (or re-take) the examinations further widened the socioeconomic divide between pupils who could advance in their studies and those who could not. Pupils from foreign schools could also take the Egyptian primary school exam, which was as challenging for them as it was for their peers from Egyptian primary schools. In 1900, only 235 such students passed the exam, out of the 522 enrolled.[48]

Alternatively, students in foreign schools could gain certificates equivalent to those in the schools' respective countries. There were different types of primary and elementary certifications. In the French educational system, for example, the primary school certificate marked the end of primary instruction, while the elementary *brevet* or patent allowed students

[45] Vice-consul in Suez Lucien Labosse, Report, 18 September 1885, CADN, Suez, n. 10.
[46] For a summary of the schooling ladder for this period, see Reid (2002, 16).
[47] Results of Primary School Certificate Exams, 1899–1902, MAE, CP, 99, 5.
[48] Ibid.

to in turn become schoolteachers. The economic condition of the students' families, as well as where they lived, could determine whether and what kind of exam they could take. In June 1903, for example, 14 girls studying in two of the French schools in Port Said, one directed by the nuns of Bon-Pasteur and the other by a woman named Madame Bourousse, took their French primary school exam. All candidates passed the exam and were awarded the primary school certificate. Mary Vallauri, reportedly from the Syrian province, got the highest score, a 65.5 out of 70. Olga Manoli, a Greek, received one of the lowest points, 51.1. The only Egyptian girl who took the exam, Khadija Mohamed, scored around the middle with a 59.5.[49] However, putting together an examination committee for the elementary *brevet* was more difficult. The two schools had been unable to find individuals 'equipped with certificates for primary education and having worked in either public or private education' to administer the exam.[50] Therefore, it was decided that the students would take the exam in a bigger city nearby, where an appropriate committee could be put together. The following year, 16 girls took the exam for the *brevet* in Cairo, 9 of them from Port Said. The number of students willing to take this patent (or who could afford the 40 Egyptian pounds and travel expenses to Cairo) was thus lower than those who had received their primary certificate. Vallauri, Mohamed, and even Manoli passed the exam and got the elementary *brevet*.[51]

In some cases, parents would send their children to school for only part of the elementary cycle, without ever taking a final exam. In the Italian colonial school in Suez during the 1896–1897 academic year, for example, there were 25 pupils enrolled in their first year, but only 11 in their last.[52] While attendance rates throughout the elementary cycle may have been more consistent in larger schools, it appears that one of the ways of taking advantage of the available educational offering was to receive only a general introduction to elementary instruction. Finally, for a number of students, being able to go to school was simply not an option. The merging of the Italian state school of Port Tawfik with that of Suez, the closing of al-Tawil's school, or the absence of an Egyptian *amīriyya* school in Ismailia show that in certain instances it was extremely difficult, if not impossible,

[49] Ibid., Primary Certificate Examinations in Port Said, 26 June 1903.
[50] Ibid., Project for the Elementary Brevet, 1 September 1903.
[51] Ibid., Brevet in Cairo Report, 31 December 1904.
[52] School Statistics, ASDMAE, 75 f. 4.

for some families to ensure that their children could access or progress in their education. While we are told by al-Tawil himself that he found a job as a correspondent for the Cairo-based newspaper *al-Muqaṭṭam* (est. 1889), we are left with no information regarding the educational prospects of his former students.

Ultimately, the context of the British occupation, extraterritorial privileges, inter-imperial competition, and uneven cosmopolitan relations variously affected the experience of students, from their initial enrollment to their final exams, depending on where they lived, their communal and legal standing, their gender, and their social extraction. For example, the fact that Italian state schools were open to all religions and nationalities and offered scholarships to some of their students meant that in areas otherwise not catered to by Egyptian state schools, local students of limited means could receive primary schooling and potentially continue on to secondary education. Still, statistical records show that Italian state schools were attended for the most part by Christians and Jews, while a larger proportion of Muslim students was found in local *ahlī* and state schools. Finally, despite the opening of an increasing number of schools for girls in this period, gender continued to represent an important axis of uneven access to education, with an enduring disparity between enrollment rates for boys and girls.[53]

Conclusion

In line with the three guiding criteria outlined in the introduction, this chapter sought to foreground the experiences of common individuals whose names may appear in the sources only fortuitously and in a scattered manner, sometimes next to an examination result and nothing more. Individuals like al-Tawil, a migrant from the Levant and resourceful school director reportedly turned newspaper correspondent, never became an acclaimed writer at the forefront of the new intellectual and cultural movements of his time. Most pupils receiving some form of elementary schooling, as well as those who were excluded from it, did not play as visible a role as their secondary and higher-educated peers in the political crises and mobilizations that were soon to erupt in Egypt in this period, chief among them the 1919 revolution.[54] Still, through their mundane interactions,

[53] Ibid., and Public and Private Schools in Egypt, 1906–1907, MAE, CP, 102, 7.
[54] Lockman (2002, 145-9).

choices, and life trajectories, these individuals also contributed to shaping the history of education and Egypt's political, cultural, and intellectual life.

In turn, by looking closely at the activities of migrants from the Arab Ottoman provinces and European nations, this study problematized the use of 'Mediterranean migration' as an all-encompassing term, and of 'cosmopolitanism' as a shared feature of a locale in its entirety. Rather, one's legal status, social and financial standing, degree of institutional support, and cultural affiliation differentiated drastically between various migrants and their respective positions in an uneven cosmopolitan and colonial field. Further research on individuals from other regions of the Southern Mediterranean and North and Sub-Saharan Africa will provide an even fuller picture of these migratory phenomena. Finally, examining the workings of education from an urban and sub-regional perspective brought to the surface some of the specificities of elementary schooling in the Suez Canal region, while simultaneously showing these schools' connections to state and inter-imperial dynamics. Localized forms of interaction and competition between multiple and oftentimes mixed forms of state, religious, and private schooling, and of local, national, and transregional initiatives were key to determining how education functioned in the region. Elementary schooling had been developing in a way that hindered the attempts of many non-European educational providers and reformers, while at times opening new avenues for professional, financial, and intellectual gain. Most crucially, it created novel spaces of interaction and reinforced preexisting axes of discrimination for the thousands of students from various linguistic, socioeconomic, religious, and communal backgrounds who lived and went to school in the Suez Canal region.

References

Archival Collections, Libraries, and Reports

ASDMAE: Archivio Storico Diplomatico del Ministero degli Affari Esteri, Rappresentanza Diplomatica Egitto, Rome.

CAD: Centre des Archives Diplomatiques, Nantes.

DKM : Dār al-Kutub al-Miṣriyya, Cairo.

DWQ: Dār al-Wathā'iq al-Qawmiyya, Cairo.

Gazzetta Ufficiale del Regno d'Italia, 1 April 1899, Rome.

MAE, CP: Archives Diplomatiques du Ministère des Affaires Étrangères, La Courneuve.

Bibliography

Belli, Mériam. 2013. *An Incurable Past: Nasser's Egypt Then and Now*. Gainesville: University Press of Florida.

Driessen, Henk. 2005. Mediterranean Port Cities: Cosmopolitanism Reconsidered. *History and Anthropology* 16 (1): 129–141.

Fahmy, Khaled. 2017. For Cavafy, with Love and Squalor: Some Critical Notes on the History and Historiography of Modern Alexandria. In *Alexandria, Real and Imagined*, ed. Anthony Hirst and Michael Silk, 263–280. London: Routledge.

Fahmy, Ziad. 2013. Jurisdictional Borderlands: Extraterritoriality and 'Legal Chameleons' in Precolonial Alexandria, 1840–1870. *Comparative Studies in Society and History* 55 (2): 305–329.

Floriani, Giorgio. 1974. *Scuole Italiane all'Estero: Cento Anni di Storia*. Roma: Armando.

Fuhrmann, Malte. 2003. Cosmopolitan Imperialists and the Ottoman Port Cities. Conflicting Logics in the Urban Social Fabric. *Cahiers de la Méditerranée* 67: 149–163.

Halim, Hala. 2013. *Alexandrian Cosmopolitanism: An Archive*. New York: Fordham University Press.

Hanley, Will. 2008. Grieving Cosmopolitanism in Middle East Studies. *History Compass* 6 (5): 1346–1367.

———. 2017. *Identifying with Nationality: Europeans, Ottomans, and Egyptians in Alexandria*. New York: Columbia University Press.

Lockman, Zachary. 2002. Exploring the Field: Lost Voices and Emerging Practices in Egypt, 1882-1914. In *Histories of the Modern Middle East: New Directions*, ed. Israel Gershoni, Hakan Erdem, and Ursula Woköck, 137–154. Boulder: Lynne Rienner Publishers.

Margaroli, Giovanni, and Giovanni Battista. 1829. *La Turchia Ovvero l'Impero Ottomano Osservato nella sua Situazione Geografica-Statistica Politica e Religiosa, non che nella sua Storia etc*. Vol. 2. Milano: Schiepatti.

Reid, Donald. 1977. Educational and Career Choices of Egyptian Students, 1882–1922. *International Journal of Middle East Studies* 8 (3): 349–378.

———. 2002. *Cairo University and the Making of Modern Egypt*. Vol. 23. Cambridge: Cambridge University Press.

Sāmī, Amīn. 1917. *Al-Ta'līm fī Miṣr fī Sanatay 1914 wa-1915*. Cairo: Maṭba'at al-Ma'ārif.

Viscomi, Joseph. 2019. Mediterranean Futures: Historical Time and the Departure of Italians from Egypt, 1919–1937. *The Journal of Modern History* 91 (2): 341–379.

Yousef, Hoda. 2016. *Composing Egypt: Reading, Writing, and the Emergence of a Modern Nation, 1870-1930*. Stanford: Stanford University Press.

CHAPTER 19

Transnational Labour in Conflict: The Italian and Greek Personnel of the Suez Canal Company and the Second World War

Angelos Dalachanis

The history of the Suez Canal Company (SCC) was replete with crises. From its foundation by Ferdinand de Lesseps in 1854 to its nationalisation by Gamal Abdel Nasser in 1956, it experienced strikes, two world wars and other local or regional conflicts. What made these events critical was that each of them could potentially prevent the SCC from conducting its basic mission: to keep the canal open to navigation in times of peace and war. A crisis is however useful to a historian because it constitutes a magnifying glass that makes apparent otherwise invisible threads that link small history with the big picture. Crises allow the historian to transcend different levels of analysis between local, regional and global, between the micro and macro scales. In this respect, a crisis related to a global company that was also universal in name, *Compagnie universelle du canal maritime de Suez*, allows us to observe how ordinary people of several ethnic and

A. Dalachanis (✉)
CNRS-IHMC, Paris, France
e-mail: angelos.dalachanis@cnrs.fr

B. Curli (ed.), *Italy and the Suez Canal, from the Mid-nineteenth Century to the Cold War*,
https://doi.org/10.1007/978-3-030-88255-6_19

national backgrounds who worked for it make decisions in the face of world-level events. Even though the history of the SCC has mostly been examined through the lens of its administrative elites, its impact on the urban context and the architecture of the canal cities, the history of its personnel remains largely understudied.[1]

This chapter contributes to the filling of this gap. It examines the personnel of what was one of the biggest private employers in the Eastern Mediterranean for more than a century, with particular emphasis on the Italian and Greek personnel, in relation to the Second World War. I argue that the romantic, rosy and often nostalgic perspective on Egypt's cosmopolitanism overplays the peaceful coexistence between foreign communities and downplays the competition among them.[2] At the same time, the Egyptian struggle for full independence and the consequent pro-colonial or anti-colonial discourse that has been very much focused on Suez obscure other social dynamics. For these reasons there is a tendency to attribute the unemployment of foreigners in immediate pre- and post-war Egypt solely to the Egyptianisation of the economy and labour.[3] In this sense, there is a tendency to see the departure of foreigners as a result of Egyptian legislation and not to the antagonism between foreign communities.

Personnel at War: Mirroring State Policies

Legally an Egyptian joint-stock company, the SCC was rather a French company, in terms of the nationality of its founder (Ferdinand de Lesseps), the majority shareholding, the administration and the majority of highly qualified employees. Britain, however, after occupying Egypt in 1882, placed the country and the canal under its protection and owned the biggest proportion of shares (around 40 percent). After the Anglo-Egyptian treaty of 1936, the British constructed their biggest military base in the world across the canal, which hosted 10,400 soldiers, officers and pilots. On the eve of the Second World War, the SCC board consisted of thirty-two directors: nineteen of whom were French, ten British, two Egyptian, and one Dutch. The associated prestige and the prospect of reduced

[1] Piquet (2008), Bonin (2010), Piaton (2009) and (2011). Additionally, Beinin and Lockman (1998) have linked the labour movements of the canal with the wider dynamics in Egypt. Curli (2014) is the only recent study on the work organization and Egyptianisation of the personnel of the Company from the 1940s until the Suez Crisis, and also offers a gender approach.

[2] For a solid critique on Egypt's cosmopolitanism, see Hanley (2008).

[3] Alleaume (1997), Karanasou (1992), Piquet (2008).

passage fees made membership of the board appealing for many countries. From 1899 to 1915 there was also a German director, but during the First World War a Frenchman replaced him. The Italians first sought representation on the board in 1922. Unsuccessful, in the late 1930s they became more insistent. By this time, their claim to a boardroom position had become more substantial because their military involvement in Libya and Ethiopia had resulted in Italy being the second-highest user of the canal after Britain.

For the British Foreign Office, the real reason for Italy's request was its desire to access vital information about the canal's security. Only a few high-ranking SCC officials were aware of top-secret plans that had been drawn up by the British navy and the company. Indeed, the British were extremely sceptical about the appointment of an Italian director because this would provide 'a very considerable advantage to the Italian government in the event of war'.[4] Within the antagonistic environment in the Eastern Mediterranean, London considered the Italian demand as the first step in a policy to undermine Britain's dominant position in Egypt, fearing it would endanger communications with the empire east of Suez. British officials also had concerns that Italian claims for representation would lead to similar demands from Germany and Japan. France and Britain did not share the same opinion as to how they should deal with the Italian demands and with Mussolini in general. For instance, when Paris agreed to discuss Italian claims on Suez, London only countenanced the possibility of reducing fees.

In August 1939, the SCC examined the possible impact on its personnel of a possible future war between Italy and France. At that time, the SCC personnel comprised 3600 persons: 600 employees (white-collar workers, engineers, navigation pilots and foremen) (364 French, 78 Italians with the remaining 158 comprising Greeks, Egyptians and other nationalities) and 3000 workers (100 French and 370 Italians, with the remaining 2530 composed of Egyptians, Greeks and other nationalities).[5] According to the company's estimates, in the event of war, 172 French employees would be called up and 78 Italians would be dismissed. As for the Italian workers, who would leave their job in the event of war, they represented 12 percent of the SCC workforce. As for the Greeks, their dependence on a state that had no military rivalry with the French or British made them an ideal workforce for critical times, according to canal

[4] Note of V. Gavendish-Bentick, 3 January 1939, TNA, FO, 371 23343.
[5] Note, 23 August 1939, ANMT, ASCC, 1995 060 1326.

administrators. In June 1938 the head of the transit service G. Douin, wrote from his base in Ismailia to the Paris SCC headquarters that he was much in favour of recruiting Greeks because '[t]hese are people who will never make war – whose patriotism will never boil over to the point of making them commit reprehensible acts – and who, by the way, are good and disciplined sailors'.[6]

In May 1940, Italy's intention to go to war was becoming increasingly evident. Paris adopted a cautious approach, suggesting that the SCC should refrain from any action that could antagonize the Italian government. At the same time, London and Cairo believed that there was no room for concessions with the Italian government and the only important matter at that stage was 'to prevent any harm being done to the canal by Italian sympathisers, especially the Italian employees against whom there were grounds for suspicion'.[7] Suspicion of Italians was not only confined to the canal but widespread in Egypt because of fascist propaganda and activities.[8] British and French intelligence provided classified information to the SCC administration about gatherings of Italians at the *Domus Italica* in Ismailia. The secretary of the local Fascio had informed his compatriots that an Italian attack on the Suez Canal zone was imminent and that they should provide assistance to the Italians parachutists. Moreover, he even gave some Italians concrete instructions about what to do in the event of an attack. Additionally, he asked them to remove the fascist insignia from their lapel and to avoid relationships with Greeks because they were unreliable and could betray them.[9] At the time, relations between Italy and Greece were rapidly deteriorating, which led to the outbreak of the Greek-Italian War in late October 1940. After the Italian defeat in Greece in April 1941, the subsequent German invasion broke Greek defence. Until late 1944, Greece would remain under tripartite German, Italian and Bulgarian occupation.[10]

Under the terms of the 1936 Anglo-Egyptian treaty, the British took control of Egypt following the outbreak of the Second World War and put pressure on the SCC to dismiss Italian personnel. The Italians were given a fortnight's leave, starting on 15 May 1940, and their full dismissal

[6] Douin to Febvre, Ismailia, 9 June 1938, ANMT, ASCC, 2000 038 0189
[7] Young to Norton, Paris, 24 May 1940, TNA, FO, 371 24601.
[8] Viscomi (2016, 66).
[9] Note of Mitteiman, Ismaïlia, 20 May 1940, ANMT, ASCC, 1995 060 1326.
[10] Mazower (1993).

followed shortly thereafter. After Italy's declaration of war on 10 June 1940, Egypt ordered the internment of Italian citizens. In total, 60 percent of Italian males living in Egypt were arrested and detained in camps, where they remained for up to four years.[11] By 1 September 1940, 331 Italian SCC staff had been dismissed: thirty-eight employees, thirteen pilots, eight foremen and 273 workers.[12] Eighty-three workers with more than fifteen years' service had the right to a proportionate pension. The others received indemnities. This was not the first time the company removed 'enemy subjects'. At the outbreak of the First World War, German and Austrian employees (who were far fewer in number) were also summarily dismissed. After the war, only those Austrians who subsequently obtained Italian or Yugoslav citizenship were reemployed.[13] As for the Germans, they were never reemployed. In 1940, however, the SCC offered the Italians pensions and indemnities. Thus, thirty-eight white-collar workers, twelve pilots, eight foremen and eighty-three workers with more than fifteen years' service received a proportional pension. Seventy workers with less than fifteen years' service and 120 workers were dismissed with an indemnity. The SCC considered this the best possible solution since the Egyptian decrees made no provision for indemnities for the Italians who would lose their jobs.

For those foreigners who were to be enlisted during the war, namely the French and the Greeks, the situation was different. The armistice of 22 June 1940 between France and Germany gave birth to the Vichy regime on 10 July. The consequent rupture in Anglo-French relations unavoidably raised questions about the security of the canal. The French were divided between the supporters of Vichy and those of de Gaulle. While the canal was under full British military control, it was administered locally under the authority of Baron Louis de Benoist, as *agent supérieur* of the SCC, who was also the leader of the de Gaulle-aligned French National Committee in Egypt.[14] De Benoist and other French officials continued their cooperation with the British military authorities during the war.[15] Apart from the French personnel of the *Ateliers généraux* who took position in favour of de Gaulle in June 1940, the majority of the French

[11] Viscomi (2016, 113).

[12] Extrait du livre vert 'Guerre 1939' questions du personnel – Tome I , p. 19 et les suivantes ANMT, ASCC, 1995 060 3017.

[13] Charles-Roux to Belardo, Paris, 17 June 1948, ANMT, ASCC, 1995 060 0976.

[14] Note of Young (s.d). TNA, FO 371 24601.

[15] 'No Fears About the Suez Canal', *Sunday Times*, 21 July 1940.

staff shared the opinions of the directors and the senior managers in France, who were 'inveterate Vichystes'. Many of the Vichy sympathizers left the SCC and Egypt in 1942 and most of them never returned.[16]

In December 1941, the Greek government in exile, which had established itself in Egypt the previous May, decided, in accordance with its British allies, the enlistment of Greek nationals residing in Egypt.[17] This decision provoked strong reactions from the leadership of the Greek community in Egypt, who feared that those who left their jobs would not be able to return to them after the war because Egyptian officials were already demanding that Egyptians filled the vacancies created as a result of the mobilization. Believing that the future of the community was being jeopardized, the community dignitaries unsuccessfully tried to exempt the over twenty-fives from enlistment.[18] Members of the Greek personnel, who constituted the biggest foreign group in the SCC and which had professional characteristics similar to the Italians', were called up and joined the Greek armed forces in the Middle East. At the same time, Maltese and Greek Cypriots with British citizenship were also enlisted in and joined the British army. Indeed, Egyptian nationals replaced those who were dismissed or enlisted.[19] This was in line with the Egyptianisation economic and employment policy, which sought for more Egyptians to be hired by the SCC but also by other private companies in the country. But the Egyptians were not the only problem the Greeks had to deal with in the labour market. What would happen after the war when Italians and other nationals who had been interned or enlisted started claiming their jobs back, or worse, if they were liberated from the internment camps or demobilized before the Greeks?

Italian Claims: Seeking the Return of Their Pre-war Life

The Italians were no less sensitive than the Greeks about what the loss of employment could entail for their post-war position in Egypt. Their pressure to be rehired by the SCC started as early as October 1943.[20] The

[16] Carelli to Saillant, Port Said, 15 June 1945, TNA, FO, 371 45954.
[17] Clogg (1979).
[18] Dalachanis (2017, 86-87).
[19] Piquet (2008, 292).
[20] Note of the Greek Embassy in Cairo, Cairo, 18 October 1943, AYE, D/2/1943-1944.

Greek embassy in Cairo closely monitored the situation, and especially Italian activities, in the country after the collapse of the fascist regime in July 1943. Greek diplomats feared that, after the armistice with Italy in September 1943, the British would announce that Italy was then considered at the Allies' side. The Greeks should be aware that at least 250 internees had signed up to the anti-fascist *Movimento Libera Italia* (MLI) and were about to be released from Fayed internment camp.[21] In an internal note, the embassy expressed its anxiety about MLI advice encouraging Italians to request their release from the camps in order to be first in line in the labour market, ahead especially of the Greeks who were still serving in the Allied army of the Middle East. According to classified information, negotiations between anti-fascist organizations and the British authorities in Egypt got underway in October with the intention of supporting the Italian personnel in returning to their pre-war jobs. Indeed, the Italians sought the moral support of the British for their demands, even if the Cairo-based *Agence supérieure* of the SCC was negatively disposed to the reemployment of the Italian workers and employees in Port Said, Ismailia and Suez. However, the Greeks did not take the SCC's refusal for granted: 'There's no doubt whatsoever that the *Agence supérieure* wants to accept the Italians back to their job. Thus, the positions of the enlisted Greeks will be fully occupied by them.'[22]

Most of the Italians would not be released from the camps until the end of the war.[23] After the war, the pressure to reemploy the Italians of the SCC became more and more systematized and concerned all levels: from the diplomatic authorities to individual claimants. As early as December 1945 the reemployment of Italians became a matter of negotiations at the highest level in France between the company administration and the French Foreign Ministry. While the ministry had no objection to the rehiring of Italians, the company considered it a very delicate issue that was 'likely to have multiple and serious impacts in various areas'. The administrators believed that a decision could not be made without a thorough examination of the situation in Egypt.[24] In Cairo, the first representative of the post-war Italian government Giovanni de Astis[25] asked de

[21] Viscomi (2016, 151).
[22] Note of the Greek Embassy in Cairo, Cairo, 18 October 1943, AYE, D/2/1943-1944.
[23] Viscomi (2016, 150).
[24] Bonnet to Bonneau, Paris, 15 December 1945, ANMT, ASCC, 1995 060 3017.
[25] Turiano (2016, 322).

Benoist to facilitate the return of the Italians to their pre-war jobs.[26] The demand mostly concerned workers and seven Italian pilots. The return of workers was not a simple matter as the Egyptians who were hired during the war could not be dismissed, especially as the government was increasingly putting pressure on the company to hire more and more Egyptians, as is confirmed by the minutes of a meeting between Hackney, a British SCC director, and King Farouk.[27] Farouk was particularly eager to see the company hire Egyptians as pilots, as it had none up to that time. In September 1946, Abdel Meguid Ibrahim Pasha, the Egyptian minister for Social Welfare, also made clear that companies and factories had the right to reemploy old employees and workers who had been enlisted during the war on the condition that it did not conflict with the interests of the Egyptian workers who had been hired during the war.[28]

Egypt's claims should be seen as part of the long-term Egyptianisation of the economy and labour market, a process that, although underway since the nineteenth century, had accelerated in the 1930s. The Egyptian workforce of the SCC rose from 30 to 40 percent in the late nineteenth century to 59 percent on the eve of the Second World War, an increase that took place mostly during the 1930s. From 1939 to 1943 the Egyptian personnel increased from 59 to 67 percent.[29] After the Second World War, Egyptianisation accelerated as a consequence of the 1947 company law and the 1949 agreement between the Egyptian government and the SCC. For the latter, the increase did not only concern workers but also employees and pilots. Especially for the prestigious corps of pilots, the Egyptians were not represented at all in 1939 and by 1949 they represented only 7 percent of the pilots.[30] Italy's ambassador in Egypt, Cristoforo Fracassi Ratti Mentone di Torre Rossano in a letter dated 24 July 1948, asked the SCC to look into the matter once again and, if possible, to help those former employees who were in a difficult situation. The director general of the company proposed to give 'personal gratuities of twenty Egyptian pounds a year to eleven workers aged fifty years or more and who were unemployed or in bad health or in a difficult family situation and a yearly allowance of twenty Egyptian pounds for nine agents

[26] F.R Hoyer-Millar, 6 November 1945, TNA, ADM, 1 19321,

[27] Hakney to Killearn, Cairo, 9 February 1946, TNA, FO, 371 53271.

[28] Triantafyllidis to Greek Ministry of Foreign Affairs, Alexandria, 23 September 1945, AYE, CS, 1946 36 3 5035.

[29] Piquet (2008, 303).

[30] Piquet (2008, 311).

having three dependent children under nineteen with a supplement of five Egyptian pounds for every additional dependent'.[31]

In the second half of the 1940s, dozens of Italians who did not leave Egypt after their liberation wrote to the company seeking their old jobs back. In their letters, many highlighted their long-term relations with the SCC, which they presented as their proper home and measured in decades the generations of their family who worked for the canal: Loreto Coppola wrote about his father, Gaetano, and his uncle, Alessandro Coppola, who 'together, gave more than 70 years of devoted service to the company'.[32] The recruitment of members of the same family was very common in the company. Its recruiting policy aimed to reinforce loyalty and secure durable settlement through employing generations of employees and workers in the canal zone. Recollections about decades of uninterrupted service served to remind the company that loyalty should be rewarded. Another set of arguments insisted on the cultural or linguistic orientation of the personnel. Xavier Croci, who was dismissed from the works department, claimed that, even though he had only lost his job because of his Italian citizenship, his loyalties were not necessary related to his nationality. He pointed out that his parents had been living in Paris since the 1930s, his younger brother was a naturalized French citizen, that his mother tongue was French and, above all, he had received a French education and French professional training in the hope of working in a French environment.[33]

There are cases of ex-workers who filed a claim in the Egyptian court against the company. Joseph Novelo accused the company discriminating against the Italians because it recruited the French who returned from the front but not them.[34] He also claimed to the court that an Egyptian joint-stock company mostly active in Egyptian territory should not favour the French over other nationals. The SCC rejected the allegations, claiming that it provided pensions and indemnities and that the French had not been dismissed, unlike the Italians. Another interesting case was that of Wilhelm Preckwinkel, a German pilot considered 'excellent' by the company, who joined it in 1929 but resigned ten years later to join the German forces. Finding himself in the British zone of occupation in Germany after

[31] Hakney to Clutton, 21 September 1948, TNA, FO, 371 69183.

[32] Coppola to the General Director of the SCC, Port Said, 22 June 1948, ANMT, ASCC, 2000 038 0035.

[33] Crocci to the General Director of the SCC, Cairo 25 January 1947, ANMT, ASCC, 1995 060 3017.

[34] Raafat to Court of Appeal, 1953, ANMT, ASCC, 1995 060 3017.

the Second World War, Preckwinkel contacted the company but, having resigned, he was not eligible even for a pension. The SCC did not discuss any possible reemployment but was willing to ask the British authorities in Germany to employ him in one of the firms under their control.[35]

The case of Rodolfo Castelli, who was an experienced long-distance captain in the Italian merchant navy before he was recruited as a pilot by the SCC on 1 January 1935, on the recommendation of the Vatican Secretariat of State, puts to the fore the role of religious institutions in the employment and reemployment of Italians.[36] This ex-pilot wrote an extensive letter from Alexandria to Giovanni Belardo, a lay official of the Secretariat of State, about the life of Italian pilots of the Suez Canal zone before the war and the necessity of getting them rehired. According to Castelli, the Italian pilots of the SCC either had the recommendation of the Holy See or were the sons or nephews of older pilots of the company. He also claimed that before the war the Italian personnel found itself in a delicate position between the fascist consular authorities, who considered them to be not patriotic enough, and the other employees of the SCC, who viewed in each of them '*un capo fascista*' or '*delle spie*' and constantly humiliated them.[37] In any case, Castelli had no chance of reemployment because British naval intelligence had advised against his reemployment on security grounds.[38]

Apart from the Vatican Secretariat of State, the Apostolic Nunciature to France also intervened on behalf of former SCC personnel. On 28 July 1949, the archbishop of *Mésembrie* (Nesebar) Roncalli wrote to the president of the SCC, François Charles-Roux, asking to employ again the Italians of the Suez Canal, who, in a precarious condition after their liberation from the camps, deserved to get their old jobs back: 'Hopefully the war is over, [the Italians] paid with the limitations to their freedom, the sequestration of their money and assets, searches, jobs, all sorts of humiliations and forced repatriations.'[39] In his response, Charles-Roux tried to defend the SCC's actions and to show that it treated the Italian personnel as fairly as possible: they received proportionate pensions or compensation, even if the Egyptian military decrees did not foresee any. Additionally,

[35] Georges-Picot to Fraser, Paris, February 1947, TNA, FO, 371 63011.

[36] Note on the Italian pilots, 9 June 1948, ANMT, ASCC, 1995 060 0976.

[37] Castelli to Belardo, Alexandria, May 1948, ANMT, ASCC, 1995 060 0976.

[38] F.R Hoyer-Millar, 6 November 1945, TNA, ADM, 1 19321.

[39] Roncalli to Charles-Roux, Paris, 28 July 1949, ANMT, ASCC, 1995 060 3017.

and in reply to the demands of Fracassi, the SCC distributed monthly aid to the Italians who were not reemployed after the war and were in real need. Roux placed the blame on the 1949 agreement between the SCC and the Egyptian government, which implemented the increase in the Egyptian personnel in the company. Of course, Charles-Roux did not explain why the SCC was reluctant to reemploy Italians even before the 1947 company law or the 1949 agreement.[40] This was a crucial question that Novelo's attorney also asked before an Egyptian court.

Prospects and Obstacles

When the SCC fired the Italians, its administrators considered that the younger people among them would easily find another job. Securing other employment, however, was not always possible because of the high degree of specialization of the SCC workforce and because of the rising unemployment in Egypt after the Second World War. Some of them managed to get rather poorly paid work in the auxiliary services of the British military base. But instead of being a viable solution to unemployment, a job in the British military services entailed several problems. First, finding a job in Egypt did not occur through Egyptian labour networks. Secondly, the collapse of the economic life of the Italian community offered limited possibilities for finding a job through the Italian networks. Thirdly, the departure of the Allied forces and the slowing down of economic activity in Egypt made it difficult for Italians, but also for other foreigners, who had been enlisted and were returning from the front to find a job in post-war Egypt. Many preferred to leave the country for good and return to their country of origin or to other destinations of their respective diaspora.

Apart from a few, mainly highly specialized staff, who were difficult to replace, the company refused to rehire most Italians. One of the lucky ones was Eduardo Esposito, who was reemployed in 1948 because he was considered young (41), because he had not been political active before the Second World War and because he still lived in Egypt at that time.[41] The SCC's attitude and reactions to Italian demands shows that it had nothing against Italians per se, but it did not want to rehire en masse a big number of dismissed employees and workers. Its official stance was that it could

[40] Charles-Roux to Roncalli, Paris, 9 September 1949, ANMT, ASCC, 1995 060 3017.

[41] Gambi to General Director of the SCC, Port Said, 11 June 1948, ANMT, ASCC, 1995 050 3017.

not fire those who had been recruited to replace the Italians during the war and that Egyptians should now fill all the vacant posts. However, the problem with the second argument is that negotiations between the company and the Egyptian government for the Egyptianisation of the personnel started in 1948 and concluded in March 1949. The resulting agreements sought the hiring of more Egyptians to the company, determining that new hires should be at a ratio of nine Egyptians for every ten administrative vacancies and four Egyptians for every five technical vacancies. Even so, Floresca Karanasou argues that the terms of the 1949 agreement allowed the SCC plenty of room for manoeuvre.[42] Vacancies came about by creating new jobs or retirement and it was not always easy to find qualified Egyptian pilots, for instance. Thus, the SCC continued to hire more non-Egyptians.[43]

We also know that companies affected by the 1947 Egyptianisation law that wanted to keep or to hire employees and workers managed to do so despite any legal restrictions.[44] The companies could increase the overall number of the personnel. Besides, it is important to keep in mind that the SCC was not an ordinary company with limited resources. It had a huge budget, which was equivalent to that of some states. It employed thousands of employees and workers and provided several benefits, with family and housing allowances and so on. During the war the personnel costs reached almost 40 percent of the turnover because of the aid being given to enlisted persons and the families of the interned Italian and because of the extra cost of the personnel who replaced those who had left.[45] In 1948 it had 3300 workers, which is estimated to have reached 4000 by the end of 1949.[46] However, after the war 75 percent of the workforce was Egyptian.

[42] Karanasou (1992, 305).

[43] This was possible also because after the war and after the 1949 agreement, the SGG partly filled the Egyptian quota by hiring Egyptian women white-collar workers, in order to keep the most qualified positions for Europeans; see Curli (2014).

[44] The Company Law 138, in July 1947. Its main aim was to secure employment for the ever-increasing Egyptian labour force, which could not be absorbed by either the public or the agricultural sector. It also sought to pass control of joint-stock companies to Egyptians. The law stipulated that companies had three years to ensure that 51 percent of the capital, 40 percent of the board of directors, 75 percent of the employees and 90 percent of the workers of these companies were Egyptian.

[45] Piquet (2008, 146).

[46] Wylie to Steward, Ismailia 10 April 1949, TNA, FO, 371 73622.

If nationality was the main qualification for finding a job after the war, had nationality been the only criterion to dismiss someone before the war? The president of the board, Charles-Roux, denied this in a confidential letter of June 1948 on the matter of the ex-pilot Castelli. He argued that when the Italians were removed from May to September 1940, the SCC took into consideration political issues and not nationality.[47] 'Please remember Radio Bari!' he noted. Charles-Roux asserted that 'after years of fascist propaganda and under the effects of the events of summer 1940, many Italians, when Graziani's army moved toward Sidi-Barrani, many Italians among those who now complain at not being rehired were not immune from celebration and did not remain calm'. For the SCC president, the Italians were dismissed because of their fascist activities. This is also confirmed by the fact that Egyptian citizens of Italian origin were also dismissed. Alfred Gambi, a Port Said resident who had obtained Egyptian citizenship in 1933, was also dismissed with the other Italians in 1940.

For Charles-Roux, Italians who could not be rehired after the war had to consider themselves 'war victims'.[48] But victims of whom? In the same letter, he stressed that the main causes for the denial was the fact that the SCC personnel was not homogenous in terms of nationality and it was absolutely necessary for the company to maintain harmony within the ranks of the personnel and the pilots in particular. He stated: 'Among our pilots in service, there are some who belong to nations which harbour rancour against the Italians more than we do in France. The French people want the Italians and they have no problem with them. But there are nationals of other countries that do not share our opinion.' Charles-Roux does not reveal the nationality of the pilots in question but the fact is that pilots in 1945 were comprised as follows: forty-two British, one Dane, one Egyptian, thirty-two French, eleven Dutch, four Norwegians, one Swedish, one Yugoslav and seven Greeks. Given that the French and the British were not against the recruitment of the Italians, it is likely that the Greeks were in opposition, especially when it is considered that behind the seven pilots were hundreds of Greek employees and workers, making them the most important non-Egyptian component in the workforce.

May we assume that the Italian occupation of Greece and hostilities between the two countries could have to do something with this stance? This hypothesis may seem exaggerated. Most Greeks of the canal zone

[47] Charles-Roux to Belardo, Paris, 17 June 1948, ANMT, ASCC, 1995 060 0976.
[48] Ibid.

originated from the Dodecanese, but they were born in Egypt and some of them had close contact with Italians. They spoke Italian and most of the few mixed marriages in the area were between Greeks and Italians. At the same time, we know that relations between the foreign communities in Egypt were not always harmonious despite the popular image of cosmopolitanism and peaceful coexistence. Competition between the foreign communities in the labour market was sometimes harsh in the immediate post-war years. In a period of increasing unemployment, all communities sought to reinforce their position in Egypt, often by turning against other communities. And this did not happen without the support of their respective state. The head of the SCC transit department, L. Lucas, admitted in 1947 that the SCC was the subject of pressing requests in favour of the Greeks of the company 'either through diplomatic channels or through their patriarch'.[49] As we have seen, before the war Greek communities were particularly anxious about possible unemployment after the war. As a result, they did all they could to return to civilian life before the liberation of the Italians from the camps. In addition, there was no question of replacing Greek personnel with other nationalities. The SCC felt obliged 'not to dismiss some important Greek captains who performed very well during the hostilities'.[50] Additionally, for the company, most Greeks had an excellent reputation as workers and a professional profile very similar to the Italians. Therefore, in the absence of the Italians, the chief of the transit department believed that it was necessary to retain Mediterranean elements in his service.[51]

But pressure did not only come from diplomats or have to do with a country's importance in the international scene. After the war, the Greeks still remained by far the largest non-Egyptian national group in the SCC. At a time when Egyptian visa restrictions were becoming ever more strict, Greeks represented a well-qualified workforce already settled in Egypt. In addition, some of them held Egyptian citizenship. These characteristics were a major advantage in the post-war environment and must have had some kind of influence on the SCC's recruiting policy. Even if we cannot take national or ethnic solidarities within a community for granted, we need to bear in mind that the Greeks of the canal zone strongly linked

[49] Lucas to General Director of the SCC, Ismailia, 18 February 1947, ANMT, ASCC, 2000 038 0189.

[50] Ibid.

[51] Ibid.

to each other through job specialization, kinship networks, common origin and residence in a rather isolated area. Besides, the SCC's recruiting policy was based on generational family ties. Evidence shows that while Italians were banned from being rehired, Greeks were recruited to the company for the first time in the late 1940s despite quotas that favoured Egyptians and pressure from Paris to this end.[52] This may explain the fact that many Italians left the canal zone after the company's refusal to rehire them. The Greeks' rather strong interconnections and feeling of belonging in the area were also illustrated some years later during the Suez Crisis, when most Greeks remained in the area even after the company itself left Egypt for good.

References

Archival Sources and Abbreviations

ANMT, Archives nationales du monde du travail, Roubaix
ASCC, Archives of the Suez Canal Company
AYE, Archives of the Greek Ministry of Foreign Affairs, Athens
CS, Central service of the Greek Ministry of Foreign Affairs
TNA, The National Archives of the United Kingdom, London
FO, Archives of the Foreign Office
ADM, The National Archives of Admiralty, Navy, Royal Marines, and Coastguard

Bibliography

Alleaume, Ghislaine. 1997. La production d'une économie 'nationale': remarques sur l'histoire des sociétés anonymes par actions en Égypte de 1856 à 1956. *Annales islamologiques* 31: 1–16.

Beinin, Joel, and Zachary Lockman. 1998. *Workers on the Nile: Nationalism, Communism, Islam, and the Egyptian Working Class, 1882–1954.* Cairo: American University in Cairo Press.

Bonin, Hubert. 2010. *History of the Suez Canal Company: Between Controversy and Utility.* Geneva: Librairie Droz.

Clogg, Richard. 1979. The Greek Government-in-Exile 1941-4. *The International History Review* 1 (3): 376–398.

[52] Homolle to Lucas, 2 April 1948, ANMT, ASCC, 2000 038 0168.

Curli, Barbara. 2014. Dames Employées at the Suez Canal Company: The 'Egyptianization' of Female Office Workers, 1941–56. *International Journal of Middle East Studies* 46: 553–576.

Dalachanis, Angelos. 2017. *The Greek Exodus from Egypt: Diaspora Politics and Emigration, 1937-1962.* New York: Berghahn Books.

Hanley, Will. 2008. Grieving Cosmopolitanism in Middle East Studies. *History Compass* 6 (5): 1346–1367.

Karanasou, Floresca 1992. 'Egyptianisation: The 1947 Company Law and the Foreign Communities in Egypt'. Ph.D. diss., Oxford University.

Mazower, Marc. 1993. *Inside Hitler's Greece: The Experience of Occupation, 1941-44.* New Heaven: Yale University Press.

Piaton, Claudine, ed. 2009. *Ismaïlia: Architectures XIXe – XXe siècles.* Cairo: IFAO.

———. 2011. *Suez : Histoire et architecture.* Cairo: IFAO.

Piquet, Caroline. 2008. *La Compagnie du canal de Suez. Une concession française en Egypte, 1888–1956.* Paris: Presses de l'Université de Paris-Sorbonne.

Turiano, Annalaura. 2016. 'De la pastorale migratoire à la coopération technique. Missionnaires italiens en Égypte: Les salésiens et l'enseignement professionnel (1890-1970', Ph.D. diss., Aix-Marseille University.

Viscomi, Joseph. 2016. 'Out of Time: History, Presence and the Departure of the Italians of Egypt, 1933-present', Ph.D. diss., University of Michigan.

PART VI

Postwar Suez

CHAPTER 20

Leaving Egypt: Rethinking 1956 Through Italian Departures

Joseph John Viscomi

In June 1946, Vittorio Emanuele III departed Italy. He arrived in Alexandria, Egypt, where he benefitted from a history of exchange between the house of Savoy and the Egyptian royal family. Around 30,000 Italian residents who held strongly to this transnational affinity welcomed the abdicated king. They were a generation of immigrants who had experienced the Fascist consolidation of the larger Italian community in Egypt—numbering around 55,000 and dispersed between Cairo, Alexandria, and the Suez Canal cities of Port Said, Ismailia, and Suez—when political propaganda emphasised that Italian emigrants were a fundamental 'asset' for the expansion of national prestige and empire. Many Italians across social strata remained loyal to the king when he arrived. In other words, he was recognised in his exile. He lived quietly in Alexandria until his death in December 1947, when flags at Italian institutions flew at

J. J. Viscomi (✉)
Birkbeck, University of London, London, UK
e-mail: j.viscomi@bbk.ac.uk

B. Curli (ed.), *Italy and the Suez Canal, from the Mid-nineteenth Century to the Cold War*,
https://doi.org/10.1007/978-3-030-88255-6_20

half-staff and then King Faruk provided a full military funeral.[1] His body was entombed behind the altar of St. Catherine's Church in Alexandria—a centre of Italian religious life—and remained there until December 2017, when his corpse was transferred to Northern Italy.

Narrating Vittorio Emanuele III's arrival in Egypt, the *Giornale della Sera* condemned the new government in Rome for showing 'disinterest' in the Italian community in Egypt: 'it is as though the Italians of Egypt are stateless, more so than the beneficiaries of the Nansen passport [for refugees] reserved for those without nationality and protected by international conventions.'[2] Indeed, diplomatic ties between Italy and Egypt had been ruptured since June 1940. During the war, Anglo-Egyptian authorities imprisoned over 5000 Italian men (and select women) in the desert alongside the Suez Canal. Freedom of movement had been restricted for Italian nationals, their bank accounts frozen, and other foreign and Egyptian subjects had been forbidden from signing contracts with Italians. Diplomatic ties would not be completely re-established until 1951. This scene illuminates extended socio-historical processes that were reshaping geopolitical constellations in the decolonising Mediterranean; processes that materialised in the connections between Italy and Egypt. The former king's recognition amongst Italians in Egypt does not simply signify a community ideologically committed to what the monarch represented. It denotes, instead, an indebtedness to the symbolic and material structures erected during the period of interwar imperial rivalry and played out in the socio-political landscapes of colonial and post-colonial Egypt. During the early-twentieth century, past and present together created a political horizon for Italians in Egypt that had vanished by 1946. Yet, the period after 1945 is often portrayed as a rupture in the history of the Mediterranean, a moment when historical actors on northern and southern shores turned their backs on the sea. 1956 is depicted as a pivotal event in this narrative break. The fragmentary categorisation of historical time in the Mediterranean, however, obscures the significance of processes that challenge strict periodisation.[3]

In this chapter, I argue that the 'uncertainty' which characterised the post-war period for Italian residents in Egypt reveals multiple, layered

[1] Telegramma n.39, ACGA, 'Vittorio Emmanuele di Savoia.'

[2] 'Dannoso disinteresse del Governo per gli italiani d'Egitto,' *il Giornale della Sera*, 26 July 1946.

[3] Viscomi (2019, 2020).

continuities and discontinuities that are concealed by conventional periodisation. Indeed, by 1953, an Italian diplomat described, the Italian communities in Egypt 'have again begun… to suffer from [chronic] uncertainty and disquiet.'[4] By reading against the grain of modern periodisation (*pre-WWI, interwar, post-war, and so forth*) and by focusing instead on the layered temporal processes of decolonisation, we gain insight into how seemingly disconnected trajectories converged to give meaning to particular conjunctures. In breaking teleologies of periodisation arranged around eventful years such as 1956, lived and remembered pasts conjoin with anticipated futures in this post-war uncertainty. Social and diplomatic imaginaries are brought closer together. 'Uncertainty' as a historical category, then, allows us to see the historical mechanisms behind transformations in Mediterranean geopolitics during the mid-twentieth century.

Orientations

The longue durée of this heterogeneous population is complicated at once by the *lack* of historiographical material available on the community and by the narratives that *do* exist of its history (which are mostly nostalgic accounts by elite members of the community or by its many visitors, or scholarly works that focus on fragmentary, synchronically isolated instances of the community's pre-fascist history and then generalise their claims based on these). The settlement of Italians in Egypt was encouraged, among other factors, by two key contingencies. First, the Capitulations: these bilateral treaties between the Ottoman Porte and European powers signed between the sixteenth and the nineteenth centuries provided extraterritorial jurisdiction and established residency rights for non-Ottoman subjects. Although they had been intended to stimulate commercial exchange in Ottoman territories, with the rise of nation-state they became a legal backdrop for relations between migrant subjects and their respective consulates. Throughout much of the Mediterranean, the safeguards tendered by these treaties had been either adapted to colonial administrations or cancelled—for example, at the establishment of the French Protectorate in Tunisia in 1881, upon Italian occupation of Libya in 1911, and through the Treaty of Lausanne in Turkey in 1923. In Egypt, they remained in effect until 1937, when they were 'abolished' through a

[4] Janelli to PCM, 21 November 1953, ASDMAE, AP, Egitto 1952, b.870.

12-year transitional period that Egyptian nationalists hoped would lead to national sovereignty. The second factor that encouraged Italian settlement was the large number of professionals, merchants, and workers who had arrived from the Italian peninsula and Sicily during the nineteenth century.

Before Italian unification, some of these 'Italians' were merchants who crossed the Mediterranean since the time of the Maritime Republics, participants in Napoleon's campaign in Egypt (1798-1801) who remained often in administrative roles under Mehmet Ali Pasha's reign (1805-1848), and political exiles of the Risorgimento (1815-1871). Until unification, these individuals lived in Egypt as subjects of their respective consuls (some French, others Tuscan or Sardinian); by 1864, all the consulates joined under one, national consul. Although numbers from this period are unclear, an 1840 estimate counts around 2000 Italians.[5] With British occupation in 1882, an event coterminous with the great emigration from Italian shores (1870s-1913), a new wave of Italian immigrants arrived in Egypt. These arrivals were typically from Southern Italy; but another large group was formed by Italian protégés--mostly Sephardic Jews--fleeing Istanbul and Izmir during the Italo-Turkish war of 1911 and subsequent tensions between Greeks and Turks. Between 1882 and 1927, the population of Italians rose from 18,665 to 52,462. This number encompassed all social strata—from government bureaucrats to Sicilians 'fleeing the volcano' as described in Bairam al-Tunsi's 1930s poem '*Al-Imtiyazat.*'

'Uncertainty' in the mid-twentieth century can be appreciated with a more substantial depiction of the horizon fashioned after the National Fascist Party took power in Rome. Beginning in the late 1920s, the fascist regime attempted to create islands of *italianità* by investing in political and cultural propaganda of Italian institutions amongst communities of Italian emigrants in and around the Mediterranean. In addition to Italy's occupation of Libya and Greece, the 'colonisation' of these Italian communities was part of the fascist vision to rechristen the Roman *Mare Nostrum*. Such efforts received inflated proportions in Egypt, as Mussolini's regime viewed emigrants living in extraterritorial jurisdiction as a strategic 'asset' that could undermine French and British hegemony in the Mediterranean. While investing in Italian cultural institutions during the 1930s, the regime poured its resources into anti-British movements among Egyptian nationalists. From Rome, the fascist government bribed

[5] Clot-Bey (1840), Iacovella (1994), Zanier (1985), Cuno and Reimer (1997).

the staff of newspapers published in Arabic, French, Greek, and English.[6] In this emerging constellation, Italian residents became central to narratives of 'friendship' between Italy and Egypt. The consequences of the 1930s are best condensed in the work of Italian historian Angelo Sammarco, a confidant of Fuad I and later Faruk. Fuad entrusted Sammarco with the task of reconstructing modern Egyptian history, giving him sole access to the Royal Family's archives. Until 1940, Sammarco, alongside other prominent members of the National Fascist Party present in Egypt, promulgated the notion that since Mehmet Ali's reforms—long before Italian unification– the Italian *colonia* and its 'timeless friendship to Egypt' helped deliver Egypt to modernity. The stories employed to substantiate this narrative referred overwhelmingly to the generation prior to the immigration of Italians that followed 1882.[7]

In this historical imaginary, Sammarco and other intellectuals knotted the politics of community-making to imperial rivalries. This approach was measurably successful, in part because the Capitulations facilitated a direct extension of the Italian state over its subjects in Egypt, one relatively free of British or Egyptian interference. What contemporaries in the early 1920s had described as a fragmented group of Italian subjects, representing a broad political spectrum—in the same Mediterranean that saw the spread of global radicalism[8]—by the 1930s had morphed into a tightly integrated national community organised around state institutions.[9] It was one that, while heterogeneous in origin, ardently participated in state-led initiatives and, when called upon to donate for imperialist causes (e.g. in 1935 during Italy's Ethiopia campaign), gave more per capita than in metropolitan Italy. The collective transformation was so extensive that upon Italy's entry into the war, Anglo-Egyptian authorities ruptured diplomatic ties, repatriated political figures, closed all Italian institutions, froze Italian bank accounts, and made it illegal for others to sign contracts with Italians. Egyptian police began to arrest and imprison in Fayed what would amount to over 5000 Italian civilians. This period of isolation would detach internees from social life in Egypt, from their families, and from the events of the war until 1945.

[6] 'Pretesa propaganda italiana in Egitto' ASDMAE, AP, Egitto 1931-45, b.16; Grange (1994, 171).

[7] Balboni (1906).

[8] Khuri-Makdisi (2010).

[9] Monroe (1938), Martelli (1938).

Reintegration

The landscapes of post-war uncertainty break through the formality of diplomatic reports, correspondences, and exchanges, as this intimate space evoked concern among various actors in the changing political constellation. After World War II and the closure of internment camps, it quickly became apparent within and around the Italian community how little post-war Egypt—and the larger realm of Mediterranean geopolitics—resembled its former incarnation. In 1944, the Italian government, based temporarily in Salerno, expressed apprehension over the situation in Egypt. The captain of an Italian ship passing through Egyptian ports observed, 'a community that just yesterday resonated with admirable and prosperous patriotism… is today taken as undesirable and lives precariously, in a state of [lethargy], on the boundaries of destitution.'[10] Juridical rights and financial accounts remained blocked, despite renewed freedom of movement for Italians.[11] The isolation and detachment imposed by the internment tore many families apart, and the severe economic hardship drove others to poverty. Rumours abounded that many Italian women, whose wage-earning family members had been interned during the war, had turned to regulated and unregulated sex-work to supplement the meagre pecuniary support they received from the Swiss Legation.[12] The implications of the community's declining conditions went far beyond the personal; they generated an array of unsettling political questions for the Italian state. Without formal diplomatic ties, it was difficult for then Secretary General of the Ministry of Foreign Affairs, Renato Prunas, to act on behalf of Italian citizens. Several requests to send a delegation to Egypt to resolve some of the growing problems went unanswered by Anglo-Egyptian authorities.[13] Italian diplomats themselves suspected that the Egyptian regime, aiming towards greater national sovereignty, sought to renegotiate the terms of the 1936 treaty before it dealt with other matters. By 1945, Prunas finally obtained permission to send a small mission to Egypt with the scope of reconnecting the Ministry of Foreign Affairs (now

[10] 'Notizie sulla collettività italiana d'Egitto,' 21 December 1944, ACS, PCM 1944-47, 15-3/11222.

[11] 'Situazione politica interna,' 22 February 1946, ASDMAE, AP, Egitto 1946-50, b.1.

[12] Memoire June 1944, SFA E2001-02#1000/111 BD:7; 'Interessi italiani in Egitto,' 1944, ASDMAE, AP, Egitto 1931-45, b.33.

[13] Memorandum 30 November 1943 (Brindisi) and 12 February 1944 (Salerno), ASDMAE, AP, Egitto 1931-45, b.33.

in Rome), to Egypt. Giovanni de Astis arrived in Egypt on 23 September 1945 and stayed until December intending to 'explore the situation.' No attempts, however, to reinvigorate diplomatic ties between Italy and Egypt were made.[14]

Drawing attention to disjointed layers of historical time, one diplomat described Italians in Egypt as 'out of sync' with the present.[15] In other words, uncertainties caused by the rupture during the war were inextricable from processes of political transformation in Italy. De Astis's brief visit, which included banquets amongst the elite circles and careful diplomacy in attempt to establish new diplomatic talks, provoked anger amongst many residents who perceived themselves as ignored. The collapse of fascism (the implications of which were largely unknown in Egypt), the loss of Italy's colonial possessions, and Egypt's acceleration towards national sovereignty punctuated Italy's 'imperial hangover' for Italian residents in Egypt.[16] They were notably 'reluctant' to break from the political structures of fascism. Emphasising their role on the imperial periphery, some residents attributed the collapse of the Fascist government to the 'cowardliness' of metropolitan Italians and the sabotage of international antifascists.[17] During a break from naval service with the allied forces, one Italian from Egypt visited family where he observed 'great contempt by the entire colony, which classified me as a traitor to the *patria* and a mercenary of the English.'[18] It was 'natural,' observed one diplomat, that Italian residents had an inflated sense of their centrality due to the image of community 'cultivated' by the Fascist regime in Rome, one in marked contrast to the perceived 'neglect of prefascist governments.'[19]

The government in Rome was alarmed by this deepening divide in the Italian residents' perception of past and present. At the end of the war, one diplomat contrasted the once 'vibrant' community—with its cultural institutions, schools, hospitals, the main daily newspaper *Il Giornale d'Oriente*—'to this recent past (passato prossimo) corresponds a miserable

[14] Rappresentanza italiana in Egitto to MAE, 4 September 1946, ASDMAE, AP, Egitto 1946-50, b.2; Pizzigallo (2008, 13-14).

[15] 'Notizie sulla collettività italiana d'Egitto,' 21 December 1944, ACS, PCM 1944-47, 15-3/11222.

[16] Seton-Watson (1980).

[17] 'Situazione collettività italiane in Egitto,' 1944, ACS, CRI servizi affari internazionali, b.144; Caccia Dominioni (1966, 269).

[18] Cuppari Annibale, 2 February 1944, ASDMAE, AP, Egitto 1931-45, b.33.

[19] Appunto 28 July 1944 and Appunto 23 June 1944, ASDMAE, AP, Egitto 1931-45, b.33.

present'[20] Institutions and schools remained closed, hospitals were occupied by the Anglo-Egyptian forces, and *Il Giornale* had been shut down since 1940. Indeed, the new government circulated a plan for the systematic 're-education' of Italians in Egypt, seeking to suture the divide between the past and the present.[21]

The conditions of possibility for Italian residents in post-war Egypt were quickly changing, however, and any ideological reintegration had to confront the material realities of everyday life. Many residents questioned whether it would be possible to transition from a community whose lifeline was state-provided assistance, as it had been in the years prior to and during the war, to an employed community. The 1947 Company Law set quotas for hiring of Egyptian employees, addressing a longer process aiming to dismantle foreign privilege in hiring practices. Unemployment was between 18-25 per cent and, when work was procured, it was often on a provisory or temporary basis. In the canal zone, following the closure of the internment camps, over 90 per cent of Italians were unemployed. Positions in the Suez Canal Company had been filled by Egyptian, Greek, and Maltese workers; the hiring anew of Italians appeared unlikely, especially given the increasing tendency to 'Egyptianise' the workplace. Maurizio Basso, a diplomat filling a temporary post in Port Said, estimated that Egyptian employees were projected to fill around 80 per cent of the total workforce in the canal zone. Attempting to understand the predicament facing the reintegration of Italians in the local workforce, Basso noted: 'It is not excluded [...] that [behind] the difficulties confronting Italians there are concealed motives linked to the Egyptian Authorities' desire to abolish the Montreux Convention [which ended the Capitulations] and then to be free to expel undesirables. Today, an unemployed foreigner is always an undesirable one.'[22]

Fuelling uncertainty, the break in diplomatic ties had excluded Italians from the benefits entitled to former capitulatory subjects in the transition away from the Capitulations. Italian residents were therefore required to renew residency permits each year (rather than every five years), and many of them speculated on an expulsion of foreigners with the approaching end of the transition period in 1949. One resident in Alexandria penned a letter to then President of the Senate, Ivanoe Bonomi, writing, 'reports,

[20] 'Interessi italiani in Egitto,' 1944, ASDMAE, AP, Egitto 1931-45, b.33.
[21] Prunas to Bounous, 17 February 1944, ASDMAE, AP, Egitto 1931-45, b.33.
[22] Basso to de Astis, 6 November 1946, ASDMAE, AP, Egitto 1946-50, b.2.

memorandums, [...] have all been written by authoritative people [...] but it all seems to finish in nothing [...] no one moves to meet the needs of (the Italians in Egypt).' The disjuncture opened by this uncertainty reveals wider temporal landscapes in the post-war Mediterranean. The angry resident continued, 'we await rescue'[23]— rescue from a collapsed state which had inflated their centrality to imperial ambitions in colonial Egypt, a detail the current government saw as a geopolitical burden noting Egypt's imminent national sovereignty. Sensing that the Italian state had abandoned them at this horizonless conjuncture Italians began to look elsewhere.

Departures

The inability to suture the past and present rendered departure as the most sensible solution to many Italian residents. Passport renewals with requests for 'repatriation' and 'emigration' streamed into the consulates in Alexandria and Cairo. Italian residents were aware that requests for new residency privileges would appear to the Egyptian government as extensions to the capitular regime. In other words, in addition to the fading past of fascist imperialism the horizon of colonial Egypt too was closing. The new Minister in Egypt expressed 'angst' over 'the future' of Italians in Egypt.[24] *Italiani nel Mondo* estimated that between 1945 and 1951 around 6000 Italians had left Egypt, searching in 'other locales the security of a future more prosperous and serene.'[25] With these departures in mind, the Italian government closely monitored the arrival in Italy of a delegation from Argentina soliciting around four million Italian immigrants. In addition to relieving the growing question of DPs in metropolitan Italy, this was seen as a preferable option to avoid an eventual repatriation en masse of Italian residents.[26] Concurrently, the locally published Italian newspaper that had emerged in Egypt began to run advertisements and articles featuring possibilities for emigrants in Belgium,

[23] Carbone to Bonomi, 1 June 1947, ASDMAE, AP, Egitto 1946-50, b.9.

[24] Zoppi to Fracassi, 9 December 1948, ASDMAE, AP, Egitto 1946-50, b.7.

[25] Raniero Giani, "Declino a Porto Said della collettività italiana," *Italiani nel Mondo*, 25 January 1951; G.P.d.F "Il passato e il presente degli italiani in Egitto," *Italiani nel Mondo*, 25 February 1951.

[26] Con.Gen.d'Italia London to MAE, 3 December 1946, ASDMAE, AP, Egitto 1946-50, b.1.

France, Switzerland, and Venezuela.[27] In 1937, Italians had represented a quarter of all foreigners in Egypt, numbering 47,706 persons. By 1947, the population had declined to 19 per cent (decreasing in size more than any other community), numbering about 28,000. Broken down further, around 17,200 Catholic Italians and 2400 Jewish Italians appear absent from Egyptian census data. These data are imprecise yet their inaccuracy is revelatory: while the decrease in the population was most likely sizable (doubtfully as high as 20,000), it might indicate hesitancy in participating in such a survey given the uncertain debates on residency status with the prospects of a community already looking elsewhere for its future.

Uncertainty encompassed more than the perceived horizon of Italian residents, it also marked the Italian government's hesitancy in taking a political position knowing that regional hierarchies of power were quickly changing. Unresolved antagonism between dominant political factions in Egypt—the nationalist Wafd, the Monarchy, and the military—and the British Authorities' forceful attempts to maintain a foothold in the Suez Canal yielded strong criticism from opposition groups between 1948 and 1952.[28] In a last-ditch effort, just before the end of the transition period, British authorities called a meeting of former capitulatory powers. Fracassi, then Italian Minister in Egypt, had been warned by the undersecretary of the Egyptian Ministry of Foreign Affairs that the British were living in an 'imperial and capitular fog' and that Italy and Greece (with the largest communities in Egypt) would only arrive at settlements through bilateral talks with the Egyptian government.[29] Neutrality was the line Italy projected in its approach to Egypt's ongoing changes, but the Italian government benefitted from interwar links between Italy and Egypt's nationalist movements to align itself with the rising military. Discussions began in the spring of 1951 for an Egyptian military visit to Italy to study Italian military and police infrastructure and receive instruction from Italian military technicians.[30] This 'advisory role' was to play a key part in Italy's emerging

[27] 'I nostri lavoratori e le possibilità d'emigrazione,' *Il Mattino della Domenica,* 7 July 1946; 'L'emigrazione nel Venezuela è aperta agli Italiani,' *Il Mattino della Domenica,* 14 July 1946.

[28] Tarchiani to De Gasperi, 9 November 1951, ASDMAE, AP, Egitto 1952, b.870.

[29] Fracassi to MAE, 23 April 1948, ASDMAE, AP, Egitto 1951, b.703; Tignor 1987, 488.

[30] Appunto per DGAP, 25 May 1951 and 'visita in Italia Ministro Commercio Egiziano,' 15 May 1951, ASDMAE, AP, Egitto 1951, b.698.

'Mediterranean vocation,' its alleged ability to navigate Atlantic, European, and Middle Eastern political spheres.[31]

As the political situation devolved after January 1952,[32] the Free Officers seized power in July. The military quickly passed laws to settle residency issues. They increased the terms of residency for foreign residents, especially those whose economic and technical skills could benefit the faltering Egyptian economy. Rather than reading this as a solution to the unresolved preoccupations of foreign communities, however, most Italians interpreted new regulations as a 'political measure' to contain the flight of foreign capital. For residents with no clear answers to long-term policies on residency, living amidst wide-spread rumours of expulsion, sudden attempts to resolve issues, wrote one Italian, 'will not eliminate the growing sense of unease… it has deep roots and depends, by now, on the uncertainty of the political and economic future of the country.'[33]

Behind the scenes, the Italian government partnered with the Egyptian military in dismantling of the interwar worlds which had connected Italy and Egypt. Faruk's expulsion was discussed and organised in talks between the Italian government and the military officers. In fact, in accord with the officers and given the refuge provided to Vittorio Emanuele III in 1946, the Italian government agreed to host Faruk to advance their 'Mediterranean' agenda.[34] Through this symbolic exchange, Italian and Egyptian governments tightened their relations. The Italian government had agreed to furnish two large police squadrons to the Egyptian military, justifying the deal to the international community as a 'non belligerent' gesture.[35] But their motives were much larger. The Egyptian military arranged for further study of the Italian police structure, which was followed by investment in training and equipment from the Italian state. Immediately after the Free Officers declared a three-year transitional period of martial law in 1953, the Italian Ministry of Defence, Randolfo Pacciardi, set out to visit Naguib in Egypt. He intended to solidify new geopolitical connections rhetorically unified around the 'defense of the Mediterranean.' The scope of the visit was broader: Pacciardi was to inquire about Naguib's

[31] Janelli to MAE, 29 November 1951, ASDMAE, AP, Egitto 1952, b.787. Consiglio (1958), Brogi (2004), (2006).

[32] Tignor (1987, 492-493), Reynolds (2012), Kerboeuf (2005).

[33] Ferrero to MAE 6 June 1952 and 8 August 1952, ASDMAE, AP, Egitto 1952, b.788.

[34] Appunto 19 September 1952, ASDMAE, AP, Egitto 1952, b.788; ACS MI PS H2 1952 B220.

[35] Appunto 22 February 1952, ASDMAE, AP, Egitto 1952, b.789.

personal insights and the military's plans for the future. Pacciardi was a major proponent of Italy's participation in NATO and this exchange marks the entry of the United States into regional affairs. Aware that Italy and Egypt had this long and historical exchange, the United States government used this channel to covertly influence regional politics.

The importance of the interwar links between fascist imperialism, the Italian community, and Egyptian nationalists was central. Pacciardi's visit took place under the guise of 'refreshing the [...] Italian colony [...] (which had been) subject to the vicissitudes of war.' He had been warned of growing opposition among Italians towards the Italian state: a strong coalition had formed in support of the Italian Social Movement (MSI), the post-fascist political party. Tellingly, just before he gave speech to an audience of Italian residents, the word 'prosperous' was removed from his description of the community so as not to unnecessarily draw attention to the perceived discrepancy between their past and the present. Aside from a quick toast to the Italian residents, who, he said, signified 'the most solid wedding band uniting our two peoples,' no other mention was made of the community. Pacciardi assured the Italian Ministry of Foreign Affairs that Naguib was united in the fight against communism in the Mediterranean, but would only join Italy once Egypt's 'national aspirations' were fulfilled. The path was paved for Italy to serve as a 'bridge' between the Middle East and Europe.[36] Italian residents in Egypt surfaced merely as an index to a vast historical narrative of national kinship, just as they saw departure as the resolution to an uncertain future.

Becoming Refugees

Italian residents and the political actors around them defined the community by embracing social and legal categories of belonging that mapped onto geopolitical histories of imperial rivalries in the twentieth-century Mediterranean. Each of these centred on the displacement of Italians from Egypt, but they differed in how they highlighted entitlements based on the relevance of the 'national' community's historical experiences. The departure of Italians had accelerated in the late autumn of 1956 and over the course of winter 1957. Because Italy and Egypt were not in a state of political conflict, Italians arriving in Italy would not qualify as refugees.

[36] Ferrero to MAE, 5 February 1953, ASDMAE, AP, Egitto 1953, b.871; Appunto de Strobel 1953, ASDMAE, AP, Egitto 1952, b.788.

They were classified instead as 'poverty-stricken' (indigenti), which entitled them to the same aid as any (national) recipient of welfare.[37] This, however, did not account for their physical and symbolic displacement; perceived inequities that were points of political contention. In February 1957, Edoardo Costa, responsible for foreign affairs for the MSI, travelled to Egypt to meet Nasser, stirring a group of Italians against the current Italian government. He claimed that it had failed to see that 'repatriates' were not 'impoverished Italians,' but rather were victims of international political relations. Costa portrayed the state's absent 'reception' as a repudiation of the 'historic role' Italian residents had played in 'developing Italy's prestige' abroad.[38]

In many ways, displacement defined this post-war conjuncture for Italy.[39] Over 3000 Italians from Egypt had requested assistance from the state or from its regional offices by early 1957. News of the 'exodus' of Italians sparked controversy in the Italian parliament. Although, during the Suez conflict, parliament voted to extend the legal parameters of 'refugee' (according to Law 137 of 1952) to include Italians from Egypt, it would take years for the legal status to take effect, doing little to resolve the sense of uncertainty.[40] The government worked with the newly established Intergovernmental Committee for European Migration (ICEM) to internationalise integration by facilitating emigration.[41] Many repatriates had left Egypt with little more than a suitcase and lacked resources or networks to travel, even when emigration had been approved. The cost of travel was above and beyond the budget for many families.[42] Displacement thus bound many to a horizonless present. One group of Italians preparing to leave Egypt wrote to the Italian Ministry of Foreign

[37] Telegramma no.28647, 2 December 1956, ASDMAE, AP, Egitto 1956, b.1059.

[38] Telespresso no.787/317, 22 February 1957, ASDMAE, DGAP, 1948-60, Uff.III, b.148, fasc.465.

[39] Ballinger (2020).

[40] Tambroni to Ministero del Tesoro, 6 February 1957, ACS PCM 1959-61 15/3/57749(3); Law 1306 of 25 October 1960, *Gazzetta Ufficiale*, n.279, 15 November 1960.

[41] 'Elementi di risposta alla interrogazione n/3209' 8 March 1957, DGE Appunto (Del Bo) 11 May 1957 and MLPS to Centri di Emigrazione, 22 October 1957, ASDMAE, DGAP, 1948-60, Uff.III, b.148, fasc.465; 'Propaganda contro l'emigrazione,' ACS, MLPS—DG Collocamento della manodopera—divisione VIII accordi verso paesi comunitari, b.383.

[42] Appunto 16 October 1957 and Prefettura to ECA & Questura, 4 June 1957, ACL, Archivio ECA, no.713, 'Profughi dall'Egitto 1957-62.'

Affairs inquiring about the possibility to transfer money in Italy, noting that some Italians feared they would abandon their limited resources in Egypt for more profound financial insecurity in Italy.[43] The restriction on exports of more than 20 pounds sterling per family and strict limitations on belongings that could be taken out of the country led to the impoverishment even of those Italians who had managed to find work upon arrival. By summer 1957, the Italian Red Cross (Croce Rossa Italiana, CRI), which oversaw the refugee housing centres in Italy, advised that departures be expedited, drawing from typical situations to illustrate the severity of the challenges facing repatriates. [44] In following months, around 350 families were sent to South America through ICEM, while the organization continued to pressure Australia to raise its quota to accommodate Italians from Egypt.[45] The marginalization of repatriates from Egypt replicated the shared sense of isolation Italian residents had experienced during World War II.

Deteriorating conditions in Italy's improvised refugee camps exacerbated a sense of having been abandoned to the past. Collectives of repatriates began appealing to politicians. In Genoa, a group formed the National Committee of Italians of Egypt (Comitato Nazionale Italiani d'Egitto, CoNIE). They looked to the past to measure their present and safeguard their futures.[46] CoNIE institutionalised links between repatriated Italians and Italian residents in Egypt, subverting state efforts to diffuse these connections. The Ministry of Foreign Affairs promptly refused any 'moral' support or financial backing of CoNIE and warned regional political offices to do the same. They feared the formation of analogous 'associations and committees' would constitute powerful political blocks. In addition to their possible mobilization by opposition parties (the MSI on the right or PCI on the left), associations like CoNIE were thought to embrace hostility towards the post-fascist state.[47] But CoNIE was not isolated. In

[43] Amb. Cairo to MAE, 8 March 1957 & 17 May 1957, ASDMAE, DGAP, 1948-60, Uff. III, b.148, fasc.465.

[44] Memorandum (Mascia), 19 October 1957, ACS—Ufficio del consigliere diplomatico, b.24 fasc. 3 "profughi"; 'Padre Governanti,' ND, ACS CRI Ufficio Madrinato, b.83 fasc. 'varie' s.fasc. 'profughi dall'Egitto', Sanfilippo (2006, 848).

[45] Appunto 4 October 1957, ASDMAE, DGAP, 1948-60, Uff.III, b.148, fasc.465.

[46] Gennaro to Istituto Italiano dell'Africa, 24 October 1957, ASDMAE, DGAP, 1948-60, Uff.III, b.148, fasc.465.

[47] Straneo to PCM, 30 November 1957, ASDMAE, DGAP, 1948-60, Uff.III, b.148, fasc.465.

Messina, 89 repatriates who had remained in an improvised camp with their families for over a year claimed they had been deprived of 'entitlements' that characterized the precise circumstances of their departures and arrivals, and of their long history of maintaining links to Italy through national institutions in Egypt. They blamed the Italian government for the hostility they faced in Egypt, arguing that the 'ambiguous' Italian position on the Suez conflict had 'assimilated' Italian residents to British, French and Jewish communities and constrained them to 'suddenly leave belongings, houses, positions, and interests in Egypt.' To frame this argument, the group drew attention to the substantial donations provided to the Fascist regime during the Ethiopia campaign and the '54-month internment' during World War II. Within one year, the Messina group joined CoNIE, which had established committees in Rome, Genoa, Milan, Livorno, Naples, Brindisi, and Messina. In Rome, around 600 individuals walked to the Ministry of Foreign Affairs demanding employment or assistance. Attilio Morini, the director of CoNIE in Rome, wrote, 'days pass [without resolution] and most refugee Italians of Egypt become increasingly oppressed by harsh daily realities.' CoNIE claimed that by mid-1957 half of the Italian community had left Egypt.[48]

Conclusion

In this story, the Suez conflict in 1956 appears not as a primary event in the history of Italian departures from Egypt, but rather as a conjuncture which shows departures as outcomes of extended processes. When Vittorio Emanuele arrived in Alexandria in 1946, he had been welcomed both by king Faruk and by the Italian community. Many residents remained loyal towards the king and continued to hold annual ceremonies at his tomb in St. Catharine's Church. With the exile of Faruk several years later, Italian residents in Egypt were twice without a king, 'stateless' once again between two emergent regimes. The collapse of fascism's nation-empire and Egypt's move towards national sovereignty left Italian residents in Egypt without a political horizon. At this historical conjuncture, the Italian ambassador described residents as suffering from 'chronic uncertainty,' a condition which evoked wider transformations in Mediterranean geopolitics. Departure was the only (partial) solution to this uncertainty—but it

[48] Morini to Gronchi, 22 May 1958, and 'I connazionali tutti, dicono e chiedono,' 11 December 1957, ACS, PCM 1959-61 15.3/57749.3; Sanfilippo (2006, 851).

came with its own set of complications, which led to the formation of a political community based on the historical experience of Italians in and from Egypt. By 1960, over 40,000 Italians had left Egypt. Concurrently, new ties would result in the establishment of oil exploration rights for the Italian oil-tycoon, Enrico Mattei, in 1954, and a series of industrial partnerships that continues to this day linking metropolitan Italy with Egypt. In the midst of the mass evacuations and departures of Italian residents in 1956/57, the Italian government implored Mattei and his team to remain in Egypt so as not to put into question, or seem disingenuous to, nascent political-economic ties with the new Egyptian regime. Looking towards 'uncertainty' as a historical category that marks this post-war experience of decolonisation among Italian residents in Egypt reveals broader processes that challenge our understanding of the periodisation of the modern Mediterranean.

References

Abbreviations

ACGA: Archivio Consolato Generale d'Alessandria d'Egitto
ASDMAE: Archivio storico-diplomatico del Ministero degli Affari Esteri
AP: Affari Politici
DGAP: Direzione Generale Affari Politici
ACS: Archivio Centrale dello Stato, Roma
PCM: Presidenza del Consiglio dei Ministri
MLPS: Ministero del Lavoro e della Previdenza Sociale
DGE: Direzione Generale di Emigrazione
SFA: Swiss Federal Archives, Bern

Bibliography

Ballinger, Pamela. 2020. *The World Refugees Made: Decolonization and the Foundation of Postwar Italy*. Ithaca: Cornell University Press.

Balboni, L.A. 1906. *Gl'Italiani nella Civiltà Egiziana del Secolo XIX*. Egypt: Tipo-litografico v. Penasson.

Brogi, Alessandro. 2006. "Competing Missions": France, Italy, and the Rise of American Hegemony in the Mediterranean. *Diplomatic History* 30(4): 741–770.

———. 2004. Orizzonti della politica estera italiana: Stati Uniti, Europa e Mediterraneo (1945-1960). *Passato e Presente* 22 (62): 13–24.

Caccia Dominioni, Paolo. 1966. *Alamein: 1933-1962*. Milano: Longanesi.

Clot-Bey, A.B. 1840. *Aperçu général sur l'Egypte*, Cairo.

Consiglio, Gregorio. 1958. Il piano Pella e l'Africa. *Africa: Rivista trimestrale di studi e documentazione dell'Istituto italiano per l'Africa e l'Oriente* 13 (3): 119–120.

Cuno, Kenneth M., and Michael J. Reimer. 1997. The Census Registers of Nineteenth-Century Egypt: A New Source for Social Historians. *British Journal of Middle Eastern Studies* 24 (2): 193–121.

Grange, Daniel. 1994. *L'Italie et la Méditerranée (1896-1911): Les fondements d'une politique étrangère*. Rome: Ecole française de Rome.

Iacovella, Angelo. 1994. La presenza italiana in Egitto: problemi storici e demografici. *Altreitalie* 6 (12): 60–69.

Kerboeuf, Anne-Claire. 2005. The Cairo Fire of 26th January 1952 and the Interpretations of History. In *Re-envisioning Egypt, 1919-1952*, ed. Arthur Goldschmidt, Amy J. Johnson, and Barak A. Salmoni, 194–216. Cairo: American University of Cairo Press.

Khuri-Makdisi, Ilham. 2010. *The Eastern Mediterranean and the Making of Global Radicalism, 1860-1914*. Berkeley, CA: University of California Press.

Martelli, George. 1938. *Whose Sea? A Mediterranean Journey*. London: Chatto & Windus.

Monroe, Elisabeth. 1938. *The Mediterranean in Politics*. Oxford, UK: Oxford University Press.

Pizzigallo, Matteo. 2008. *La diplomazia italiana e i paesi arabi dell'oriente mediterraneo (1946-1952)*. Milano: Franco Angeli.

Reynolds, Nancy. 2012. *A City Consumed: Urban Commerce, the Cairo Fire, and the Politics of Decolonization in Egypt*. Stanford: Stanford University Press.

Sanfilippo, Matteo. 2006. Per una storia dei profughi stranieri e dei campi di accoglienza e di reclusione nell'Italia del secondo dopoguerra. *Studi Emigrazione/Migration Studies, XLIII* 164: 835–856.

Seton-Watson, Christopher. 1980. Italy's Imperial Hangover. *Journal of Contemporary History 15* (1): 169–179.

Tignor, Robert. 1987. Decolonization and Business: The Case of Egypt. *The Journal of Modern History 59* (3): 479–505.

Viscomi, Joseph John. 2020. Pontremoli's Cry: Personhood, scale, and history in the Eastern Mediterranean. *History and Anthropology* 31 (1): 43–65.

———. 2019. Mediterranean Futures: Historical Time and the Departure of Italians from Egypt, 1919-1937. *The Journal of Modern History* 91 (2): 341–379.

Zanier, Claudio. 1985. I fondi non inventariati delle legazioni e dei consolati degli stati pre-unitari all'archivio storico del ministero degli Affari Esteri: la rappresentanza di Sardegna ad Alessandria d'Egitto (1825-1861). *Oriente Moderno* 1-2: 49–57.

CHAPTER 21

Italian Catholics and the Suez Crisis: Between Neo-Atlanticism, Pacifism, and Third Worldism

Massimo De Giuseppe

Introduction: The Suez Crisis at the North-South, East-West Crossroads

> Dear Mattei, the real solution for Egypt is the dam: only the country that will be able to build it will have the solution to the current issues of the Egyptian and Arab world. Discuss it with Count della Torre: on this subject of the construction of the Aswan dam you can set up a great policy that concerns the peace in the world.[1]

With these words, the mayor of Florence and Christian-democrat Giorgio La Pira invited the president of Ente nazionale idrocarburi (ENI), Enrico Mattei, to break the delay which caused the international crisis

[1] La Pira to Mattei, 24 August 1956, ALP, 3, Egypt, 2, Aswan Dam, d. 7.

M. De Giuseppe (✉)
Università IULM, Milan, Italy
e-mail: massimo.degiuseppe@iulm.it

B. Curli (ed.), *Italy and the Suez Canal, from the Mid-nineteenth Century to the Cold War*,
https://doi.org/10.1007/978-3-030-88255-6_21

following the suspension of multilateral funding for the construction of the Aswan Dam and the consequent nationalisation of the Canal Company that ignited the events in the summer of 1956. The letter was written on 24 August, the day after the closing of the conference 'of the Twenty-two', held in London from 16 to 23 August to find a negotiated solution to the crisis. In the letter, La Pira made also reference to Count Della Torre, the director of the newspaper of the Holy See *L'Osservatore Romano*, who closely followed the entire Egyptian game,[2] in order to get possible support from the Vatican. That strategy revealed the peculiar Lapirian idea that Italian foreign policy should develop wide triangulations, in terms of technical cooperation and cultural diplomacy, taking advantage of the Vatican's moral weight, but also of the new horizons that the North-South, East-West crossroads were opening up in an enlarged Mediterranean in the global world.

The year 1956 opened with the birth of three new independent African states: Sudan (1 January), Morocco (2 March) and Tunisia (20 March), but it were the two crises of Budapest and Suez that would have shaken the two Cold War blocs. The 'double trauma', as Guido Formigoni defines it,[3] deeply intertwined the bipolar logics with the post-colonial countries. In 1956, the crisis matured in spring to explode in autumn, under the noise of the Soviet tanks that occupied Budapest and the cockpits of the French-British planes that dropped paratroopers on the port of Alexandria.

Indeed, the events of 1956 would have shown that the neutralist bloc, despite its heterogeneity, could play a new role between the two superpowers and their respective European and non-European allies. At the beginning of 1956, Moscow was undoubtedly opening new horizons of dialogue with the Afro-Asian world, although China seemed more rapid and unpredictable in its movements to some observers. Washington began to evaluate carefully, even from influential think tanks such as the World Peace Foundation, the capacity of the pro-communist pacifism to penetrate the Third Bloc, in particular through the dialogue established between the Indian Premier Jawaharlal Nehru and the new Egyptian President Gamal Abd el-Nasser, estimating also the effects on the European allies. In this perspective, the position of European Catholics, and in

[2] Ferrari and Zanini (2010, 204–230).
[3] Formigoni (2016, 236–243).

particular French and Italians, potentially open to Third World instances, began to represent an original element.

When the president of Yugoslavia, Josip Broz Tito, invited Nasser and Nehru to his summer residence on the Croatian island of Brijuni, on 18 and 19 July 1956, the interest for the constitution of a hypothetical Third Bloc became more spasmodic. Nasser was then officially president for less than a month (since June 23). The summit and the declaration had a strong media impact. The historiography is still divided on the outcome of the summit, which according to some scholars would have pushed the Indian leader to move away from the platform of the Colombo group focused on Afro-Asian neutralism to adhere to a new global non-alignment with uncertain outcomes (as for relations with Moscow). In May, Nasser had officially recognized the People's Republic of China, and Tito, in the previous months, had visited Ethiopia and Egypt, where he had self-proposed to Nasser as a potential mediator in the crisis with Israel, informing about the project (without raising too much enthusiasm) the US administration.[4]

As it emerged later, it was precisely on the sidelines of the trilateral summit that Nasser received the news of the US decision to suspend the financing of the Aswan Dam project. This was born on the sidelines of the 1954 Cairo meeting between Nasser and the US Secretary of State, John Foster Dulles, and resulted in an impressive credit plan that continued even after the turn of summer 1955 and the trade agreement signed in September between Egypt and Czechoslovakia (which also involved supplying weapons in exchange for cotton and wheat). The financing plan presented at the end of 1955 amounted to about 270 million dollars (56 million dollars in bilateral aid from Washington, 14 million dollars from London and the others guaranteed by the multilateral structure of the International Bank for Reconstruction and Development). The situation had therefore worsened rapidly between the spring and summer of 1956, first due to the US irritation at the Egyptian recognition of China,[5] then because of Moscow's provocative offer to finance the dam with a loan of

[4]Tito to Eisenhower, Belgrade, 30 November 1955. FRUS, 1955–1957, Central and South Eastern Europe, vol. XXVI, 269. Tito wrote to the US president during his journey from Addis Ababa to Cairo: 'Of course I have no claim to be a mediator in the conflict between Israel and Egypt, but I will try, as I have already told Mr. Dulles, during my talks with the Egyptian leaders and with the President Nasser to promote a reassuring action, according to the principles of our policy.'

[5]Zhou Enlai promoted a strategic agreement, "dramatically increasing the volume of trade to maintain the balance in favor of Egypt", Campanini and Di Donato (2021, pp. 66–67).

1.2 billion dollars and at a rate of 2%. The announcement of the suspension of Washington's aid helped to accelerate the crises.[6]

In those weeks, Washington was looking with increasing concern at the southern shore of the Mediterranean as a whole, as it emerges from the FRUS (Foreign Relations of the United States) documents, in particular from a memorandum prepared by the British envoy to Cairo, John Russell, for the Secretary of State, on the eve of the Brijuni summit, on a possible reopening of the dialogue between Egyptians and Israelis. The objective was to solve the Palestinian question, after the failure of the peace plan proposed by Eisenhower's special envoy, Robert B. Anderson, by dropping it from other Middle Eastern issues, fearing a destabilizing effect for the region. This attempt, according to the Secretariat of State, could have found an intermediary in the UN Secretary-General, Dag Hammarskjöld, who however seemed to have 'insufficient authority'.[7] The main issue remained to avoid a dangerous Egyptian isolation that could have pushed Nasser toward a true alliance with Moscow. Russell added in the memorandum's final notes that in his view the most solid hypothesis for a rapid resolution of the Palestinian question, at least between Egyptians and Israelis, was to rely on the 'Third Bloc', possibly including a European country, but still a Mediterranean one: 'I agree that neither Hammarskjöld, nor the tripartite powers', he wrote 'would be suitable for this latter role. I would prefer a group consisting of India, Pakistan, Burma, a Latin American state (perhaps Uruguay), and perhaps Italy.'[8]

The mediation plan would have been dismissed even before it could be seriously considered; certainly the Egyptian informal support to the Algerian National Liberation Front (FLN) militiamen contributed to the growing concerns of the French Prime Minister Guy Mollet, in the months in which the colony entered a spiral of blood, increasing the anxiety of the British prime minister, Anthony Eden, for a possible destabilization of the whole Middle Eastern chessboard. However, the suspension of funding for the Aswan Dam project would have quickly polarized the situation. Nasser would have suddenly changed the scenarios, taking the world by surprise with a radical decision.

[6] Dougherty (1959, 21–45).

[7] Russell to Foster Dulles, 9 July 1956. FRUS, 1955–1957, Arab-Israeli Dispute, January 1–July 26, 1956, Vol. XV, 433. Memorandum from the Secretary of State's Special Assistant (Russell), Washington, 1956.

[8] Ibidem.

A week after the Yugoslav summit, on 26 July, the Egyptian president, in front of a crowd in Alexandria, announced that Egypt would unilaterally proceed with the nationalisation of the Suez Canal Company. The move triggered an immediate controversy around the legal juniper, also because the *Compagnie universelle du canal maritime de Suez*, established in 1858 by Ferdinand de Lesseps, apparently taken aback by the Egyptian move, was a multinational corporation with headquarters in Paris and with international shareholders, mostly French and British.[9] However, Nasser's decision assumed above all a surprising political character, striking a symbol of the age of empires that still played a crucial strategic role in the global oil game. The nationalisation, the international crisis, and the war that followed indicated how the intersections between the Cold War and post-colonial dynamics became more and more unpredictable and dangerous. Even Nehru, surprised by Nasser's move, and worried by the destiny of the non-alignment project, decided to publicly support Egyptian position, as a common cause against colonialism, while initiating a tacit negotiation action.[10]

Therefore, the Suez crisis brought to light all the latent tensions within the Western alliance and contributed to reinforcing the idea of the existence of a Third World given by Sauvy four summers before,[11] whose neutrality was not passive. Although defeated on the field by the Israelis, Nasser, with the dispute over the Aswan Dam and the nationalisation of the Canal Company, had emerged as a hero in the nascent Afro-Asian front.

Suez and Italy: The Role of Catholics

The consequences of the double crisis of 1956 inevitably resonated in Italy, both in the internal political debate and on the lines of foreign policy. From this reshuffling of the strategic and political weight of the various forces in the Mediterranean, it seemed that the country could carve out 'new diplomatic opportunities'.[12] In those years, relations with some Arab countries were growing in importance, even within a pro-Atlantic approach that never seemed to be seriously questioned. But if the economic collaboration promoted by Enrico Mattei, the head of ENI (Ente

[9] Deney (1967, 293–299).

[10] https://casi.sas.upenn.edu/iit/swapnakonanayudu.

[11] De Giuseppe (2011, 29–52).

[12] Di Nolfo (1992, 26).

nazionale idrocarburi, the state-owned oil company)—returned from Egypt a day before the announcement of the nationalization of the channel (on 24 July he had inaugurated the Suez-Cairo pipeline with Nasser)—would later come on a collision course with the interests of Western oil companies, in relation to the Suez crisis, the attitude of the Italian government seemed perfectly aligned with that of Washington, in open opposition to the Anglo-French initiative. All this led to a strengthening of the bond between the two governments and to the Italian role as a privileged partner in the Mediterranean. The Italian ambassador in Washington, Manlio Brosio (2008), wrote to Foreign Minister Gaetano Martino: 'As we are determined and very loyal to the West,—this is what we are and will be—, we have the duty to preserve the sense of proportion and to distinguish between the common and vital causes of the Western world and the last remains of colonialism. For us, as for America and other countries, the freedom of navigation of the channel counts.'[13]

As reconstructed from the historiography, the 'neo-Atlanticist' position of the president of the Republic, Giovanni Gronchi, was more enigmatic.[14] The Suez crisis, the Arab-Israeli war, and the subsequent rapprochement of Tel Aviv to Washington would have contributed to the consolidation of a 'neo-Atlanticist' line within Christian Democracy (DC), which, albeit with different visions, would have brought closer together Gronchi, Fanfani, and Mattei. In search of spaces of movement for Italy's Mediterranean policy within North Atlantic Treaty Organization (NATO), they would have kept open the dialogue both with the Arab world and with Israel. By the way, in the first months of 1956, during the presidential trip to the United States, the diplomat Egidio Ortona commented: 'The president said exactly what the Americans did not want to hear. We wonder if it is not a subtle calculation to be seen as the enfant terrible and to strength in Italy his image as privileged national interlocutor of foreign powers.'[15] The diplomat's concerns, however, were solved thanks to a substantially positive judgment of the results of the trip, both for renewed economic aid from the United States and for a strengthened Italian position within NATO. The inclusion of the Atlanticist Italian Minister of Foreign Affairs Gaetano Martino in an expert committee which was

[13] M. Brosio, 9 September 1956, ASDMAE, AAPP, III, b 1063.
[14] Varsori (2017, 77–86).
[15] Ortona (1986, 153–167).

entrusted with the review of Article 2 of the North Atlantic Treaty was the most immediate consequence.

By the way, at the time of the Suez crisis, the interest in Italy for Mediterranean countries seemed to arouse cross-cutting interest. During the first phases of the crisis, on 9 August 1956, there was a summit with Gronchi, Prime Minister Antonio Segni, Martino, and the Social Democrat Giuseppe Saragat. It emerged the need to point out to the Allies that, according to Article 4 of the Treaty, Italy should have been consulted in case of military operations. According to Galante, it was from this moment on that 'DC and PCI wanted to be less and less passive recorders of international relations established by others'[16]; this coincided with the resumption of the dialogue between Togliatti and Tito, while Dulles recommended Fanfani to be cautious about 'late conversions'.[17]

More generally, after the nationalisation of the Suez Canal Company relationships with the South began to change rapidly. During the months of the Suez crisis, while still a stranger to the Italian debate, the expression 'Third World' had begun to circulate again. For the first time, after Suez, there was also the problem of the relationship between scholars and militants of the Third World idea, in an ideal comparison between different intellectual approaches: the development line of the demographer Alfred Sauvy and the revolutionary one of the psychiatrist Frantz Fanon.

In the Italian perspective, the crisis was strictly linked to the nascent Third World issues, where the Catholic component would have been crucial. A gradual process of rediscovery of the 'paesi nuovi' (new countries), the expression then in vogue among the Catholic missionaries, was then accompanied by what Forno called 'the nightmare of nationalism and communism',[18] which in some cases would be associated also to the awakening of traditional religions. The publications of Edizioni Missionarie Italiane (EMI) are revealing in this regard: it was an editorial project which made its debut right in 1956 with an emblematic text: *Il risveglio dei popoli di colore* (The awakening of the colored peoples) by the Lombard missionary Father Piero Gheddo. The text associated the urgency of evangelization with the need to prevent the advance of communism in the former colonies. In the preface to the book, Vittorino Veronese, former president of the Italian Catholic Action, at the time chairman of the UNESCO

[16] Galante (1990, 95–96).
[17] Ortona (1986, 224).
[18] Forno (2017, 68).

Executive Committee, wrote: 'The big gauges of international communism are now focused on Japan, South-East Asia, India, the Middle East, and Africa.'[19] Gheddo, in his reconstruction of the previous three years, added to the emerging dangers the state secularism and nationalizing programs promoted by Nasserist pan-Arabism and Nehru's socialist reformism. 'Faced with this massive ideological offensive', he continued, 'we, the Catholics of the West, are almost exclusively worried about defending ourselves ... Yet, it would be necessary, today, to counterattack, to wrest from communism the ideological dominance over the pagan masses.'[20]

Beyond these battle tones, the text also offered a series of operational proposals. First of all, Gheddo recognized the novelty of Bandung and that of Suez (this latter considered much more dangerous), which grasped the sense of urgency of the fight against hunger and underdevelopment as decisive challenges for international politics. The priest also expressed all his suspicions about neutralist formulas and rejected Nehru's proposal for a non-aligned front, reiterating: 'After Suez, the Afro-Asian nations, besides being separated from these historical, cultural, or racist elements, do not even agree on the political line to follow.'[21] His map of the world was in any case now divided into three blocs: the communist world (defined as 'aggressive'), the Western world ('defensive'), and the Third World, 'made of colored peoples, in revolt against everything and everyone'.

This defensive approach of the Catholics, however, took on different nuances, as emerges from the 1956 initiatives of the Italian section of Pax Christi, which suited to the national context the experiences coming from France, where Archbishop of Paris, Maurice Feltin, had accepted the *Tiers monde* formula since the Louvain Congress of 1955 on underdevelopment. In the months of the Suez crisis, the movement promoted in all its public meetings the study by former Food and Agriculture Organization (FAO) president, the Brazilian Josué de Castro (1954), *Geography of Hunger*. This text, written in 1946 to denounce the impact of the breakdown of the ecological system from a case study in the Brazilian Noreste, served to sensitize the Catholic world in terms of a reception of various issues coming from the developing countries. In some groups this new approach helped to justify the choice of Nasser in favor of the

[19] Veronese (1956, XIII).
[20] Ibid., XV.
[21] Gheddo (1956, 21).

nationalization of the channel. This new climate, of which FAO (based in Rome) was an original actor, also marked Italy's first steps into the UN after formal admission in 1955. Veronese, then chairperson of UNESCO's executive board (and UNESCO Director-General from 1958 to 1961), wrote on 18 December 1956 an in-depth analysis of the IX General Conference of UNESCO in New Delhi. In his view, the summit was deeply transformed in the wake of the Egyptian crisis, which had changed the attitude of Third World countries. 'The choice of the Indian capital', he wrote, 'was not exempt from the deliberate purpose or at least from the hope of the Afro-Asian countries to force the European and American states, even if this caused openly controversial positions. And the events of the Egyptian crisis lent themselves to the polemics, which at that time reached peaks of greater harshness.'[22]

The Role of La Pira and the Idea of a New Mediterranean Peaceful Area

In this complexity of voices, that of La Pira was undoubtedly, within the Catholic world, the most open to experiment the political and cultural innovations produced by the Suez crisis and to envisage wide-ranging collaborations that would strengthen the weight of Italy in the region and promote peace initiatives. La Pira wrote to Mattei on 7 November 1956, in a very delicate moment of the crisis: 'We therefore agree: the attention must be increasingly oriented towards the East: Italy today must grasp the evolution of events in neighboring regions. Weaving with the peoples of the Middle East, and then also of the most distant East, economic, social, political and cultural relations: being an evangelist in the history of tomorrow.'[23]

After Amintore Fanfani's appointment as secretary of the DC party and the arrival of Gronchi at the Quirinale, in the wake of Bandung, ENI had been able to take advantage of the renewed interest in the Mediterranean dialogue.[24] With the creation of the Nuovo Pignone, the technological potential of a company that was active through its holding-style affiliates (AGIP, SNAM, ANIC, ENM, and RASIOM) had grown in several sec-

[22] Cable by V. Veronese to MAE; 18 December 1956, ASILS, Veronese, UNESCO, 1974/973, 1956.

[23] La Pira to Mattei, 7 November 1956, ALP, 21, 1, 1952–1958, d. 64.

[24] Bini (2017, 3–22).

tors, from the production of materials for drilling to refineries. While the political-diplomatic potential of the company was consolidating, Mattei's initiatives met with growing resistance not only by the front of competitors in the hydrocarbon sector, but also on the part of Confindustria and some officials of the Ministry of Foreign Affairs. The ENI Mediterranean policy was undoubtedly linked to the ability of Mattei and his group of young public managers and of his political and cultural interlocutors (including La Pira) to grasp the opening of a third front as a great possibility of mutual growth. In this perspective, Mattei's rhetoric concerning the legitimate ambitions of the poor country crushed between the great international consortia seemed to be functional to that narrative which certainly did not have the objective of opposing Italian Atlanticism, if anything to articulate it.

In May 1955, a month after the Bandung Conference, a collaboration was signed that allowed Azienda Generale Italiana Petroli (AGIP) to acquire a stake in the Egyptian oil company IEOC (International Egyptian Oil Company), experimenting with an unprecedented formula. Mattei visited Nasser in Cairo on 24–25 July 1956, on the eve of the nationalization of the canal and at the height of the dispute with the World Bank on the suspension of credits for the construction of the Aswan Dam. Only eight months later, on 14 March 1957, by strengthening the Egyptian formula, the famous partnership between ENI and the National Iranian Oil Company would be announced, marking a turning point destined to break the traditional 50/50 formula that international companies stipulated with national governments. ENI offered instead the producer country a higher share (the so-called Mattei formula) between 75% and 85%, through investments, supplies, or other forms of assistance. The formula would be applied in the following months also to the contracts stipulated with the governments of Morocco and Tunisia. Mattei's activism came into conflict with French interests in the region: ENI even secretly financed the cause of the Algerian FLN, in exchange for future concessions.

La Pira represented an original political, diplomatic, and cultural referent of the dialogue with developing countries, fully supported by Mattei. His peculiar vision, however, went beyond the horizon of the public entrepreneur, opening up a new concept of the role of non-state bodies in promoting international dialogue and an original pacifism, connected to ecumenism. Since the organization of the meeting of the cities of the world, held in Florence in October 1955, La Pira had overcome a purely Mediterranean vision to look at the emergent bloc of non-aligned countries in a comprehensive way, beginning to evoke the creation of a historical planetary axis to contrast with the rigid schematic nature of bipolarity. La Pira embraced enthusiastically the thesis of the 'geography of hunger', but declined it in his original terms, so as to bring it closer to the ideal 'path of Isaiah', as he called it, of Abrahamic religions (Judaism, Christianity, and Islam).[25] Along this path he deemed necessary a strong ideal presence of the Holy See, identified as the historical-teleological epicenter of the dialogue, from which Italy could have drawn a diplomatic weight of great importance, exploiting its own unique geopolitical position. However, the role of cities emerged as mediation tools, and therefore the historical mission attributed to Florence and those multilateral institutions that should have to contribute over time to the construction of an 'integral UN of peoples'.

La Pira's proposal, which was refined in the conciliar years, remained semantically and culturally quite distinct from Sauvy's Third World formula, while absorbing several elements of it: the weight and potential of a Third Bloc, the centrality of the human dimension of the institutions, and the idea of integral development. An internationalization of aid to development was proposed, inspired not so much by Rostow's thesis on development, but by a mix of Beveridge and the social doctrine of the Church, a theme he had reiterated to Fanfani, already the previous year, with an open letter invoking an Italian commitment to promote a Marshall Plan for 'depressed countries'.[26]

From this point of view, the Suez crisis, in its various evolutions, represented a decisive passage for the mayor of Florence and was intended as a powerful political sign, in view also of an innovative strategy of action that would have developed in the following years. This would have led to the reappointment of La Pira at the Palazzo Vecchio, but at the head of a new alliance that also included an agreement with the socialists.[27] In April 1956, La Pira intervened in the Tenth Assembly of Catholic intellectuals, celebrated in Beirut, with a speech focused on the dialogue between three 'fundamental cultures: that of the West, of Islam and of India', invoking a dialogue between the societies of exasperated technicality and those societies still marked by transcendent and spiritual values, so as to 'mutually

[25] La Pira (1956, 6).

[26] La Pira (1955, 1).

[27] The local elections of 27 May saw a growth of the DC from 36.24% to 39.27%, and the preferences of La Pira passed from 19.132 in 1951 to 33.907.

integrate'. The mayor invited to learn the lessons coming from the new post-colonial countries and spoke of a spiritual fortitude to preserve, invoking, 'in the atomic age of impossible war … the only allowed war: a war of ideas, of technical, economic, social, cultural, spiritual, and religious elevation of the peoples'.[28]

A month before the announcement of the nationalization of the canal, Florence hosted the Fifth Convention for Christian Peace and Civilization, *History and Prophecy*, which saw among others the greetings of President Gronchi, the interventions of the French Jesuit Jean Daniélou, the Italian pacifist Igino Giordani, and the bishop of Jericho, Alberto Castelli. Participants included representatives from Egypt, India, Indonesia, Iraq, Israel, Libya, and Pakistan. In the introductory speech, La Pira recalled 'the great processes of human promotion' that were 'working on those classes and peoples who history is revealing in all continents', appealing to a principle of mutual solidarity and responsibility.[29]

The 'double' crisis in Budapest and Suez seemed to reinforce La Pira's conviction of Italian responsibilities in the new international context of an expanding bipolar system. With the support of Fanfani and interacting with Gronchi and Mattei, La Pira actively followed the evolution of the events that marked the Suez crisis between July and November. On 23 August 1956, he wrote to Nasser, offering Florence as the venue for a neutral conference of all the Arab peoples, through the Arab League and the UN, to 'start a life and fraternal voice to rethink a new agenda and face the problems affecting peace, the dignity of the Arab world and all other parts of the world'.[30]

Despite the impossibility after a failed attempt to initiate a real mediation action through former Egyptian Culture Minister Taha Hussein, La Pira intensified his initiatives for peace, as emerges from the messages written in those months to Eisenhower, the Soviet Foreign Minister Alexander Bogolomov, Nehru, and Hammarskjöld, whom he thanked for a work which, in his view, could have opened a new era for the UN and multilateralism in general.[31] The correspondence on the Suez crisis preserved in the La Pira archive also includes letters to the ambassador to Cairo, Massimo Magistrati, and the Egyptian diplomats and politicians Ahmed

[28] La Pira (1989, 13).

[29] La Pira (1957), opening speech, 11.

[30] La Pira to Nasser, 23 August 1956, ALP, 3, 'Egypt', 3, 2, d. 1.

[31] La Pira to Hammarsköld, 4 December 1956, ALP, 10, 'UN/1', d. 45.

Naguib Hashim, Mahmoud Fawzi, and Edine Hassan. As major of Florence, he also sent official protest notes to London and Paris (a copy of these were also sent to Nasser and to the governments of Lebanon, Syria, Jordan, Iraq, Iran, and Israel) and established contacts with the Brussels-based Association International des Juristes Democrates.[32]

Though his action was considered by many (even within the DC) to be idealistic, unrealistic, utopian, if not counter-productive, La Pira was convinced that Italy could make its contribution to Egypt's hopes of recovery and development, becoming a privileged partner in the dialogue with the Afro-Asian countries; he also openly advocated an economic intervention by Mattei in the Aswan Dam game, to present a new model for Euro-African cooperation and dialogue. According to his reading, the facts of Hungary and Suez were not at all separated indeed. It was therefore necessary to seize the opportunities offered by the historical situation, to open up to a third way of acting as an element of conciliation, in line with the calls of 'the spirit of Bandung'.[33]

The most interesting passage emerges, however, from the speeches given by La Pira in the Florence municipal council (in the sessions of 29 and 30 November 1956) to respond to a provocative interpellation of the communist councilors who denounced the passivity of the Christian Democratic government in relation to the Suez crisis. After stigmatizing the Soviet armed intervention in Budapest, and expressing strong disappointment with the Moscow political actions, the mayor stated: 'The sad and bloody events in Egypt deserve a condemnation: and we, with the whole civilized world, have not been afraid to condemn them: but how do you raise a voice of protest for them as you carry the blood of an unheard and painful heart of Abel scattered in the land of Hungary?' He continued: 'What have we done for Egypt and for peace? What we did for Hungary: we have without hypocrisy expressed our condemnation of what happened to the Egyptian people and to Egyptian cities: no one for any reason has the right to infringe the essential norms—legal and ethical—which govern the community of peoples and nations: no one, then, and for no reason, has the right to initiate war operations which by their very nature not only sow destruction and death, but often irreparably upset the very structures of society and the community of peoples. This is the reason why we have intervened in all phases of the Egyptian story: from the initial

[32] Giovannoni (2006, 80–81).

[33] De Giuseppe (2008, 11–32).

one in Suez to the final one of the bombardments of Port Said and other Egyptian cities.'[34] La Pira's idea was to organize diplomatic initiatives to support the UN effort to restore the troubled juridical order. Then, after the bombings, he wanted to support the reconstruction of infrastructures and to promote a symbolic act of peace, recalling an analogy between the events of 1956 and those of 1939. On 1 November, Taha Hussein sent a telegram to thank the mayor and the city of Florence for the commitment to solve the crisis. The most interesting point, however, concerned La Pira's attempt to seize the opportunity to relaunch the dialogue between Egypt and Israel, coordinating a broad multilateral intervention plan that would enhance the role of the UN General secretariat.

On 23 December, the day after the signing of the agreement between Nasser and Hammarskjöld that sanctioned the control of the Suez Canal by the Egyptians, La Pira resumed part of those speeches in his first radio message to the Florentine citizens, recalling: 'the great ferments—good and not good—that animate the immense space of Asia and Africa must be seen in this light: they are produced, so to speak, under the mysterious, but true and irresistible action of this divine yeast that is in bloom on the whole front of peoples and nations.' The future of the Afro-Asian Third Bloc should have been at the center of the Sixth Congress for Peace and Christian Civilization, which however failed to materialize due to the 1957 crisis of the Florence municipal council.[35]

Conclusion

Six years later, recalling Mattei on the third anniversary of his death, La Pira would have returned to that fateful 1956 to affirm: 'The historical and political (social, economic and cultural) emergence of the Mediterranean peoples (Israel and the Arabs), of Africa, Asia, and Latin America (of the so-called underdeveloped peoples): the historical and political emergence of the Third World.'[36] In a Lapirian perspective, in short, the Middle East had to become a radiating center of a policy of dialogue (instead of war) as an example to the rest of the developing world; that should also offer a lesson to the two blocs to address the strategic problems of the region as a crucial area of world policy in terms of economy and diplomacy. La Pira wrote it again to Nasser in March 1957,

[34] La Pira (1989), *Interventi sulle vicende ungheresi ed egiziane*, 265–282.
[35] La Pira, ALP, Conferenze, 1956–1957.
[36] La Pira (1962, 7).

inviting him to consolidate 'that timid dawn of peace that sprouts over the Mediterranean and, therefore, over the whole world', but also manifesting a new fear and a latent concern for the widespread misunderstanding of the peace vanguards.[37] La Pira seemed increasingly convinced that the Middle East had become a seismic fault at the intersection of East-West and North-South interests, and that without integral peace in the region it was increasingly difficult to promote peace on a global level.

In 1958, the project of the *Colloqui mediterranei*[38] (Mediterranean Colloquies) would start in the fragile hope of overcoming the rigid bulkheads of a bipolarity which, after Bandung and Suez, opened up definitively to the prospects and uncertainties of global complexity.

References

Abbreviations

ASDMAE	Archivio Storico Diplomatico del Ministero degli Affari Esteri, Rome
AAPP	Affari Politici
ALP	Archivio della Fondazione Giorgio La Pira, Florence
ASILS	Archivio Storico dell'Istituto Luigi Sturzo, Rome
FRUS	Foreign Relations of the United States

Bibliography

Bagnato, Bruna. 2005. La Pira, de Gaulle e il primo Colloquio Mediterraneo di Firenze. In *Giorgio La Pira e la Francia. Temi e percorsi di ricerca. Da Maritain a de Gaulle*, ed. Pierluigi Ballini, 99–134. Florence: Giunti.

Bini, Elisabetta. 2017. Oil for The Free World: US Public Relations as Cultural Diplomacy in Post-World War II Italy. *Ricerche di storia politica* 1: 3–22.

Brosio, Manlio. 2008. *Diari di Washington, 1955–1961.* Edited by Umberto Gentiloni Silveri. Bologna: il Mulino.

Campanini, Massimo, and Marco Di Donato. 2021. *Il canale delle spie, storia della crisi di Suez.* Roma: Salerno Editrice.

Castro, Josué de. 1954. *Geografia della fame.* Italian Translation with a Preface by Carlo Levi, Bari, Leonardo da Vinci (or. ed. *Geografia da fome: a fome no Brasil*, Rio de Janeiro, O Cruzeiro, 1946).

De Giuseppe, Massimo. 2008. La Pira, Firenze e il Terzo mondo. *Quaderni della Biblioteca Ballestrieri* 1: 11–32.

[37] La Pira to Nasser, 16 March 1957, ALP, 3, 'Egypt', 2, d. 2.

[38] Bagnato (2005, 99–134).

———. 2011. Il Terzo mondo in Italia. Trasformazioni di un concetto tra opinione pubblica, azione politica e mobilitazione civile (1955–1980). *Ricerche di storia politica* 1: 29–52.

Deney, Nicole. 1967. L'Égypte et la Compagnie universelle du canal maritime de Suez. *Revue française de science politique* 2: 293–299.

Di Nolfo, Ennio, ed. 1992. *Power in Europe? Vol. II, Great Britain, France, Germany and Italy and the Origins of the EEC, 1952–1957*. Berlin and New York: Walter de Gruyter.

Dougherty, James E. 1959. The Aswan Decision in Perspective. *Political Science Quarterly* 1: 21–45.

Ferrari, Silvio, and Paolo Zanini. 2010. La questione del Medio Oriente. In *Singolarissimo giornale: i 150 anni dell'Osservatore romano*, ed. Antonio Zanardi and Vian Giovanni, 204–230. Turin: Allemandi.

Formigoni, Guido. 2016. *Storia d'Italia nella guerra fredda (1943–1978)*. Bologna: il Mulino.

Forno, Mauro. 2017. *La cultura degli altri. Il mondo delle missioni e la decolonizzazione*. Rome: Carocci.

Galante Severino. 1990. *Alla ricerca della potenza perduta*. Manduria. Lacaita.

Gheddo, Piero. 1956. *Il risveglio dei popoli di colore*. Milan: EMI.

Giovannoni, Marco Pietro. 2006. *Il grande lago di Tiberiade. Lettere di Giorgio La Pira per la pace nel Mediterraneo (1954–1977)*. Florence: Polistampa.

La Pira, Giorgio 1955. Lettera aperta all'on. Fanfani del prof. Giorgio La Pira. *Il Giornale del Mattino*, 28 August, 1.

———. 1956. Per combattere la geografia della fame. *Il Focolare* 14: 6–7.

———. 1957. *Storia e profezia*. Florence: Ufficio Stampa del Comune.

———. 1962. *Enrico Mattei a Firenze: discorso commemorativo in occasione del Trigesimo della morte, 27 novembre 1962*. Florence: Giuntina.

———. 1989. *Giorgio La Pira sindaco, vol. 2, 1955–1957*. Edited by Giorgio Giovannoni, Gianni Giovannoni, and Ugo De Siervo. Florence: Cultura, Comune di Firenze.

Ortona, Egidio. 1986. *Anni d'America. Vol. II. La diplomazia*. Bologna: il Mulino.

Varsori, Antonio. 2017. La politica estera italiana durante la presidenza di Giovanni Gronchi. In *Giovanni Gronchi e la politica estera italiana (1955–1962)*, ed. Antonio Varsori and Mazzei Federico, 77–86. Pisa: Fondazione Piaggio/ Pacini Editore.

Veronese Vittorino. 1956. Presentazione, In Gheddo Piero. *Il risveglio dei popoli di colore*, Milan, EMI.

CHAPTER 22

The Oil Route: Bernardo Bertolucci's Filmic Journey Through the Suez Canal

Giuseppe Natale

ENI's Bridging Culture Project

'The culture of energy, the energy of culture' is ENI's motto, intended to encapsulate the company's philosophy and its synergic harmonization of seemingly dissimilar elements. There is probably no better symbolic representation of the company's culture and technology integration than the black-and-yellow AGIP logo that ENI adopted as its trademark. Featuring the famous six-legged fire-breathing dog, the logo was the result of a nationwide competition instigated by Enrico Mattei, ENI's first president from 1953 to 1962. Conceived by sculptor Luigi Broggini, the eye-catching image hybridizes man and machine (the four legs of the dog combined with the two legs of the driver) and also bridges the modern and the archaic, spanning across distant countries and cultures. While the fire-spewing dog directly evokes the energy burst produced by burning oil, it also calls to a learned viewer's mind a host of virtual linkages: the

G. Natale (✉)
University of Nevada, Las Vegas, NV, USA
e-mail: giuseppe.natale@unlv.edu

B. Curli (ed.), *Italy and the Suez Canal, from the Mid-nineteenth Century to the Cold War*,
https://doi.org/10.1007/978-3-030-88255-6_22

mythical Tarantasio dragon of Lombardy, the Persian lion of ancient Iran, the Etruscan chimera of Arezzo, which, in Homer's and Hesiod's narration, was a fire-breathing monster that terrorized Lycia. In short, the logo functioned as a trans-Mediterranean icon, a cultural pipeline connecting north and south, west and orient.[1]

Mattei's reliance on modern visual art reflected his wider strategy to create public support for Italy's nascent state-owned oil and gas enterprise entrusted to his management. As Frescani points out, under Mattei ENI became a sort of experimental lab for new forms of communication aimed at sharing corporate information both internally and externally. Endowed with keen acumen for marketing techniques, Mattei employed a plethora of advertising forms and tools to foster AGIP and ENI brand names, from commercials to billboards.[2] Mattei firmly believed that the media could play a fundamental role in raising industry awareness and exercising soft power, or, as his critics would say, strong-arming the Italian politicians. At his initiative, the newspaper *Il Giorno* was founded in 1956. One of its main functions was precisely to create support for a national policy favorable to the oil-producing Arab countries which, in turn, could allow Italy to break free from the multinational petroleum cartel.[3] ENI communication strategies, however, went way beyond mere corporate propaganda through the founding of an in-house news magazine, *Il Gatto Selvatico*, and a cinema production department.

Published between 1955 and 1964, *Il Gatto Selvatico* was slated to be a virtual meeting space for the 'large ENI family'. Behind the rhetoric of corporate familism, however, there hid Mattei's combative stance toward his international competitors and domestic doubters—stance which transpired from the magazine's title, *Gatto Selvatico*, the Italian translation of w*ildcat*, the moniker for an exploratory well drilled in difficult or unproved territory. Mattei fancied being a sort of *wildcatter* himself, an emulator of Colonel Drake and the first Italian to drill successfully for oil. The magazine title also evoked his favorite parable, the hungry little cat attacked and killed by big dogs while trying to feed from their large bowl full of broth. As Mattei often intimated, the cat represented Italy, an energy-starving

[1] ENI official website https://www.eni.com/en_IT/company/eni-history/the-six-legged-dog-history-eni-logo.page.

[2] Frescani (2015, n.p.), Bini (2013, 131–36).

[3] With a similar strategic purpose—that of backing the growing interest of ENI in foreign markets—the Agenzia Giornalistica Italia (AGI) was created in 1950. AGI became an official ENI subsidiary in 1965, after Mattei's death. https://www.agi.it/.

country that was denied access to plentiful oil resources by big international companies. The wild cat, conversely, stood for Italy's endurance and ability to find her own energy resources in uncharted areas.

While officially targeting the group employees, *Il Gatto Selvatico* aimed at a larger audience. Consistent with Mattei's belief in the sharing of technology and free circulation of knowledge, the magazine sought to provide corporate information and disseminate culture. To this end, Mattei entrusted its editorship to the poet Attilio Bertolucci, father of film director Bernardo. As editor, Bertolucci ended up designing a magazine quite modern in its high-low target, readable by everyone 'from the President of Italy to the ENI oil driller farthest abroad'.[4] Except for two specific requirements—a recurrent section on corporate news and a cover picture illustrating ENI's positive presence in the world—Bertolucci had the greatest latitude in editorial policy, with his magazine covering a wide array of subjects, including art history and literature. In the wake of Leonardo Sinisgalli, the 'engineer poet' who in *Civiltà delle macchine* had grafted humanism and science, Bertolucci sought the collaboration of those writers and artists open to a discourse of modernity, a critical prerequisite given the nature of the magazine.[5] As evidenced by Elio Vittorini, traditional literati saw industry as an unmitigated cause of alienation and were unable to respond to it appropriately by creating a new type of scientific humanism.[6] Likewise, as film director Antonioni evidenced, contemporary cinema also suffered from a chasm separating technology and its visual representation. The industrial landscapes that were starting to dot Italy called for a new figurative description, a terse style void of frippery, a streamlined form to showcase the object *per se*—be it a mechanical equipment or the architectural structure housing it.[7]

The need for a new aesthetic was all the more indispensable in techno-films, that is, films having industrial process and technology as their main subject.[8] Yet, in describing the novel sphere of oil and hydrocarbons, documentarians often defaulted into an outmoded aesthetic sensibility. Such was the case of Ubaldo Magnaghi in the AGIP-produced *Le vie del metano*

[4] A. Bertolucci, 'Così ho addomesticato il Gatto', *Il Sole 24 Ore*, 4 September 2011.

[5] A thorough assessment of the magazine's import can be found in Frescani (2016, 161–79).

[6] Vittorini (1961, 13–20).

[7] Antonioni (2007, 285–86).

[8] Verdone (1962, 27). For a thorough classification of techno-films see Latini (2011, 136, n.27).

(1952). Introduced by an overly dramatic musical score, the documentary illustrated the discovery of Cortemaggiore gas fields with turgid narration, indulging in such a lyrical celebration of the landscape: 'the ancient rustic world is altered by the iron castles of the drill rigs, high like the obelisks of a new world'. The film script regularly filtered technological precision through poetic metaphors and similes, as with the central power station and its grid network being likened to 'a mechanical heart' and 'an arterial system' respectively.[9] Even more disturbingly, the film's celebratory tone was reminiscent of Luce newsreels from the fascist era, like them aiming to lull viewers into an acceptance of technological progress. Other ENI documentaries from the same period displayed a drier, more scientific informative style, such as 'Le Ricerche del Metano e del Petrolio' (*Methane and oil explorations*) (1951) by Virgilio Sabel, who through direct narration and ingenious use of special effects and graphics successfully illustrated soil stratigraphy and deposit exploration to the layperson.[10]

In the 1950s, ENI's production of documentaries increased in parallel with its expanded search for oil and gas resources. With the unstated goal of spreading its brand nationwide, ENI invested more and more in the film sector so as to produce documentaries to be shown in regular cinema theaters, beyond the strict ENI circuit. Despite an apparent lack of dedicated distribution, the viewing of documentaries was facilitated by state incentives for the theatrical screening of educational short films on art, folklore, tourism, agriculture, and industry. Even though ENI's techno-films were more concerned with function than aesthetics, a decision was made to recruit skilled professionals who could elevate their artistic level. A point in case was Carlo Capriata, brought in to direct *Pozzo 18 profondità 1650* (Well 18 1650 depth) (1955), a film chronicling AGIP's success in locating oil deposits in Northern Italy. Capriata was assisted by Alessandro Blasetti, one of many intellectuals who, after some initial reluctance, agreed to cooperate with a large-scale industry like ENI, which, for its part, seemed willing to grant intellectuals freedom of expression without effacing their cultural identity. As ENI's patronage of documentary filmmaking increased, in 1957 Mattei decided to create an in-house film office. He appointed as its director Pasquale Ojetti, journalist, writer and film critic, entrusting him with overseeing techno-films production.

[9] *Le vie del metano* (1952), produced by SNAM and Istituto Nazionale Luce, film script by Vittorio Zincone.

[10] Latini (2011, 57).

The decision to invest more heavily on cinema was based on a multi-pronged approach. Besides their vocational function, techno-films also had a promotional and political purpose. According to Charles Dechert, an early Mattei chronicler, 'the most significant of ENI's policies was its effort to "politicize" the petroleum industry, to take it from the sphere of private initiative and cast it into the public arena'.[11] Mattei saw film as a means to strengthen ENI's image both domestically and internationally and to foster its strategy of oil exploration in Africa and Western Asia. To this end, ENI's print media and cinema worked synergistically. Aggressive media campaigns were employed to create a narrative of energy self-sufficiency in Italy based on the discovery of oil and methane deposits in the Po Valley and off the Sicilian coast; concurrently, an anti-colonial narrative was advanced so as to favor ENI's new partnerships with the recently independent oil nations.

In 1957 a number of articles on ENI's policy toward the third world appeared in *Il Gatto Selvatico*. In recounting the anti-colonial turmoil in countries like Morocco and Tunisia, which had recently gained independence, the magazine emphasized Mattei's public support for their struggle. ENI's stance, showcased by its multiple media, facilitated the forging of alliances with several North-African and Western Asian countries. The 1957 agreement between AGIP Mineraria and the National Iranian Oil Company was particularly meaningful in that it served a twofold purpose. On the one hand it provided ENI with the ability to conduct oil explorations in areas outside the control of the Anglo-American multinationals, commonly referred by Mattei with the mythological moniker Seven Sisters; on the other hand it introduced an innovative accord based on a 75:25 proportional division of revenues between the producer state and a Western oil company. This type of agreement, known as the 'ENI formula', served as an inroad into other oil countries.[12] The accord further solidified Mattei's ongoing discourse on the positive role of enlightened capitalism: the discovery and exploitation of energy sources was presented as both an opportunity for commercial profit as well as a progress propeller.

[11] Dechert (1963, 68).

[12] Contrary to the prevalent opinion that ascribes the creation of the 'ENI formula' to Mattei, Bucarelli maintains that such division of revenues was first proposed by the Iranian NIOC. Mattei's merit was that of accepting such a revolutionary concept, thus breaking the established oligopolistic norms. The agreement with Iran had a great political import, in that it positioned ENI as a player in the oil business despite the company having been excluded from the 1954 International Oil Consortium. Bucarelli (2010, 3–11).

Mattei's liberal ideology was echoed in the pages of *Il Gatto Selvatico*, which, in covering the working conditions of the Italian oil field laborers in ex-colonial countries likened them to 'legionaries' engaged in a peace building mission. In the late 1950s and early 1960s, the magazine placed increased emphasis on the nations with which ENI purported to entertain equitable partnerships. Particular attention was given to the education of the young African managers trained at the ENI schools of Metanopoli in San Donato Milanese and IROM in Venice, their technical education seen as a form of empowerment and viaticum toward decolonization.[13] *Il Gatto Selvatico* did not limit itself to training issues. The magazine editors, in fact, believed that the emancipation of colonized nations was inseparable from bringing knowledge of their culture to its readers. This gave rise to a series of informative articles on African arts and literature. The parity relationship between ENI and African nations also demanded that the cultural exchange go both ways, so much so that in 1960 a special issue of *Il Gatto Selvatico* was published in Arabic. In some regards, ENI's policy was ahead of its time. By reneging on the discoverer's hegemony of technological knowledge it anticipated a postcolonial discourse based on bi-directional interaction; by fusing together the economic and ethic notions of value it anticipated the tenets of axiology.[14]

The national and transnational complexities of oil exploration also ended up permeating ENI's cinema production. Under Ojetti, ENI's cinema office transited from documenting the most practical aspects of oil technology to creating culturally layered films that at times seemed to use corporate activity as a social springboard. The cultural turn was facilitated by the expressive freedom granted to film directors, both during Mattei's patronage and after his demise in 1962.[15] In the time span between the first documentary by Giulio Briani (1950) and the one by Bernardo Bertolucci (1966), ENI techno-films underwent an evolutionary trajectory by increasingly reflecting the director's identity. A glaring example of the cultural shift is *L'Italia non è un paese povero* (Italy is not a poor country, 1960) by Joris Ivens. In Mattei's intentions, the documentary was to celebrate Italy's strides toward energy independence. The director, however, in line with his communist conviction, instead of producing an all

[13] Prattico (1962, 7–9), Bini (2013, 191–214).

[14] For a discussion of the ethical dimension of science see Agazzi (2004).

[15] During Mattei's presidency sixteen documentaries were filmed by established filmmakers, plus an unknown number of documentaries by non-professionals directors.

rosy picture on the merits of democratic capitalism spotlighted the persisting poverty in parts of Italy that the energetic revolution had yet to eradicate.

Notwithstanding the sporadic social denunciation, ENI cinema projected a self-image that was generally positive and proactive, especially in the international films meant to show the company's political openness towards those countries with which it had reached equal-level agreements, such as Egypt and Iran. ENI international documentaries hinged on five main points: a) respect for diversity; b) cooperation with the producing country; c) technological research; d) industrial strategy; e) advertising and propaganda.[16]

The Oil Route

Like previous ENI documentaries, also Bernardo Bertolucci's *La via del petrolio* (The Oil Route, 1965) was to be animated by trans-Mediterranean intentions. Commissioned by ENI and coproduced by RAI, the documentary was filmed in 1965–1966. Before *La via del petrolio*, Bertolucci had tried his hand at directing two feature films, *La comare secca* (The Grim Reaper, 1962) and *Prima della rivoluzione* (Before the Revolution, 1964), both receiving disappointing critical and commercial reception. Unable to secure the funds for a new feature film, Bertolucci accepted to direct the ENI documentary on the black gold route, and taking advantage of his relative artistic freedom, he actually turned the techno-film into a semi-fictional work.[17] Ojetti and Bertolucci together developed the idea of a three-part documentary, each 50-minutes long. Part one, titled 'The Origins', was to describe the points of oil extraction, the Zagros Mountains in southwestern Iran, a bare 4000 meter high range left untapped by the

[16] Frescani (2014, 21–28).

[17] Reflecting on the origins of *La via del petrolio*, Bertolucci stated, 'I remember saying to Godard, whom I met some days before leaving [for Iran]: "I'm so unhappy, I can't make a film, so I'm going to Iran to make a documentary on oil." He [Godard] answered that this was a wonderful opportunity to discover unusual, fascinating places. Today I'm still grateful for that job, as it triggered within me the pleasure of traveling. It was my first real journey: I discovered that other cultures existed and I immediately fell in love with them… I always thought that all cinema is cinéma vérité…When I was shooting *La via del petrolio* I wanted the documentary to turn into fiction, just as I later turned fiction into cinéma vérité'. *La via del petrolio, intervista a Bernardo* Bertolucci, 21 September 2013. http://www.centrostudipierpaolopasolinicasarsa.it/molteniblog/la-via-del-petrolio-intervista-a-bernardo-bertolucci/.

Oil Consortium due to its difficult access, and the off-shore platforms (*wildcats*) in the Persian Gulf; part two, titled 'The Journey', was to document the transport of the unrefined oil on the tanker AGIP Trieste, from the Persian Gulf to the Red Sea, through the Suez Canal to the Mediterranean Sea and the port of Genoa; part three, titled 'Across Europe', was to follow the oil's journey in a pipeline from Genoa to the Ingolstadt refineries.

While in Egypt, instead of navigating the Suez Canal on the tanker, Bertolucci descended to film the isthmus and its surrounding areas. The sequences shot in Suez ended up forming a separate 12-minute short, aptly titled *The Canal*, and were eventually included in part two of *The Oil Route*. In these sequences, Bertolucci treats the canal not just as a geographic limen between continents and seas but also as a nexus connecting multiple dichotomies—cultural, historical, and philosophical. For Bertolucci the canal is responsible for holding the adjacent countries in a state of contradiction, suspended between their archaic past and dawning progress. 'Contradiction' is indeed the key word in both films.

The *Oil Route* opens with the black and white image of an esplanade before a modern mosque, with a child lying still on ground. The camera then zooms on a columned portico and a flight of geometric steps leading into an ancient looking cuneiform arch. The balanced still image is suddenly animated by the rising child, soon joined by other children who run into the dark mosque. Contrast of light and shadow, static and motion, past and present displayed in the very first images is a recurring motif that Bertolucci extends to the entire Iran. In fact, the voice-over immediately introduces incongruous information on the country's profile, a veritable mix of encyclopedia factoids and personal impressions:

> Year 1344. A calendar 626 years behind ours. Country surface: 1,621,860 square km. Form of government: monarchy. The children are the future. Country name: Iran. Teheran, Tabriz, Isfahan, Mashhad, Abadan, Shiraz, Qom, Ahvaz. Persepolis. The Tomb of Darius, the Tomb of Cyrus. The *One Thousand and One Nights*. Petroleum. Persia. Sixth place in the world for oil production. This film is dedicated to the Persian children.

The hesitation by which the narrator wavers between past and present transpires from the country's dual name—Persia and Iran—Persia deriving from the ancient kingdom of Parsa, Iran being the contemporary political entity with which ENI had recently reached a commercial agreement. A temporal displacement also informs the next sequences. The voice-over announces that we are in Persia in the year 1344, 626 years earlier than the viewers' time, thus laying the foundation for an alternative, parallel narrative. The next scene shows a rig worker shooting a rocket into the night sky in order to ignite the gasses separated from crude oil—an action clearly inconsistent with our western perception of 1344. Drawing on erudition, the commentator immediately cites Zarathustra, recalling the relationship that the Persians have entertained with fire rituals since time immemorial. The reference to fire worship and Zoroastrianism implicitly belies the agency of technology and deflects the awe induced by oil processing. This is not an isolated instance. Bertolucci's documentary, in fact, not only does constantly shift its focus from the actual present to the distant past, but also moves away from the technological to the humanistic. Even the oilrig is given a human character through a comparison of the pumps' sound to the muezzin's voice, which are 'answered by the workers' prayers'. The six-minute introductory sequence provides no account of oil machinery; rather, it delves into history through the display of art ruins and the recitation of Aeschylus' tragedy, *The Persians*. The advent of the new 'oil age' announced in the next sequence is also framed in historical terms.

Through a series of aerial shots, the camera finally takes the viewer from inland to coastal waters to open sea, where some offshore oil rigs finally appear. The voice-over intimates that under land and sea there lies an invisible oil empire, a more lasting one than Darius and Xerxes' ancient empire because it will not succumb to hubris. The camera finally moves into a collection tank and then pans back at the light circle formed by its entrance, very much reminiscent of a cathedral oculus. Here Bertolucci forgoes erudite verbal description in favor of modernist visual minimalism, showing the building from within, viscerally, and not from without as an intellectualized abstraction. The structure inside the tank is presented as free of ornamentation, a geometric space echoing with hollow metallic sounds.[18] Here the camera, through an approach consonant with modernist art, lingers admiringly on pure forms.

Depiction of contrasts resumes in the ensuing scenes shot in Tehran's souk, a magical site brimming with carpet sellers and piles of spices. As the script suggests, the bazaar is the place where the 'little sequence' of the 'great contradiction' is filmed. The director's comment refers to the image of a young girl in a black chador who first reveals her face to the camera

[18] Toffetti (2010, 16).

then coyly hides it; her behavior mirrors that of the young boys who, in the opening sequence, came tantalizingly close to the director's car, mimicking his filming them, while keeping some distance at the same time. Through these images Bertolucci implicitly demystifies the power of the cinema machine, showing that its technology is unable to fully merge with the outside reality. Like the oil rigs, also his camera appears to be a modern object cast within an archaic surrounding. Through additional images of the bazaar, the viewer eventually comes to understand that the 'great contradiction' is the entire country's inability to own the progress brought by the 'stone oil', as the ancients named the petroleum. The oil revolution, Bertolucci reflects, will not change the essence of this country—which he now calls Iran—it will only change its revenues:

> Between its hydrocarbons and the endless traffic of the bazar, Iran lives at the same time in the 20th century and the past centuries. […] Oil was discovered at the beginning of this century, almost at the same time in Iran and in Texas. But oil produced an endless economic boom in Texas, in a society prepared to receive it, while in a society still being in medieval times it only produces deferred utility capital. Oil can be absorbed by the machine civilization, while for the pre-industrial societies it is little more than a commodity to be entrusted to the ships, like the ancient spices.

After a detour to Mount Rig in the inland, which allows the director to carry out his corporate task by interviewing the Italian crew, in which they discuss their life abroad, their relations with the local hands, and some technical details about their work and equipment, Bertolucci returns to what interests him the most, the diachrony brought by the stone oil's exploitation. Thus part one aptly ends with a dual reference, one historical one fictional, to colonel Drake and Captain Ahab, two men haunted by the search for oil and sperm oil respectively. The precious crude, now transformed into a commodity as whale oil once was, is entrusted to the bowels of the tanker that will carry it to the more developed parts of the world.

Before beginning the journey back to Europe, after skirting the Sinai peninsula, the director descends the tanker, carving out some time to explore the Suez Canal area. Here he uncovers the existence of an even greater dichotomy that the canal itself, acting like a catalyst, seems to have precipitated.

The Canal

As in the *Oil Route*, also the initial sequence of *The Canal* is placed under the sign of the children. The first images of Suez city present an Egyptian shoeshine boy, described as 'the younger brother of the Neapolitan *sciuscià*… cheerful as all puppies of the world are'. This is the first of many comparisons by which the director depicts Suez in supranational terms. As he continues his exploration, it becomes clear that Bertolucci has no interest in extolling or even describing the technical achievements that led to the canal's realization. Except for a fleeting comment about its 161 km. total length, the director fails to provide any meaningful technical data. The documentary script, in fact, is an open invitation to transcend the canal's physicality: 'the most absolute, the most pure, the most river of all is not a river: it is an artificial channel dreamed up by a visionary Italian and realized by a determined French, Mr. de Lesseps'. Not surprisingly, then, Suez's attributes are intangible. In witnessing the uncanny silence enfolding the city, Bertolucci offers a psychological diagnosis: 'Suez suffers from a ship complex: men, animals, goats and children are intimidated by the appearances of the gigantic tankers'. After lingering on groups of men at work, of goats grazing on piles of dirt, of children playing quietly, he finally directs the viewer's attention at the waterway itself. The stationary camera is placed on a narrow city street, between symmetrical buildings, with the channel visible in the distance. The static image is soon enlivened by a visitation, a giant cargo ship that glides across the visual space at excruciating slow speed. No comment is provided to the viewer, at least not until the ship is about to disappear out of sight:

> This is the real time a ship takes to get past the intersection down the road. She came silently at engine idle speed and no man on deck. A moment that lasts forever and the ghost has turned the corner. This apparition down the road repeats itself every six minutes, all day and all night long; Suez has not yet got used to it and each time baits its breath.

If the canal represents the nexus of adjacent countries, by virtue of its central position in the film this sequence represents the nexus of the perceived temporal dichotomy. By dilating static and motion, this ninety-second long sequence offers a non-scientific, near-metaphysical view of the canal, in which the accelerated time of modernity is forced to decelerate.[19] Uninterested in the technical data behind the canals' construction or use, Bertolucci veers away into artistic imagination, as if the latter could reveal relationships between man and environment precluded to the scientific eye.

The very next images show a group of Muslim faithfuls who pray facing Mecca. These images are followed by graffiti depicting a pilgrimage to the Holy City, where a large ship has replaced the camel as a means of transportation, thus proving that the canal has entered the modern imagination of Suez' inhabitants. As further proof that cargo ships mark the transition to a new modern age, the camera then focuses on a procession of gigantic vessels navigating the canal. Notable among them is a large tanker with the six-legged dog logo on its funnel, advancing against the backdrop of land and sand dunes. Instead of seizing this opportunity to illustrate ENI's presence in the area and discuss the commercial benefits of the waterway, the director once again veers away from the technical, counterpointing the ships' images with literary references to Rimbaud's poem, *Le Bateau ivre*. The verses' symbolic imagery of the drifting and sinking boat lost at sea serves him as a springboard for conveying his impression that the Suez Canal is actually an insubstantial causeway collapsed under the desert sand. The canal, the voice-over states, has been in existence for only 100 years, the blink of an eye in earth history. The desert, Bertolucci tells us in *The Oil Route*, exists in the sacral unity of time and place. We now come to understand that the canal's construction has infringed upon that unity, and that the ships navigating it may be subject to being subsumed into the desert. As the ships keep gliding along the channel, they appear and disappear behind the dunes. It is no coincidence, then, that the director compares their traveling through the desert to the *Voyage à travers l'impossible* (The Impossible Voyage, 1904) by illusionist and film director Georges Méliès. A series of tracking shots gives the viewer the impression that the vessels are suddenly buried underneath the dunes, only to be resurrected right away. In this new 'game of ships', as the voice-over defines it, the funnels follow one upon another, while the seemingly eternal landscape, provided by the banks, stays unchanged.

Leaving the illusionistic 'game of ships' behind, the camera moves on to a roundup of fishing boats lying idly on the channel's banks, left by Port Said's fishermen. These 'empty shells' not only do symbolize Suez's past

[19] An abridged version of this sequence, with some minor script variations, is also used in Part II of *The Oil Route*.

but also provide a transnational dimension to the modernization process that Suez just started to experience. Through their 'sun-seasoned wood', their 'shriveled sails', their 'red, green, yellow, and blue colors', Suez is immediately connected to all the fishing boats around the 'great Mediterranean lake'.

As the director finally arrives at Port Said, the northern end of the Suez Canal and gateway to the Mediterranean Sea, the image of two playing children in front of a Nord Bar sign takes the viewer to the end of a journey come full circle. The notion of circularity is indeed made explicit by the voice-over, which likens 'the smells, colors, and noises' of Port Said to those of 'Crete, Athens, Marseille, Barcelona', that is those of 'our civilization'.

After a quick exploration of the city's old houses with grand balconies, 'still immersed in their colonial dreams', the documentary brings the viewer back to Bertolucci's core vision of the canal. Through a reversal of the central sequence described earlier, the camera pans back along a narrow street, keeping its focus on an immobile cargo ship visible in the distance, past symmetrical buildings, and it leaves the canal to its seemingly eternal present. It is with this prolonged sequence-shot, inspired by the stylistic dictates of the nouvelle vague, that Bertolucci concludes *The Canal*.

While the *Oil Route* strives to meet some expectations of ENI's international documentaries, such as the portrayal of Italian workers abroad, the cooperation with the producing country, the respect for diversity, *The Canal* can be considered the director's cut of his Middle Eastern personal vision. Leaving aside the desiderata of his industry commissioner and the needs of corporate film, Bertolucci is free to concentrate his attention on the 'modern' landscape that the canal has created. Yet, as we watch this short film, we can see the director's actual attempt to unveil the semiotic suture of past and present and future that holds the landscape together. As he weaves his poetic imagination with a plethora of literary reminiscences drawn from the past, from Aeschylus to Xenophon, from Marco Polo to Melville, Bertolucci breaks with the present of corporate documentary style and realism. With his continuous detour queries, exploring images, ad hoc lyrical script, he creates a counter narration which, to use Banita's words, 'tells the "magical" story of petroleum in reverse, as a transcendent agent of cultural separation and disunion'.[20]

[20] Banita (2014, 155). Bertolucci rejected the commentary script written by journalist Alberto Ronchey, an official film consultant, on the ground that it was too aseptic and pre-

In conclusion, both *The Oil Route* and *The Canal* are arguably the least 'techno' of all ENI's techno-films. With them, Bertolucci creates a veritable poetic document akin to lyric criticism which implicitly, and likely unintentionally, questions ENI's narrative of modern progress.

References

Accorinti, Giuseppe. 2008. *Quando Mattei era l'impresa energetica. Io c'ero*. Matelica: Hacca.

Agazzi, Evandro. 2004. *Right, Wrong, and Science. The Ethical Dimension of the Techno-Scientific Enterprise*, Poznan Studies in the Philosophy of the Sciences and the Humanities. Vol. 81. Amsterdam: Rodopi.

Antonioni, Michelangelo. 2007. *The Architecture of Vision: Writings and Interviews on Cinema*. Chicago: University of Chicago Press.

Archivio Storico ENI, https://archiviostorico.eni.com/aseni/it/.

Banita, Georgiana. 2014. From Isfahan to Ingolstadt: Bertolucci's *La via del petrolio* and the Global Culture of Neorealism. In *Oil Culture*, ed. Ross Barrett and Daniel Worden, 145–168. Minneapolis: University of Minnesota Press.

Barrett, Ross, and Daniel Worden, eds. 2014. *Oil Culture*. Minneapolis: University of Minnesota Press.

Bini, Elisabetta. 2013. *La potente benzina italiana. Guerra fredda e consumi di massa tra Italia, Stati Uniti e Terzo mondo (1945–1973)*. Roma: Carocci.

Bonifazio, Paola. 2014. United We Drill: ENI, Films, and the Culture of Work. *Annali d'Italianistica* 32: 329–350.

Bucarelli, Massimo. 2010. All'origine della politica energetica dell'ENI in Iran: Enrico Mattei e i negoziati per gli accordi petroliferi del 1957. *Nuova rivista storica, XCIV* 2 (2): 465–500.

Bertolucci, Bernardo 2013. La via del petrolio, intervista a Bernardo Bertolucci, Pier Paolo Pasolini, Centro Studi Casarsa, http://www.centrostudipierpaolopasolinicasarsa.it/molteniblog/la-via-del-petrolio-intervista-a-bernardo-bertolucci/.

Dechert, Charles R. 1963. *Ente Nazionale Idrocarburi: Profile of a State Corporation*. Leiden: E. J. Brill.

cise; instead, Bertolucci injected the original script with a more poetic and creative text of his own: 'Alberto Ronchey was supposed to write the spoken text. The things he was saying were dry, precise, and informative. However, I was so clinging to the idea of poetic cinema that it seemed to me almost impossible to keep Ronchey's text; so I managed to persuade him to shorten it and added a text that I wrote on my own. ENI never gave me any directive on how to give the movie, maybe through its soundtrack, a more celebratory or informative tone'. Paolo Mereghetti, 'Conversazione con Bertolucci', in Toffetti (2010, 32).

Frescani, Elio. 2012. 'L'Italia non è un paese povero. Le vicende del film di Ivens commissionato dall'Eni di Enrico Mattei, *Meridione. Sud e Nord nel mondo* XII (2–3): 86–102.

——— 2016a. Cultura e oro nero. Strategie comunicative e intellettuali nell'Agip e nell'Eni di Enrico Mattei. *Storia e Futuro*, 38. www.storiaefuturo.eu.

———. 2016b. 'Una rivista per tutti «Il Gatto Selvatico» di Attilio Bertolucci nell'Eni di Enrico Mattei. *Memoria e Ricerca* 1: 161–179.

Invernizzi, Emanuele, and Stefania Romenti. 2014. *Progetti di comunicazione per la reputazione aziendale*. Milano: Franco Angeli.

Latini, Giulio. 2011. *L'energia e lo sguardo: il cinema dell'Eni e i documentari di Gilbert Bovay*. Roma: Donzelli.

Prattico, Franco. 1962. Con gli allievi africani alla Irom. *Il Gatto Selvatico*, VIII, 12, December, 7–9.

Toffetti, Sergio, ed. 2010. *La scoperta dell'altrove*. Milano: Feltrinelli.

Verdone, Mario. 1962. *Cinema del lavoro*. Roma: Realtà editrice.

Vittorini, Elio 1961. Industria e letteratura. *Il Menabò* 4, rpt. in Rodondi, Raffaella 2008, *Letteratura arte società. Articoli e interventi 1938–1965*, 955–962. Torino: Einaudi.

CHAPTER 23

Afterword: Nations and Empires in a Connected and Unequal World

Frederick Cooper

Barbara Curli and her colleagues have given the last word in this volume on Italy and the Suez Canal to someone who has done no research on Italy or on the Suez Canal. Moreover, they have chosen a confirmed skeptic about the utility in historical analysis of some of the key concepts behind the study of Italy and the Suez Canal–globalisation and national construction.[1] However, the conference held in Turin in May of 2019 and this collection of essays that has emerged from it are of at least as much interest to the skeptic as to a historian working within the frameworks of globalisation and national construction. The chapters of this book reveal the gap between expectations of the benefits of a more connected world and the continued unevenness of who, in fact, benefits. They point to the fragility of the politics of building a nation in a world where economic and political

[1] Cooper (2001, 2011).

F. Cooper (✉)
New York University, New York, NY, USA
e-mail: fred.cooper@nyu.edu

B. Curli (ed.), *Italy and the Suez Canal, from the Mid-nineteenth Century to the Cold War*,
https://doi.org/10.1007/978-3-030-88255-6_23

power is concentrated—very unevenly—in different sorts of units, imperial and corporate as well as national. This is a book about the spatial dimensions of economic and political power—on the effects of a change in the geography of commercial connections—but it underscores the impossibility of spatial determinism. Italy's apparently privileged geographical position in relation to the new routes of maritime commerce turns out not to be the blessing many thought it would be.

As I was reading the draft chapters of this book in the last week of March 2021, the Suez Canal suddenly became the focus of attention in the news media. A giant container ship, the *Ever Given*, had managed to wedge its bow on one side of the canal and its stern on the other. Traffic was completely blocked for nearly a week. Hundreds of ships were backed up, and the press warned of impending economic catastrophe. The vessel was freed through great effort using well-known maritime technology—tugboats and dredges—but, in the end, through the decisive intervention of something even older, an especially high tide during a full moon.[2] The episode—and the anxieties it unleashed—were a splendid illustration of the point that Valeska Huber made in her book *Channelling Mobilities: Migration and Globalisation in the Suez Canal Region and Beyond 1869–1914*: that the canal was not just a link that enhanced connectivity around the globe, but a choke point.[3] Western newspapers relished the image of the modern leviathan entrapped alongside the dilapidated houses, dry fields, and wandering goats of the village adjacent to the canal—a twenty-first century echo of the imagery of the steamships navigating the canal against sand dunes and Arab villages that were part of the iconography of the canal in the 1950s and 1960s.[4]

The chapters of this book illustrate simultaneously the importance of the new channel to the movement of people and goods and to the exercise of political power over space, the gap between the actual patterns of movement and the confident assertion of the 'modern' world of commerce and civilization that would flow from the new channel, the social consequences of the uneven impact that canal had on the region in which it was located, and the importance that potential—or actual—closure of the canal or the

[2] "Ship Is Freed After a Costly Lesson in the Vulnerabilities of Sea Trade", *New York Times*, March 29, 2021.

[3] Huber (2013).

[4] "A Tiny Village Finds Itself an Onlooker in the Scramble to Free the *Ever Given*", *New York Times*, March 29, 2021; Giuseppe Natale, this volume, Chap. 22.

exclusion of certain categories of people from using it had on world politics. In the end, the Suez Canal was a narrow line connecting two oceans and usable by only a select set of actors—corporations, states, individuals—not a web-like set of interconnections. It, as Huber makes clear, channeled economic, social, and political connections in quite specific ways—a point that does more to undermine than reinforce the image that advocates of globalisation have tried to promote.[5]

The story told in these pages is above all about uneven and unequal relationships. The coming together of an Italian state was not a step toward a Europe of equivalent national forms, even if Italian statesmen—Cavour and many others—were striving for something like that. The canal helped to make the richest and most powerful empire—Great Britain—even more so. Instead of fostering ship traffic across most of the Mediterranean, it encouraged Europe's leading shipping lines—based in Northern Europe—to bypass once-active ports in the eastern Mediterranean.[6] It made it easier for European colonizers to get to their colonies, which accentuated the subordination and exploitation of the latter rather than integrate them into a web of global commerce.

The contributors to this volume provide illuminating accounts of different forms of political imagination that were in play among backers of the canal. The conflicts among these visions are consistent with recent scholarship that goes against the old empire-to-nation-state metanarrative of the eighteenth to twentieth centuries and presents a much more complex picture of the late nineteenth and early twentieth century, in which different forms of political organization are overlapping and in unresolved tension with each other.[7]

At first glance, there is reason to privilege the national framework, particularly since the early decades of the Suez Canal were the time of the consolidation of an Italian nation—or at least an Italian state. Certainly, this is an essential part of the story. Giovanni Cristina refers to the 'patriotic enthusiasm' for the canal project.[8] But not all the figures promoting the canal fit within the national framework. Pietro Paleocapa was of Greek

[5] Cooper (2001).

[6] Kevin Tang, Chap. 6. While facilitating sea links, the canal blocked land routes long used by Bedouins. Huber (2013, 33–34, 141–71).

[7] Berger and Miller (2015), Leonhard and von Hirschhausen (2011), Burbank and Cooper (2010).

[8] Barbara Curli, Chap. 2, Andrea Giuntini, Chap. 3, Giovanni Cristina, Chap. 4; Filippo Gattai Tacchi, Chap. 8.

origin, lived part of his early life in the Austrian Empire, and was a minister of the Venetian Republic before he was exiled to the Piedmont, where he became a leading figure in the promotion of transportation linkages. He served on behalf of Piedmont-Sardinia in the commission planning the canal in the 1850s, alongside members from the French, British, Habsburg, Spanish, and Dutch empires. Barbara Curli describes him as a 'leading figure in the European technocracy'. Luigi Negrelli was another engineer who played a leading role. Sometimes referred to as 'Italian', he was a quintessentially imperial figure—a loyal citizen of the Habsburg Empire—and a quintessentially cosmopolitan one, whose principal language was German and who worked on railroads and other communications networks in several different parts of Europe.[9] Baron Pasquale Revoltella, Venetian born, was a leading entrepreneur and a promoter of the imperial interests of the Habsburgs and the urban interests of Trieste.[10]

These imperial and cosmopolitan figures, the *Commission internationale pour le percement de l'isthme de Suez*, and the multiple sources of finance, contributed to making 'Europe,' as much as any state, into a collective actor. This dimension was enhanced by the treatment of much of the world on the far side of the Sinai Peninsula as the object of the endeavor and not a participant. But the picture of European empire-building was complicated by another empire that straddled the Europe-Asia divide, the Ottoman. In the 1850s and 1860s, the Ottoman Empire was still in a position of geopolitical strength in the eastern Mediterranean but compromised by indebtedness, the ambiguity of its relationship with Egypt, and the power of local rulers in the Red Sea region.[11] The extent to which the Ottoman rulers were included or merely placated during the planning and construction of the canal is a question that deserves more attention on the basis of Ottoman sources.[12] At a minimum, the Ottomans were able to insist on diversification of financing to avoid too much influence going to France.[13] The papers collected here show that the canal contributed to swinging the balance of power in the region toward the European imperial

[9] Barbara Curli, Chap. 2; Andrea Leonardi and Alice Riegler, Chap. 7.

[10] Giovanni Modaffari, Chap. 9.

[11] Giampaolo Conte, Chap. 12.

[12] Andreas Guidi, Chap. 13, and Giorgio Ennas, Chap. 15, make the most use of Ottoman sources.

[13] Recent scholarship has complicated, if not overturned, an earlier view of the Ottoman Empire as experiencing an inexorable decline in the nineteenth century. For a pioneering contribution to the trend, see Deringil (1998).

powers, a club Italy was striving to join, but that they could not simply act any way they wished.[14]

One of the most important contributions of this volume is to an inter-empire story, a tale of jockeying for power among European states with imperial ambitions.[15] It is a story with twists. Great Britain at first opposed construction of the canal, not surprisingly, given that it was the premier imperial power under the status quo. But as France joined by Piedmont-Sardinia, the Habsburg empire, and other potential challengers seemed to be onto a good thing, Britain went along with the consortium. And it—and shipping companies within its orbit—were among the biggest winners of the accelerated commercial exchange that the canal produced. Much of the tonnage going through the canal in the late nineteenth century was under a British flag, and the vastly improved sea connection between England and India was a commercial boost, and by reducing the time necessary to bring military force from Great Britain to South Asia, it reduced British dependence on intermediaries and enhanced the security of the empire that had been severely challenged by the Indian revolt of 1857.[16]

Italy, which hoped its geographic position and newfound unity would propel it to a position of centrality and importance among the empires, was instead largely bypassed by the new trade routes. Perhaps the illusion of economic gain helped promote national consolidation and some infrastructural reform that benefitted internal commerce. But as Kevin Tang puts it, 'while the Suez Canal was intended to remake Italy and other Mediterranean countries as centres for global trade, the direct route to Asia actually made the region more peripheral, port activity more localized, shipping lines more direct, and consequently less able to benefit from the transshipment business that had historically made the Mediterranean Sea the centre of global trade'.[17] Ironically, Italian shippers continued to find a niche in which they could play a leading role not via the 'modern' world of steamships but through sailing ships that carried low-priced bulk

[14] For a broad look at naval power in Italy's rivalry with other empires, see Fabio De Ninno, Chap. 14.

[15] On the inter-empire perspective, see Hedinger and Heé (2018). On Italian empire in relation to other imperial powers in the 1920s and 1930s, see Pergher (2017).

[16] Kevin Tang, Chap. 6. For a related argument, about the importance of the telegraph to promoting imperial control, see Wenzlhuemer (2013). There is a large and growing interest in connections across the Indian Ocean. See for example Bose (2006), Alpers (2014).

[17] Tang, Chap. 6.

commodities, but which could not use the canal and instead took the old routes around Africa.[18]

Italy would eventually make good use of the canal for its imperialist ventures. Those projects had boosters like Giuseppe Sapeto, whose book of 1865 called for Italy to bring commerce, civilization, and Christianity to the part of Africa adjacent to the Red Sea. His vision was not purely Italian—he had earlier tried to persuade Emperor Napoleon III of France to take an interest in Ethiopia before he tried to make a similar case to Piedmont-Sardinia and then to Italy. The ensuing project in Assab, in which he cooperated with the Baron Pasquale Revoltella with his Habsburg connection, was a less than stellar showpiece for Italian colonialism.[19]

This venture was a forerunner to the colonization of Eritrea and to the first attempt to colonize Ethiopia, which ended in failure in 1896. Mussolini built on the fantasies of a Roman-scale imperial development centered on the Mediterranean, and he made good use of the Suez Canal in his brutal conquest of Ethiopia in 1935–1936, which he was able to pull off only because the British let him get away with sending soldiers and military supplies through the canal until the outbreak of World War II. For Africans, the failure of Britain and the League of Nations to stop the fascist invasion of Ethiopia showed that Europeans were in the imperialist game together—there were no 'good' imperialists—and the entire edifice had to be confronted directly.[20]

How Italy was able to use the Suez Canal in its conquest of Ethiopia is one of a number of empire stories told in this book.[21] The imperial perspective was particularly important to the Habsburgs, whose Vienna-based, German-speaking monarchy hoped to look beyond its Central European predominance and to use its connection through the port of Trieste to join the inter-empire competition in the Mediterranean-Red Sea-Indian Ocean sphere.[22]

The 'technoscientific' dimension of the Suez project, to use Barbara Curli's expression, was part of a discursive move that heightened the

[18] Leonardo Scavino, Chap. 5.

[19] Filippo Gattai Tacchi, Chap. 8, Luca Castiglioni, Chap. 11.

[20] Steven Morewood, Chap. 16. On the importance of the invasion of Ethiopia to the development of black internationalism, see Srivastava (2018).

[21] Georgio Ennas, Chap. 15, Steven Morewood, Chap. 16, Joseph John Viscomi, Chap. 20, Massimo De Giuseppe, Chap. 21.

[22] Giovanni Modaffari, Chap. 9.

distinction between Europe and 'the rest'.[23] Both secular and religious figures saw the canal as unifying the world, but it was not unifying it as a world of equals. European science and the Christian religion would deepen European empires' belief in their own superiority and—at least for colonizers with some sense of higher purpose—enhance their role in bringing civilization to peoples in need of uplift.[24]

While the story told here is most often too big to fit in national (or even inter-*national*) containers, it highlights the role of cities—and their relation to the spacial configuration that the canal reshaped—as well. Genoa, Brindisi, Venice, and other Italian ports perceived that they had something to gain and perhaps something to lose by whatever sea routes across the Mediterranean would be encouraged by the canal. Changing sea routes could change the relation of these cities to each other and thereby affect the process of Italian unification, as well as linkages by land to other parts of Europe.[25] The canal region itself and the Red Sea area beyond it were potential or actual sites of hubs in networks linking sea routes to hinterlands (Assab, Port Said), producing complex and often conflicting relationships between fledgling loci of European power and commercial interests and local potentates who were, in the late nineteenth century still forces with which the various imperial actors had to reckon.[26]

If at first glance, the canal appears to be a force for integration, much in this volume suggests it was a new element in an already fragmented set of political relationships, and it increased the asymmetrical nature of those relationships. The overlap and ambiguity of political jurisdictions—Khedival Egypt, Ottoman Empire, British officials and their regional clients, the Christian Empire of Ethiopia, local 'sultans' in the canal region, different European states claiming extraterritorial privileges in parts of the Mediterranean—brought together people operating with a variety of political visions. The conventional metaphors of 'integration', 'cosmopolitanism', and 'network' don't embrace the complexity of the dialectic of connection and fragmentation described in the chapters of this book. Nor

[23] Barbara Curli, Chaps. 1 and 2.

[24] On religious ideas and endeavors, see Carlo Bovolo, Chap. 10, and Paola Schellenbaum, Chap. 17. Olga Verlato, Chap. 18, points to the mix of secular and religious efforts in bringing educational resources to the Suez region.

[25] Andrea Giuntini, Chap. 3, Giovanni Cristina, Chap. 4.

[26] Luca Castiglioni, Chap. 11. Serels (2017) points out that it was well into the twentieth century that European powers came to exercise truly effective control over the various political entities in the Red Sea region.

does the invocation of 'imperialism' or 'colonialism' tell us all we need to know about political relationships that were asymmetrical, sometimes extremely so, but in which there was enough uncertainty and enough possibilities of cross-cutting relationships to render fragile the exercise of colonizing power in particular places and times. The Khedive of Egypt had to sign the agreement; the Ottoman empire pressured the sponsors to diversify their finances; local rulers along the Red Sea had to be placated; Italian power a few kilometers from a port like Assab or Massawa was not necessarily compelling.[27] So while the power behind the project of building a canal came from European empire-builders, other actors could push back and at the margins have influence on how the project was executed.

The Suez Canal and the port cities on both ends share with other critical nodes in communications networks around the globe the combination of being essential to world commerce and vulnerable to disruption. The ports and the Canal Company made use of polyglot labor forces recruited from around the Mediterranean world, with the usual vulnerabilities of migrants with limited roots in their place of work but strength deriving from the possibility of interrupting the movement of ships. Labor to build and maintain the canal and the ports on either end had initially come from a corvée—with all the tensions that accompany compulsion—then from people from different locations around the Mediterranean whose precarious situations pushed them to take up work under precarious conditions in the ports.[28] The paradox of the valuable and the vulnerable was typical of port labor in much of the world in the late nineteenth and early twentieth centuries: dockers were usually employed by the day, organized in work-groups that could become gangs. Port cities have reputation as dangerous places—sites not just of strikes and economic disruption but of prostitution and various forms of criminality.[29] Why was such disorder tolerated in places on which the capitalist economy depended? This situation did provoke anxiety—in Victorian London for example—but it was mainly in the years after World War II that efforts were made in both colonial and European ports to 'stabilize' dock labor on the grounds that a continuously employed, regularly paid work force was less dangerous than workers moving in and out of employment and other urban activities. But by the 1970s, there was an evident trend in much of the 'Third World' toward

[27] Luca Castiglioni, Chap. 11.

[28] In addition to Angelos Dalachanis, Chap. 19, see Huber (2013, 28–29, 105–38).

[29] Davies et al. (2000), Cooper (1987).

'informal'—that is unregulated—labor, and the move away from regular, protected employment reversed the trend of previous decades so much that scholars now refer to a current trend toward 'precarious labor'.[30] In the case of the Suez Canal zone, Huber describes labor employed by the Suez Canal Company, coaling companies, and other employers as 'rootless and disposable', often 'unruly' and frequently engaging in strike action.[31]

The canal's situation outside of the domain of any one state was, however, the vulnerability that eventually came to haunt the process of inter-empire cooperation that had run it for nearly a century. Even when Egypt became self-governing, the canal zone was not under its control. This situation was overturned by an upstart nationalist, President Nasser of Egypt. His seizure of the canal in 1956 resulted in an invasion by France, Great Britain, and Israel to reverse the takeover, an intervention that failed largely because the United States—which did not want defense of a colonialist project to get in the way of the struggle against communism—pulled the plug on its erstwhile allies. That the US acted this way and the operation of retaking the canal failed was a sign of a profound change in world politics. Deference to the legitimacy and normality of European powers doing what they willed in Africa, the Middle East, and Asia had been coming apart since World War II.

Influential Italians tried to play a mediating role in the Suez crisis and its aftermath: the remarkable mayor of Florence, Giorgio La Pira, brought Catholic ideas of social cohesion to an effort at reconciliating North and South, and Enrico Mattei, the iconoclastic head of the Italian oil company ENI, not only developed a personal relationship with Nasser and commercial ties to other Middle East oil producing states but promoted a view of post-imperial politics that promised to bring about harmony and mutual economic progress. But neither the government of Italy nor such would-be diplomats had much room for maneuver in a world divided by tensions among what was becoming known as the first, second, and third worlds. The travails of Italian citizens who had settled in Egypt was from a global perspective a minor sidelight to the drama of fragmentation and polarization as colonial empires lost their grip.[32]

[30] Cooper (2017).

[31] Huber (2013, 112, 123).

[32] Massimo De Giuseppe, Chap. 21, Giuseppe Natale, Chap. 22. On the fate of Italians who had—over a long period of time—settled in Egypt and found themselves caught in the aftermath of the Suez Crisis, see Joseph John Viscomi, Chap. 20.

To conclude: The boosters of the Suez Canal in Italy and the internationally minded engineers of the time had insisted that the canal would contribute both to the expansion of commerce and to the unification of Italy. Indeed, in the half century after the canal's opening, global commerce did expand and Italy did become a coherent national entity. The chapters of this book take apart the connection of these two processes. Most of the traffic that the canal stimulated bypassed the ports of Italy and the eastern Mediterranean. The most important users of the canal were those who were already the largest-scale participants in global commerce.

The impact of the canal is best analyzed less in terms of 'globalisation' or 'state-building' than in terms of unevenness and fragmentation, and specifically imperialism. The British government had been reluctant to support the canal when it profited from the previous status quo, but when the canal was built it was in a position to make the fullest use of it: to enhance its control over India, to increase the rapidity and volume of the shipping its flag carriers transported, and—especially after 1882—to make Egypt into a hub of British power.

As a technical achievement and as an instance of inter-empire cooperation, the building of the canal contributed to Europe's sense of its civilizational superiority. Italy's imperialist ventures (along the Red Sea in the late nineteenth century, vainly in Ethiopia in the 1890s, with transient success in the 1930s) would later take advantage of the canal, but only because the British and the international partners administering the canal let them do so. Inter-empire cooperation, however, was a fragile phenomenon, breaking down in 1914–1918 and 1939–1945. Italy would try to reposition itself in relation to the world south of the Mediterranean as colonial empire came under attack in the 1950s, but its statesmen's attempts at mediating the North-South divide could not overcome the depth of that divide as well as that between East and West. The Suez Crisis of 1956 was a sign both of how much the taken-for-granted nature of an imperialist world order had by then eroded, and also how much uncertainty and conflict accompanied the opening of a post-imperial order. The canal would continue to be simultaneously a valuable link in global commerce and hence a vulnerable location in the global political economy—a potential chokepoint as well as channel. The people who lived on both sides of the canal were the least likely to enjoy its benefits and the most likely to suffer from the conflicts that the volatile politics of the region entailed.

References

Alpers, Edward. 2014. *The Indian Ocean in World History*. New York: Oxford University Press.

Berger, Stefan, and Alexei Miller, eds. 2015. *Nationalizing Empires*. Budapest: CEU Press.

Bose, Sugata. 2006. *A Hundred Horizons: The Indian Ocean in the Age of Global Empire*. Cambridge, MA: Harvard University Press.

Burbank, Jane, and Frederick Cooper. 2010. *Empires in World History: Power and the Politics of Difference*. Princeton: Princeton University Press.

Cooper, Frederick. 1987. *On the African Waterfront: Urban Disorder and the Transformation of Work in Colonial Mombasa*. New Haven, CT: Yale University Press.

———. 2001. What Is the Concept of Globalization Good For? An African Historian's Perspective. *African Affairs* 100: 189–213.

———. 2011. Alternatives to Nationalism in French West Africa, 1945–60. In *Elites and Decolonization in the Twentieth Century*, ed. Marc Frey and Jost Dülferr, 110–137. Houndmills: Palgrave Macmillan.

———. 2017. From Enslavement to Precarity? The Labour Question in African History. In *The Political Economy of Everyday Life in Africa: Beyond the Margins*, ed. Wale Adebanwi, 135–156. Oxford: James Currey.

Davies, Sam, Colin Davis, Lex Heerma Van Voss, Lidewij Hesselink, David de Vries, and Klaus Weinhauer, eds. 2000. *Studies in the International History of Dock Labour, c. 1790s–1970s*. London: Ashgate.

Deringil, Selim. 1998. *The Well-Protected Domains: Ideology and the Legitimation of Power in the Ottoman Empire 1876–1909*. London: I. B. Tauris.

Hedinger, Daniel, and Nadin Heé. 2018. Transimperial History: Connectivity, Cooperation and Competition. *Journal of Modern European History* 16 (4): 429–452.

Huber, Valeska. 2013. *Channelling Mobilities: Migration and Globalisation in the Suez Canal Region and Beyond 1869–1914*. Cambridge: Cambridge University Press.

Leonhard, Jörn, and Ulrike von Hirschhausen, eds. 2011. *Comparing Empires: Encounters and Transfers in the Long Nineteenth Century*. Göttingen: Vandenhoeck & Ruprecht.

New York Times. March 29, 2021.

Pergher, Roberta. 2017. *Mussolini's Nation-Empire: Sovereignty and Settlement in Italy's Borderlands, 1922–1943*. Cambridge: Cambridge University Press.

Serels, Steven. 2017. Early European Colonial Rule and the African Red Sea Littoral. *Northeast African Studies* 17 (1): 1–23.

Srivastava, Neelam. 2018. *Italian Colonialism and Resistances to Empire, 1930–1970*. London: Palgrave Macmillan.

Wenzlhuemer, Roland. 2013. *Connecting the Nineteenth-century World: The Telegraph and Globalization*. Cambridge: Cambridge University Press.

Index[1]

[1] Note: Page numbers followed by 'n' refer to notes.

B. Curli (ed.), *Italy and the Suez Canal, from the Mid-nineteenth Century to the Cold War*,
https://doi.org/10.1007/978-3-030-88255-6

Printed in the USA
CPSIA information can be obtained
at www.ICGtesting.com
LVHW021448221123
764620LV00001B/6

9 783030 882549